An Introduction to Engineering Systems

PUES-9

Pergamon Unified
Engineering Series

Pergamon Unified
Engineering Series

An Introduction to Engineering Systems

Samuel Seely
University of Connecticut

Pergamon Press Inc.
New York · Toronto · Oxford
Sydney · Braunschweig

PERGAMON PRESS INC.
Maxwell House, Fairview Park, Elmsford, N.Y. 10523

PERGAMON OF CANADA LTD.
207 Queen's Quay West, Toronto 117, Ontario

PERGAMON PRESS LTD.
Headington Hill Hall, Oxford

PERGAMON PRESS (AUST.) PTY. LTD.
Rushcutters Bay, Sydney, N.S.W.

VIEWEG & SOHN GmbH
Burgplatz 1, Braunschweig

Copyright© 1972, Pergamon Press Inc.
Library of Congress Catalog Card No. 75-161591

All Rights Reserved. No part of this publication may be reproduced, stored in a retrieval system or transmitted in any form, or by any means, electronic, mechanical, photo-copying, recording or otherwise, without prior permission of Pergamon Press Inc.

Printed in the United States of America
08 0168213

Pergamon
Unified Engineering
Series

GENERAL EDITORS

Thomas F. Irvine, Jr.
State University of New York at Stony Brook

James P. Hartnett
University of Illinois at Chicago Circle

EDITORS

William F. Hughes
Carnegie-Mellon University

Arthur T. Murphy
PMC Colleges

William H. Davenport
Harvey Mudd College

Daniel Rosenthal
University of California, Los Angeles

SECTIONS

Continuous Media Section
Engineering Design Section
Engineering Systems Section
Humanities and Social Sciences Section
Information Dynamics Section
Materials Engineering Section
Engineering Laboratory Section

Preface

Most existing introductory systems engineering texts are concerned with the "hand solution" of systems response problems. With the advent of the digital computer with its great power and versatility, it is no longer necessary to restrict the classes of problems that may be studied, and many of the constraints imposed by the hand-solution requirement can be relaxed. It is desirable, therefore, that the methods of analysis appropriate to hand solution be supplemented by numerical methods that are better adapted to the computer.

This book is designed to provide the student with an introduction to many important aspects of systems analysis. It seeks to provide a sufficient background to both the classical methods, which, at least in simple cases, are appropriate to hand-solution and the matrix approach that is more appropriate to computer solutions. Because numerical methods are essential to an understanding of computer oriented discussions, and because such methods can accomodate nonlinear as well as linear systems, the general scope of the study has been broadened accordingly. Thus the Kirchhoffian and the state variable formulations are considered in about equal detail and emphasis, and numerical as well as analytic methods are essential parts of the study. Furthermore, owing to their importance in system studies, discrete time systems, and continuous time systems under sampled conditions, also receive attention. The book will provide the student with an understanding of the presently important methods of systems study, and it likewise seeks to show the correlation that often exists in alternate formulations.

The book is divided into four parts of unequal length; Part I, Models and Modeling; Part II, Interconnected Systems; Part III, Systems Response a. Time Domain, b. Frequency Domain; Part IV, Special Topics. Chapter 1 is an introduction to models and modeling. It contains a discussion of the means for describing the elements which constitute the systems under study.

Chapters 2, 3, 4 contain the methods for describing the dynamics of the interconnected system. Chapter 2 details the means for describing the system dynamics by a Kirchhoff description relating the input-output variables. It also discusses in considerable detail an equilibrium description that involves input, output and state. Both the Kirchhoff and the state formulations receive general treatment. Also considered are discrete time systems, since these are in a form particularly appropriate for digital computer study. Chapter 3 introduces the signal flow graph, a graphical-analytical representation that not only has interest in itself as a tool in analysis, but also serves as a convenient step in the preparation of analog computer programs. It is also found to be of considerable aid in relating the Kirchhoff to the state formulation. Chapter 4 is a somewhat more formal approach to equilibrium equation formulation than in Chapter 2. It introduces the idea that an interconnected group of elements is conveniently studied by a two step process involving the topological description of the interconnection and the algebraic description of the elements in the interconnection.

Chapters 5 and 6 are concerned with system response and behavior in the time domain. Chapter 5 contains a discussion of classical methods of finding the response of simple circuits to different excitation functions. This chapter introduces certain important systems concepts, but it also serves to show the rather poor match to numerical problem solving, except in the simplest cases. Also included is consideration of a number of important features of numerical methods of solution, plus a discussion of the important aspects of machine methods in the solution of the dynamic equations of interconnected systems. Chapter 6. extends this study to systems which have been described in state form, with particular emphasis on the digital computer considerations, and the time-constant barrier and sampling time problem.

Chapters 7 through 11 contain a discussion of the analysis of systems response in the frequency domain. The material in these chapters is largely limited to linear systems, since transformation techniques are essential to most of these studies. Chapter 7 introduces the essential features of the Laplace transform, and a general introduction to problem solving. For linear systems, these techniques provide a better match between problem and problem solver, since the transformation allows many of the steps in the solution to be accomplished algebraically. Chapter 8 is a detailed study of features of systems in the transformed domain. Chapter 9 examines the response of systems in the steady-state to sinusoidal excitation functions, a form that is of particular importance in many areas of endeavor. Chapter 10 considers many special theorems and forms that are useful in problem analysis and solution. Chapter 11 extends considerations to the periodic nonsinusoidal function, and shows the place of the Fourier series in systems studies. These considerations are extended to the case of nonperiodic excitations, with an introduction of the Fourier integral and the features of the Fourier transform. Attention is given to the numerical solution of Fourier series components, and also to the features of the Fast Fourier transform.

Chapters 12 and 13 are interested in certain time-domain problems. Chapter 12 directs attention to the sampled data problem and to the use of the Z-transform in sampled data systems. The relation between the difference equation and the sampled data Z-transform techniques are examined. Chapter 13 considers the question of system stability from a number of viewpoints, a matter of particular interest in feedback system studies.

There are many important topics that are not discussed, including time varying systems, and systems with stochastic inputs. It is expected that the content will provide the student with a background in modern systems analysis on which he can build further in his subsequent studies. Parts of the content are rather sophisticated, but have been included for the sake of completeness. An instructor can eliminate that material which he considers to be outside of the objectives of his course.

This book is a direct outgrowth of discussions with Professor E. Folke Bolinder of Chalmers University of Technology, Gothenburg, Sweden, concerning an introductory course in systems analysis. These discussions have set the technical level and content of the book. Acknowledgement is also made of ideas that come from the reports published under the Discrete System Concepts Project funded by the National Science Foundation under Grant GE-2546. Particular note is made of Monograph No. 6, "Discrete System Techniques at the Sophomore Level," by John G. Truxal.

University of Connecticut SAMUEL SEELY

Contents

PART I MODELS AND MODELING

Chapter 1 Modeling of System Elements   3
    1-1 Introduction   3
    1-2 Model Characteristics   5
    1-3 Model Approximations   7
    1-4 Signals and Waveforms   10
1-a Electrical Elements   13
    1-5 Introduction   13
    1-6 The Capacitor   13
    1-7 The Inductor   16
    1-8 Mutual Inductance – Transformers   19
    1-9 The Resistor   22
    1-10 Sources   24
    1-11 Duality   28
1-b Mechanical Elements   29
    1-12 The Ideal Mass Element   29
    1-13 The Spring   30
    1-14 The Damper   31
    1-15 Rigid Linkage (Mechanical Transformer)   32
    1-16 Independent Mechanical Sources   34
    1-17 Mechanical Elements – Rotational   35
1-c Fluid Elements   39
    1-18 Liquid Systems   39
    1-19 Liquid Resistance   40

1-20	Liquid Capacitance, Inductance, Sources	41
1-21	Gas Systems	43
1-d	Thermal Elements	45
1-22	Thermal Systems	45
1-e	N-Port Devices	49
1-23	Transducers	49
1-24	Active Networks	51
1-25	Modeling of Complicated Situations	53
1-26	Summary	55

PART II   INTERCONNECTED SYSTEMS

Chapter 2   Interconnected Systems: Equilibrium Formulations — 65

2-1	Interconnected Elements	65
2-a	Kirchoff Formulation	66
2-2	Operational Notation	67
2-3	Through-Across Equilibrium Laws	69
2-4	Node Equilibrium Equations	69
2-5	Loop Equilibrium Equations	77
2-b	State Variables and State Equations	82
2-6	Introduction to the State Formulation	82
2-7	State Equations for Linear Systems	85
2-8	Differential Equations in Normal Form	90
2-9	State Variable Transformation	93
2-10	Discrete and Sampled Time Systems	95

Chapter 3   Signal Flow Graphs — 108

3-1	Properties of SFG	108
3-2	Graphing Differential Equations	110
3-3	Simultaneous Differential Equations	112
3-4	The Algebra of SFG-s	116
3-5	State Equations and the SFG	122

Chapter 4   System Geometry and Constraint Equations — 130

4-1	Interconnected Elements	130
4-2	Graph of a Network	131
4-3	The Connection Matrix	134
4-4	General Form of Topological Constraints	135
4-5	Node Pair and Loop Variables	139
4-6	Branch Parameter Matrixes	141
4-7	Equilibrium Equations on a Node-Pair Basic	144

4-8	Equilibrium Equations on the Loop Basis	146
4-9	The Canonic LC Network	149
4-10	The General LC Network	151
4-11	The Canonic LC Network Containing R	153
4-12	The General RLC Network	157
4-13	Duality	159

PART III SYSTEM RESPONSE

a Time Domain

Chapter 5 System Response — 169

5-a Classical Differential Equations — 169
- 5-1 Features of Linear Differential Equations — 169
- 5-2 General Features of Solutions of Differential Equations — 171
- 5-3 The Complementary Function — 174
- 5-4 The Particular Solution — 178
- 5-5 Variation of Parameters — 182
- 5-6 Evaluation of Integration Constants — Initial Conditions — 185
- 5-7 The Series RL Circuit and its Dual — 187
- 5-8 The Series RL Circuit with an Initial Current — 190
- 5-9 The Series RC Circuit and its Dual — 193
- 5-10 The Series RLC Circuit and its Dual — 198
- 5-11 Switching of Sinusoidal Sources — 206

5-b Numerical Methods — 207
- 5-12 The Newton-Raphson Method — 209
- 5-13 Numerical Solution of Differential Equations — 210
- 5-14 Difference Equation Approximation — 212
- 5-15 Nonlinear Systems — 216
- 5-16 Various Methods for Numerical Integration — 219

5-c Machine Solutions — 221
- 5-17 The Operational Amplifier — 222
- 5-18 Computer Simulation of Differential Equations — 228
- 5-19 Introducing Initial Conditions — 230
- 5-20 Time and Magnitude Scaling of Analog Computers — 231
- 5-21 Simulation Languages for the Digital Computer — 233
- 5-22 Problem Oriented Languages — 234

Chapter 6 General Time Domain Considerations — 243
- 6-1 Singularity Functions — 243
- 6-2 Superposition Integral — 246

6-3	Convolution Integral	249
6-4	Convolution Summation	254
6-5	State Equations	255
6-6	Numerical Solution of Continuous Time Systems	261
6-7	Discrete Time Systems	262
6-8	Continuous Time Systems with Sampled Inputs	263
6-9	Steady-State Output to Periodic Inputs	265

b Frequency Domain

Chapter 7 The Laplace Transform — 275
- 7-1 The Laplace Transform — 276
- 7-2 Laplace Transforms of Elementary Functions — 277
- 7-3 Properties of the Laplace Transform — 279
- 7-4 Inverse Laplace Transform — 290
- 7-5 Problem Solving by Laplace Transforms — 291
- 7-6 Expansion Theorem — 293
- 7-7 Linear State Equations — 297
- 7-8 Initial Conditions and Initial State Vectors — 303

Chapter 8 s-Plane: Poles and Zeros — 310
- 8-1 The System Function — 310
- 8-2 Impedance and Admittance Functions — 312
- 8-3 System Determinants — 316
- 8-4 The s-Plane — 320
- 8-5 $T(s)$ and its Pole-Zero Constellation — 322
- 8-6 Step and Impulse Response — 326
- 8-7 Step and Impulse Response of a System with One External Pole — 327
- 8-8 State Models from System Functions — 328
- 8-9 System Function Realization using Operational Amplifiers — 336

Chapter 9 System Response to Sinusoidal Functions — 344
- 9-1 Features of Sinusoids — 344
- 9-2 Steady-State System Response to Sinusoidal Excitation Functions — 350
- 9-3 Power — 353
- 9-4 Phasor Diagrams — 355
- 9-5 Q-Value and Bandwidth — 359

9-6	The $T(j\omega)$ Plane	368
9-7	Magnitude-Phase and Bode Plots	369

Chapter 10   Special Topics in Systems Analysis — 382
- 10-1   Thévenin and Norton Theorems — 382
- 10-2   Maximum Power Transfer Theorems — 385
- 10-3   Source Transformation — 388
- 10-4   Two-port Passive Networks; y-System Equations — 388
- 10-5 z-System Equations — 391
- 10-6 T and Π Equivalent Networks — 394
- 10-7   Hybrid Parameters — 395
- 10-8   Cascade Parameters, $abcd$ Coefficients — 397
- 10-9   Input, Output, and Transfer Impedances — 400
- 10-10 Active Networks — 401
- 10-11 Tellegen's Theorem — 407

Chapter 11   General Excitation Functions — 414
- 11-1   Periodic Excitation Function — Fourier Series — 414
- 11-2   Effect of Symmetry — Choice of Origin — 419
- 11-3   Complex Fourier Series — 423
- 11-4   Properties of Fourier Series — 428
- 11-5   Numerical Determination of Fourier Coefficients — 431
- 11-6   The Fourier Transform and Continuous Frequency Spectrums — 435
- 11-7   Properties of Fourier Transforms — 437
- 11-8   Frequency Response Characteristics — 441
- 11-9   The Discrete Fourier Transform — 443
- 11-10 The Fast Fourier Transform — 446

PART IV   SELECTED TOPICS

Chapter 12   The Z-Transform and Discrete Time Systems — 457
- 12-1   Time Sampling and the Z-Transform — 457
- 12-2   The Z-Transform — 461
- 12-3   Properties of the Z-Transform — 462
- 12-4   Discrete Time System Function — 464
- 12-5 Z-Representation of Differentiation — 465
- 12-6   Difference Equations and the Z-Transform — 472
- 12-7   System Description by Difference Equations in Normal Form — 477

Chapter 13 Stability — 482

- 13-1 Pole Locations and Stability — 482
- 13-2 Properties of Driving Point Functions — 486
- 13-3 Routh-Hurwitz Test — 487
- 13-4 The Nyquist Criterion — 489
- 13-5 Discrete Time Systems — 493
- 13-6 Controllability and Observability of Linear Systems — 494
- 13-7 Observing the State of a System — 496
- 13-8 Stability in the Sense of Liapunov — 498
- 13-9 The Direct Method of Liapunov — 498
- 13-10 Generating Liapunov Functions — 501

References — 509
Appendix A Matrixes — 512
Index — 521

Part I *Models and Modeling*

1
Modeling of System Elements

In undertaking a study of the behavior of an engineering system, or when undertaking the design of a new system, it is usually necessary to employ a suitable manipulable representation of the system, often referred to as a model of the system. This chapter will be concerned with the modeling process.

1-1 INTRODUCTION

System models are of three basic types—physical, analytical, and descriptive. The traditional and perhaps most familar type of model is a physical representation of the system. Such models have the advantage that experiments can be conducted on them; they have the disadvantage that they may be very costly. Analytical models are mathematical expressions of the characteristics of the system, usually in the form of sets of simultaneous equations. Descriptive models, and these may bridge the entire scale of abstraction, may be simple word pictures of the processes that make up a system. The description will cover the interactions among the variables, the attributes of the variables, and the constraints which limit the performance. Such models can be prepared in a minimum of time, are relatively low in cost, and their development requires few special skills. Descriptive models can be applied to solve a wide variety of problem situations requiring decisions. Models of business systems, for example, are usually most easily expressed in descriptive terms.

The need for a wide variety of model types is evident when one considers that the description of an electronic circuit that carries out a specified function, and which is made up of many resistors, capacitors, inductors, transistors, diodes, etc., would be undertaken in a manner that is quite different from that of a chemical processing plant, with its oxidation reactors, heat exchangers, distillation columns, and compressors. Both of these would have rather different descriptions from a highway network with its exits, access points, interchanges, and the random introduction and removal of vehicles, or from a business concern that manufactures rubber products. But in spite of their differences, to undertake a study of these or almost any other system requires that some convenient model be developed. Further, it may not be possible to represent fully all quantitative aspects of the physical system. Clearly, therefore, the validity of a study can certainly be no more accurate than the quality of the model that is used.

As noted, a system study is usually done by developing an equivalent model of the actual system and then studying the model. This is especially true in carrying out a design, since no actual system exists, but is ultimately to be developed. The system is often represented by the interconnection of a collection of appropriate components, each of which is individually modeled. For carrying out mathematical studies, a mathematical model is usually necessary; for physical studies, physical models would be used. Thus, for example, to ascertain the performance characteristics of a chemical plant, a pilot plant is built as a physical model for test and study. With the advent of the digital computer, it is quite common to model the chemical process by mathematical relationships that fully describe the interconnected components of the system, and then ascertain the performance characteristics from computer studies.

In carrying out the modeling, the procedure may be rigorously analytical and may be accomplished from the basic laws that describe the components. This is often possible for some physical devices, and much of our subsequent studies will be of systems composed of such elements. Often the components are so involved that the models are approximations that involve empirical data correlations.

We shall be concerned in this chapter with a study of the mathematical modeling process of the elements which may comprise the components of an extensive physical system. We shall discuss a variety of elements, but the systems that we shall study will be composed largely of electrical or mechanical elements.

1-2 MODEL CHARACTERISTICS

Our study will be concerned only with *lumped* elements. A lumped element is one that can be isolated and treated as one that possesses the features of idealized elements. Actually in many cases the real elements are not lumped. Clearly, it might appear quite unreasonable to consider that a long section of pipe or a long electrical transmission line could be considered as being lumped, yet in many cases this is a valid approximation. Also, clearly discernable springs exist, and if these are not too large, it seems reasonable that they would well satisfy any reasonable criterion of lumpiness. However, a solid rod also possesses compliance or spring qualities, and such a rod under loading does stretch, and with the removal of the load, the rod will return to its original length, provided that it had not been stressed beyond its elastic limit. An easily stated criterion for lumpiness does not appear possible. Thus one must be careful that in any critical case that the assumption of lumpiness, if it is made, is indeed valid.

Often it is assumed that the elements are *linear*. This assumes that there is a linear relation between cause and effect, or in engineering terms, between excitation and response. A casual inspection will not indicate linearity—both lumped or distributed elements can be linear. We shall find that systems analysis can often be carried out completely in closed mathematical form for systems composed of linear elements. This is rarely possible for nonlinear systems, and nonanalytic methods (graphical or numerical) must be employed in studying such systems. In an effort to deduce a closed-form solution, one often may assume linearity for a nonlinear system. Extreme care is necessary in such cases, since the assumption of linearity may completely negate the important features of the system. The use of a digital computer in analysis often makes unnecessary such linear approximations, since the computer can handle nonlinear systems with about the same ease as a linearized description.

Most elements that are considered linear are, in fact, nonlinear. However, there is usually a reasonable range over which their response-excitation characteristics are linear, to a reasonable engineering approximation. Where the approximation is not valid, then the behavior of the real system will probably not agree with that predicted for the modeled system. The discussion is clarified somewhat by reference to Fig. 1-1 which shows graphically a response-excitation relationship. If the excitation were limited to the small range *a–b* (this is often referred to as the small-signal or *incremental* operation), then the assumption of linearity of

6   Modeling of System Elements

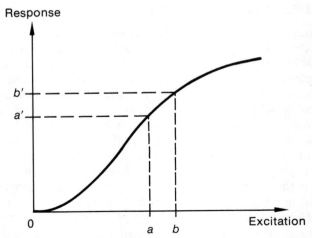

Fig. 1-1. A representative nonlinear response-excitation relationship.

response over the associated range $a'-b'$ is reasonably valid. If the excitation were to extend over the range $O-b$, then, of course, the assumption of a linear response would not be valid.

There is an implication in the foregoing that must be discussed. It is that upon reversing the sense of the excitation that the response-excitation relation will remain unchanged. Elements for which this is true are known as *bilateral* elements. There is a wide range of elements which are not bilateral. Figure 1-2 illustrates a nonbilateral characteristic. In the extreme

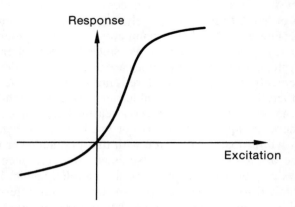

Fig. 1-2. A nonbilateral response-excitation characteristic.

case for which there is zero response for all excitations upon the reversal of the excitation, the elements are said to be *unilateral*. Figure 1-3 shows a representative diode characteristic (this is typical of the semiconductor diode, and is also closely approximated by vacuum diodes).

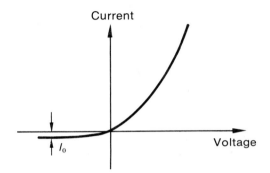

Fig. 1-3. The response characteristic of a unilateral element.

1-3 MODEL APPROXIMATIONS

When attempting to carry out a study of a system that may include nonlinear nonbilateral elements, one is faced with the problem of providing a description of the elements. An exact description is not often possible and one must employ an approximate representation. The representation is usually chosen in a form that is most convenient to the method of analysis that is to be employed. For example, if the analysis is to be done numerically or by a digital computer, a tabular numerical description with appropriately chosen intervals of the independent variable may suffice. Specifically, refer to Fig. 1-3, which is now to represent the terminal characteristics of a silicon diode. A theoretical description of this diode characteristic is possible which is based on semiconductor theory. The result is the semiconductor diode equation,

$$I = I_0(e^{qV/k\theta} - 1), \tag{1-1}$$

where q, V, K, θ, are respectively; electron charge, voltage, Boltzmann's constant and absolute temperature.

The diode equation is not a particularly convenient expression to use in hand numerical calculations, although it is an analytical description of the device characteristics. In the case of numerical calculations, it might

8 Modeling of System Elements

be more convenient to prepare a table giving the currents appropriate to selected values of the independent variable, which is voltage in this case. In other cases it might be found convenient to approximate the curve by a number of linear segments. A simple *piecewise linear* approximation for the diode characteristic will be of the form shown in Fig. 1-4. For the

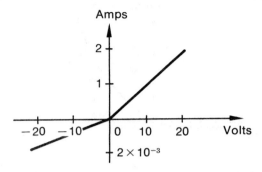

Fig. 1-4. A piecewise linear approximation to the diode characteristic of Fig. 1-3.

particular values which are shown on the figure, the characteristic is given by the expressions,

$$i = 10^{-1}v, \quad v > 0$$
$$i = 10^{-4}v, \quad v < 0$$

Let us now consider a characteristic such as that shown in Fig. 1-5. This is a fairly general nonlinear nonbilateral characteristic which is assumed to represent the terminal characteristics of a device for which an exact theoretical description is not possible. Almost any single valued nonlinear curve of this type can be approximated by a polynomial of the form,

$$i = a_0 + a_1 v + a_2 v^2 + \cdots + a_n v^n. \tag{1-2}$$

We shall not consider the general techniques involved in fitting a polynomial to a given characteristic.

An alternate piecewise linear approximation to the nonlinear curve of Fig. 1-5 is provided by the use of straight line segments, as shown in the figure. The choice of the number of linear segments to be used depends on the range of variation in the component terminal variables and the accuracy desired. It is cumbersome to write the equations for the straight line sections. For this reason, this type of representation is not particularly attractive in analytical studies. However, it is a very practical representa-

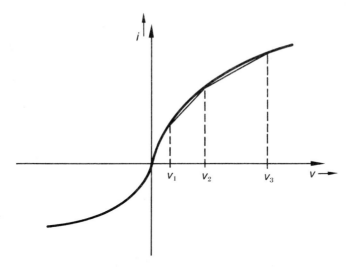

Fig. 1-5. A general nonlinear curve; to define piecewise linear approximation.

tion to use when an analog computer is used in the study of the system performance since one can simulate such a piecewise linear curve by means of a diode function generator. In this case, the simulation almost exactly duplicates the piecewise linear approximation.

A form of approximation that we shall use on a number of occasions is the *stepwise constant* or *staircase* representation that is illustrated in Fig. 1-6. It is possible to extend the stepwise constant and the piecewise

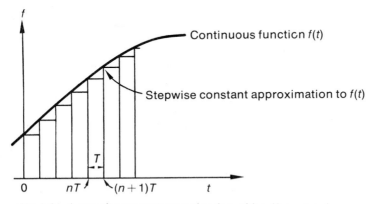

Fig. 1-6. A stepwise constant approximation, with uniform step size.

linear approximations to piecewise quadratic and piecewise higher order variations between sample points. Some of these ideas will be used in Chapter 1 when we study detailed methods of sampled data systems.

These general ideas of function approximations will occur at a number of points in the text, and further discussion will be deferred until these needs arise. Initially we shall consider lumped, linear and nonlinear, and bilateral elements, since such elements are particularly important in our studies.

1-4 SIGNALS AND WAVEFORMS

We shall consider a system to be a collection of interconnected components or elements with an input set of dynamic variables, often called excitations, and another set which are the output variables or responses. Systems analysis is concerned with the determination of the responses for specified excitations.

We shall talk about observable physical quantities, such as: voltage, current, force, velocity, and others. These are dynamic variables which will be designated respectively by the symbols v, i, f, u. These specify the time description of these quantities—more explicitly, they should be written $v(t)$, $i(t)$, etc. These may be constants, they may vary sinusoidally with time, they may denote pulses in a recurring train, or they may be of rather special shape and duration. We shall refer to these dynamic variables as *signals*, even though they are not necessarily concerned with messages (in the sense of communications). In the sense here being used, they denote any time dependent observable associated with some system. Electrical signals with almost any reasonable waveform can be generated electronically. Mechanical signals with complicated waveform are much harder to produce. We shall limit our considerations to those waveforms which are possible by reasonably straightforward methods.

Signals, in the sense here being used, can be represented graphically, and can also be described analytically, although the analytical description may be complicated. Refer to Fig. 1-7 which shows a number of representative signal waveshapes and their analytic representations. Observe that in Figs. 1-7a and 1-7b the description is a continuous function of the time, whereas in Figs. 1-7c and 1-7d the waveshapes are specified by piecewise descriptions in time. Note further that these waveforms may extend over all time, or they may be of finite duration. Thus for the waveforms shown in Fig. 1-8 which are time limited versions of the waveforms of Fig. 1-7, we have, respectively, a constant source of finite duration, a sinu-

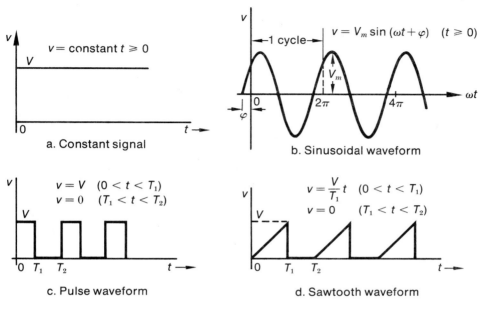

Fig. 1-7. Signal waveforms.

soid of prescribed number of cycles, a finite pulse train, a single sawtooth wave. Note particularly that even though Figs. 1-8a, c, and d illustrate signals that do not become negative at any point in time, this is not an essential restriction, and more generally any part or all of the waveshape can be negative.

Signals of all types can be generated, and the peak amplitudes can range over enormous values. For example, readily available commercially produced electrical signal generators are possible for producing voltage signals from millivolts to kilovolts. Current sources may range from microamperes to amperes. Mechanical force sources are possible over very wide ranges, from micronewtons to meganewtons; mechanical velocity sources are limited in amplitude and in recurrence period.

Our discussion will proceed in general terms, and will be appropriate to any waveshape. Later in specific applications, the waveshapes will be designated. It is emphasized that the development is general and is based on general laws, even though much of the detail will proceed for prescribed waveforms.

12   Modeling of System Elements

The signal sources with which we shall be concerned are of two general classes: one class is called *across* sources, the other being *through* sources. Correspondingly, the associated dynamic variables are across and through variables. In water flow, for example, pressure is an across variable, and flow rate is a through variable. The across variable, whether it is water pressure, temperature, electric pressure (voltage), mechanical

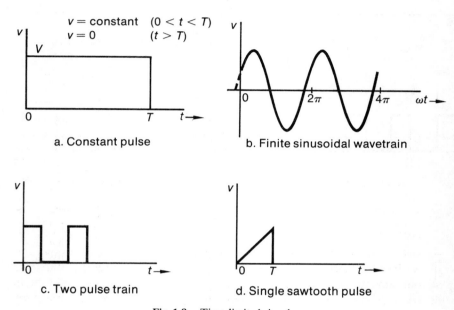

Fig. 1-8.   Time-limited signals.

velocity, or rotational velocity, denotes the variable relative to some specified or implied reference or datum, e.g., the translational velocity of a body relative to the fixed earth. Conversely, the through variable denotes a flow or transfer quantity, be it water flow, heat flow, electric flow (current), mechanical force, mechanical torque. In these cases the direction of flow must be specified as well as the magnitude. One can distinguish through and across quantities by the types of instruments that would be required and how they are connected in the system in making through and across measurements. Other aspects of source representations will be discussed later, as the needs arise.

1-a ELECTRICAL ELEMENTS

1-5 INTRODUCTION

Electrical elements are described in terms of two dynamic variables, current (through variable) and voltage (across variable) or a simple functional form of these. Either can be chosen as the dependent or as the independent variable in any particular case. We wish first to discuss these quantities.

Here we shall only consider the conduction current that exists in the connecting wires to the terminals of the electrical elements, that is, it is here implied that the current into and out of the element terminals is under survey. By restricting our discussion to the terminals, we avoid any need for considering the precise mechanism that may underlie the operation of the element. In this way, for example, we avoid the question of displacement current in the dielectric of a capacitor; we avoid the fact that in a bipolar transistor both electrons and holes are involved in the conduction process; or that in a gaseous conduction device, electrons, positive ions and negative ions are charge carriers, and so are involved in the current.

The voltage is the difference of electrical potential across the terminals of the element. For a specified voltage across the element, it is possible to specify the potential of each terminal relative to a reference datum. In electric circuits and apparatus, the reference datum is often the "ground" or "earth." This means, for example, that an element that has a 5 volt difference of potential across the terminals may have one terminal at 1000 volts with the second terminal at 1005 volts above ground potential. For most network problems the considerations do not concern themselves with any datum or reference level. The situation is somewhat different with mechanical systems in linear translation since inertial forces are related to the acceleration of the mass elements with respect to the earth as a fixed frame of reference.

1-6 THE CAPACITOR

The capacitor is the idealized circuit element in which energy may be stored in electric form. In its most elementary form the capacitor consists of two closely spaced metallic plates which are separated by a single or multiple layers of a nonconducting dielectric material (air, glass, paper,

14 Modeling of System Elements

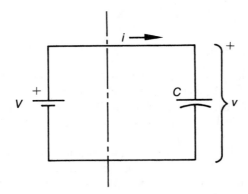

Fig. 1-9. A capacitor across a battery.

oxide). The schematic representation of the capacitor is shown to the right of the broken line in Fig. 1-9.

Suppose that an initially uncharged capacitor is connected to a battery, as shown in Fig. 1-9. There will be a flow of electrons from the capacitor terminal that is connected to the positive terminal of the battery through the battery to the capacitor terminal that is connected to the negative terminal of the battery. The loss of electrons from one plate causes it to become positively charged, whereas the plate that receives the excess electrons will be negatively charged. The amount of charge transfer will be such that the capacitor will be charged to the voltage of the battery to which it has been connected. Observe that in the charging process no electrons have passed through the dielectric material between the plates of the capacitor. Because of the changing potential across the plates, or more precisely, because of the changing electric field across the dielectric material, there is set up through the dielectric what is called a *displacement* current. Thus one speaks of a *conduction* current in a conductor and a displacement current in an insulator. For the situation illustrated in Fig. 1-9 the displacement current at every instant is equal to the conduction current, which thus insures continuity of current through the circuit.

The terminal properties of the capacitor are described graphically by a charge-voltage relationship of the form shown in Fig. 1-10. In the case where a linear q, v relationship exists, the charge is proportional to the voltage. The proportionality constant is called the capacitance C. Thus by definition,

$$C = \frac{q}{v}. \text{ (Coulombs/Volt)}(= \text{Farad}) \tag{1-3}$$

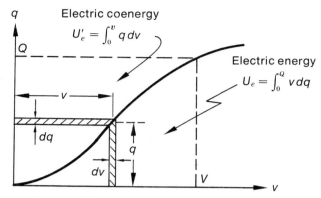

Fig. 1-10. The charge-voltage relationship of a capacitor.

For the linear capacitor, C is a geometric factor which can be deduced from elementary considerations of the theory of electricity. For a simple parallel plate capacitor, C is given by the expression,

$$C = \frac{\epsilon A}{d}, \quad \text{(Farads)} \tag{1-4}$$

where ϵ is the *permittivity* (sometimes called the dielectric constant) of the dielectric material, A is the area, and d is the separation of the plates. In general $\epsilon = \epsilon_r \epsilon_0$ where $\epsilon_0 = (36\pi \times 10^9)^{-1}$ Farads/meter, the free space permittivity, and ϵ_r is the *relative* permittivity. In the case when the q, v relationship is nonlinear, the simple concept of capacitance is no longer valid, although one might deduce an analytic expression that describes the nonlinear q, v relationship given graphically in Fig. 1-10.

Another feature of Fig. 1-10 that sometimes is of interest is the area below and above the curve. These are known, respectively, as the energy stored in the capacitor and the corresponding coenergy. As shown in the figure, these functions are,

$$\text{Coenergy} \quad U'_e = \int_0^v q \, dv,$$
$$\text{Energy} \quad U_e = \int_0^Q v \, dq. \tag{1-5}$$

This distinction is not essential to our present studies, but the distinction becomes important when nonlinear circuit analysis is under consideration. Note in particular that when the q, v curve is linear the coenergy and the energy functions have equal values, since the area above is equal to that below the curve.

The important through-across variable relationships of the capacitor are readily obtained. In general, we may write that,

$$i = \frac{dq}{dt} = \frac{d(Cv)}{dt}. \tag{1-6}$$

In the linear case, this equation becomes,

$$i = C\frac{dv}{dt}, \tag{1-7}$$

where v is the independent variable. Correspondingly, we may write,

$$v = \frac{1}{C}\int i\,dt, \tag{1-8}$$

where i is the independent variable. Both of these linear relations are of basic importance in electric circuit theory. We shall often employ these relationships in our subsequent studies.

The practical range of physical capacitors extends from several picofarads ($= 10^{-12}$ Farads) to perhaps 1000 microfarads ($= 10^{-6}$ Farads). The physical size is determined by the material and thickness of the dielectric, the total capacitance, and the maximum allowable voltage across the dielectric without rupture. Whatever the material or design, a 1 Farad capacitor would be very bulky and would not be a practical circuit element, although it is used extensively in our subsequent studies, which are usually on a normalized scale. A 1 Farad 1 Volt capacitor has been built; it measures roughly 3 in. diameter and about 12 in. long.

1-7 THE INDUCTOR

As a rough parallel to the capacitor, with energy storage associated with the fixed charges in the electric field, energy may be stored in the magnetic field produced by moving charges, or current. To discuss this matter in detail would require considerations of the work of Oersted, Ampére, and Biot-Savart. We summarize the situation by noting that a wire carrying a current produces a magnetic field. The magnetic flux that is produced will link the wire, (see Fig. 1-11) for the simple solenoid. The important parameters for describing the inductor are flux linkages ψ and current i.

The terminal properties of the inductor are described graphically by a flux linkage-current relationship, which, in general, will be of the general form shown in Fig. 1-12. For the case when a linear ψ, i relationship exists,

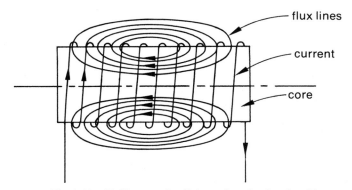

Fig. 1-11. To illustrate flux linkages in a simple solenoid.

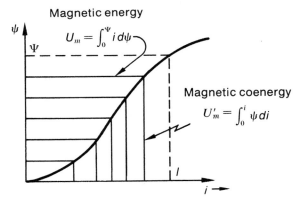

Fig. 1-12. The flux linkage-current relationship of an inductor.

the flux linkages are proportional to the current. The proportionality constant is called the inductance L. This is,

$$L = \frac{\psi}{i}. \quad \text{(Henrys)} \tag{1-9}$$

For the linear inductor, L is a geometric factor which can be deduced from basic considerations of the theory of magnetism. For example, the inductance of a uniformly wound torous (a doughnut-shaped device) of N turns on a core having a constant permeability μ is given by the approximate relation,

$$L = \frac{\mu N^2}{2\pi R} A, \quad \text{(Henrys)} \tag{1-10}$$

18   Modeling of System Elements

where μ is the permeability of the core on which the torous is wound (here assumed to be a constant), N is the number of turns, R is the mean radius of the torous and A is the cross-section of the core. In this expression there is the implicit assumption that the radius of the winding is small compared with the radius R of the torous. In general $\mu = \mu_r\mu_0$, where μ_0 is the permeability of free space ($= 4\pi \times 10^{-7}$ Henrys) and μ_r is the relative permeability. Usually the geometry of the configuration does not lend itself easily to ready calculation of the inductance. Nevertheless, L is a function of the geometry and the properties of the material.

When the ψ, i relationship is nonlinear, and such nonlinear relationships are very common with inductors owing to the fact that μ is a sensitive function of the magnetic field and hence of the current, the simple concept of inductance is no longer valid. The situation may be particularly complicated by the fact that magnetic materials usually display magnetic hysteresis, with B,H characteristics ($B \propto \psi$, and $H \propto i$) as shown in Fig. 1-13.

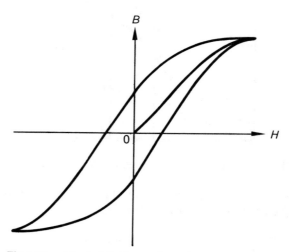

Fig. 1-13.   A typical B,H curve for an iron core inductor.

To describe the inductor as a circuit element, we require a relation between the current i and the voltage v. This is given through Faraday's law, from which we write,

$$v = \frac{d\psi}{dt} = L\frac{di}{dt}. \tag{1-11}$$

This is one of the fundamental circuit relations for the linear inductor. The corresponding relation, when v is the independent variable, is,

$$i = \frac{1}{L} \int v\, dt. \tag{1-12}$$

The graphical representation for the inductor fully marked as a circuit element is given in Fig. 1-14.

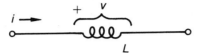

Fig. 1-14. Network representation of the inductor.

The practical range of physical inductors extends, for coils of a few turns of wire in air with inductances measured in microhenrys, to coils with special magnetic cores to several hundred henrys.

1-8 MUTUAL INDUCTANCE – TRANSFORMERS

When more than a single inductor is present in a given region, mutual interaction of the magnetic fields of the inductors can exist. To examine this matter in some detail, refer to Fig. 1-15 which illustrates two coils which react magnetically. As drawn, the two coils are shown on a common core. This is just a schematic representation since in many cases no clearly discernable core exists (two coils in air would be an example). The advantage of indicating the presence of the core is that it aids in visualizing the mutual flux that links both coils.

We call specific attention to the intimate relation of the distribution of the windings on the core in Fig. 1-15a and the dots in Fig. 1-15b. The meaning of the dots is the following: when progressing along the winding from the dotted terminal, each winding encircles the core in the same sense. This means that if the currents enter the dotted terminals, the component fluxes in the core will be in the same direction, and will add. An alternative though equivalent description follows by considering a variable source of voltage to be applied to one set of terminals, with an oscilloscope connected to the second set of terminals. The dotted terminals rise and fall together in voltage, or equivalently, the dotted terminals have the same instantaneous polarity.

It is clear from the foregoing discussion that the algebraic sign to be

20   Modeling of System Elements

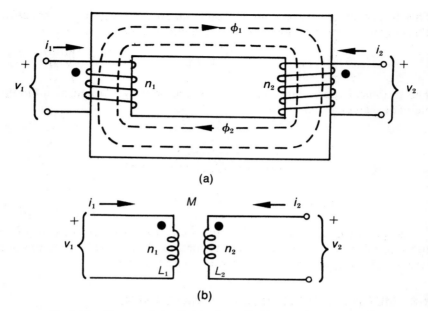

Fig. 1-15. Two coupled coils (a) schematic, (b) network representation.

assigned to the mutual flux, and so to the mutual inductance, is intimately associated with the assumed current directions. Consider the case of *unity* coupled coils, which means that all of the flux ϕ_1 produced by the current in coil 1 links coil 2, and correspondingly, all of the flux ϕ_2 produced by current i_2 in coil 2 links coil 1. For the coupled inductors illustrated in Fig. 1-15a we write, by Faraday's law,

$$v_1 = n_1 \frac{d}{dt}(\phi_{11} + \phi_{12}),$$
$$v_2 = n_2 \frac{d}{dt}(\phi_{22} + \phi_{21}),$$

(1-13)

where ϕ_{ij} denotes the flux coupling coil i that arises from a current in coil j. In the manner of Eq. (1-9), we write the following expressions,

$$L_1 = \frac{\psi_1}{i_1} = \frac{n_1\phi_{11}}{i_1}, \qquad L_2 = \frac{\psi_2}{i_2} = \frac{n_2\phi_{22}}{i_2},$$
$$M = \frac{n_1\phi_{12}}{i_2} = \frac{n_2\phi_{21}}{i_1}.$$

(1-14)

It follows from these that,

$$M = \sqrt{L_1 L_2}. \qquad (1\text{-}15)$$

More detailed calculations for the case of nonunity coupling shows that M is related to L_1 and L_2 by the expression,

$$M = k\sqrt{L_1 L_2}, \qquad (1\text{-}16)$$

where k, the coefficient of coupling, may vary from zero to one. Achieving unity coupling is very difficult, but can be approximated closely in tightly wound coils on a high permeability core.

By Eqs. (1-14) we write Eq. (1-13) in the form,

$$v_1 = L_1 \frac{di_1}{dt} + M \frac{di_2}{dt},$$
$$v_2 = M \frac{di_1}{dt} + L_2 \frac{di_2}{dt}. \qquad (1\text{-}17)$$

Note further that if we take the ratio of the two equations of Eq. (1-13) we have that

$$\frac{v_1}{n_1} = \frac{v_2}{n_2}$$

which we write,

$$v_1 = \frac{n_1}{n_2} v_2 = nv_2, \qquad (1\text{-}18)$$

where n is the turns ratio n_1/n_2. This shows that for unity coupled coils the voltage per turn is the same for the two windings, and that with different number of turns, the voltage v_2 can be greater or less than v_1.

Another feature of the unity-coupled transformer may be deduced from considerations of the power to the coils (power $p = vi$). For zero power losses in the transformer, the total power to the system, by the principle of conservation of energy, is,

$$v_1 i_1 + v_2 i_2 = 0, \qquad (1\text{-}19)$$

where due account has been taken of the reference conditions for voltage and current in Fig. 1-15b. Combine this expression with Eq. (1-18) to get,

$$-\frac{i_2}{i_1} = \frac{v_1}{v_2} = \frac{n_1}{n_2} = n, \qquad (1\text{-}20)$$

which relates the current in winding 2 with that in winding 1. Observe therefore that the electrical transformer imposes constraints on the through and across variables simultaneously.

For practical air-core transformers, the coupling is usually considerably less than unity, it is nearly unity in specially built iron-core transformers. Often the resistance of the wires of the transformer must be considered. It might also be necessary in some cases to take account of the power dissipation that occurs in the transformer core, if it is iron. Consequently, our simple coupled coil model may be somewhat in error.

1-9 THE RESISTOR

The capacitor and the inductor store energy respectively in the electric and the magnetic fields but do not dissipate energy. The resistor will dissipate energy but will not store electrical energy. Refer to Fig. 1-16

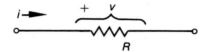

Fig. 1-16. Network representation of the resistor.

which shows a resistor in its schematic form, with current and voltage indicated by the reference states. A plot of the i,v characteristics of a resistor may, in some cases, be of the form shown in Fig. 1-17. We here

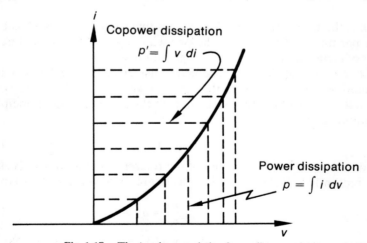

Fig. 1-17. The i,v characteristic of a nonlinear resistor.

show a nonlinear characteristic that is characteristic of a diode (*see* Fig. 1-3). If the characteristic were linear, then there would be a simple linear relationship between v and i, with the factor of proportionality,

$$R = \frac{v}{i}. \quad \text{(Ohms)} \tag{1-21}$$

For the nonlinear case, this equation really defines the slope of the straight line from the origin to any point on the curve. Such a resistance value has some significance as an average value for a particular voltage and current, but usually has little analytical interest. We can, of course, define an *incremental* resistance which specifies the slope of the curve for each value of v. Such an incremental resistance concept is used extensively for active circuit elements: diodes, tubes, transistors.

From Eq. (1-21) we may write the through-across variable relationships of the resistor in the form,

$$i = \frac{v}{R} = Gv, \tag{1-22}$$

where $G = 1/R$ is called the conductance of the resistor. When i is chosen as the independent variable, then we write,

$$v = Ri. \tag{1-23}$$

In both forms, these are referred to as Ohm's law. Clearly, of course, these are exactly valid only for linear elements.

For homogeneous materials (metals) the resistance is substantially constant, although there is a small temperature dependence. For a resistor that consists of a material of uniform cross-section A and length l and with conductivity σ ($= 1/\rho$. resistivity), the resistance is given by the expression,

$$R = \frac{l}{\sigma A} = \rho \frac{l}{A}. \quad \text{(Ohm)} \tag{1-24}$$

This expression is not valid for a material with a nonuniform cross-section. It also is not valid for a material for which ρ is not a constant, e.g., a semiconductor material.

We may speak of the power in the resistor, and the energy dissipation in the resistor. That is, the instantaneous power to the resistor is

$$p = vi = \frac{v^2}{R} = i^2 R, \quad \text{(Watts)} \tag{1-25}$$

and the total energy dissipated during the time t is expressed as,

$$W = \int_0^T p\,dt = \int_0^T vi\,dt. \tag{1-26}$$

It follows from this expression that the energy, for constant v and i, is

$$W = viT. \quad \text{(Watt-sec)(Joules)} \tag{1-27}$$

1-10 SOURCES

We have implied in the foregoing discussion that sources exist which produce currents or voltages, since if work is to be done or if energy is expended in dissipative elements, this energy must have been supplied by energy-producing or energy-converting sources. Both current and voltage sources are physically realizable, but practical current sources are less common than voltage sources. The schematic symbols for designating sources are given in Fig. 1-18. Here the arrow associated with the current source and the + sign associated with the voltage source express the *reference conditions*.

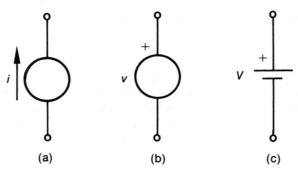

Fig. 1-18. Symbols for ideal independent electrical sources: (a) current, (b) voltage, general, (c) voltage, constant or d-c.

The significance of the reference conditions stems from the fact that measurable electrical quantities are *physical entities*. Their algebraic representations (i, v) are *algebraic quantities*. Physical entities change their direction or polarity while their algebraic representations change their sense (algebraic sign).

Refer first to the notation for current, which is the flow or through variable, as shown in Fig. 1-19. According to our system of notation, the

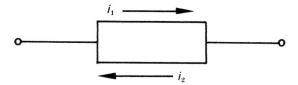

Fig. 1-19. Indicating the reference direction for current.

reference direction is indicated by the arrow, which is an essential part of the notation. The direction of the through variable when the symbol that represents it is positive is called its *reference direction*. A negative algebraic sign is the equivalent of a reversal of the reference direction, so that,

$$i_1 = -i_2. \tag{1-28}$$

Voltage exists between two points, one of which is the reference point. Such across variables exist between pairs of points or nodes. Refer to the isolated element with a pair of terminals given in Fig. 1-20. The reference

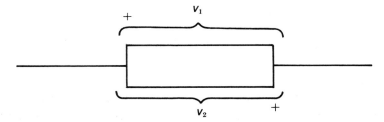

Fig. 1-20. Reference polarity for across variable.

polarity is indicated by $+$ and $-$ signs at the nodes. Since the $-$ sign is implied if the $+$ sign is given, it is sufficient to show only the $+$ sign. The polarity of an across variable, when the symbol that represents it is positive, is called its *reference polarity*. Since a reversal of the algebraic sign amounts to a reversal of a reference polarity, then,

$$v_1 = -v_2. \tag{1-29}$$

The general term *reference condition* is used to imply either a reference direction or a reference polarity. Thus in general terms, the reference condition of a physical entity corresponds to the positive sense of the algebraic quantity that represents it. It is important to note that for

26   Modeling of System Elements

electrical sources that the + sign is used to denote both the reference polarity and the algebraic sense. The distinction is important, but no confusion should result from this double use of the + sign.

Electrical sources are not ideal, in the sense that practical sources are not completely independent of the external load. Consider for example a chemical battery. The output voltage remains substantially constant with current, although a slight, and almost linear, decrease does occur, as illustrated in Fig. 1-21. This means, as we shall later understand, that the

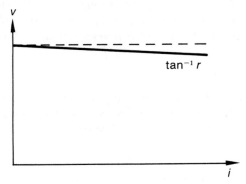

Fig. 1-21.   Load characteristics of a chemical storage battery.

practical source may be represented as an ideal source plus a resistor, as shown in Fig. 1-22. In this diagram r denotes the equivalent internal resistance of the source. This response characteristic can be described

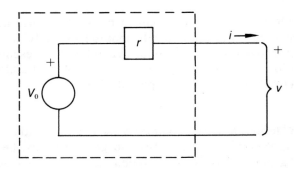

Fig. 1-22.   Representation of a practical source.

analytically by the straight-line formula,

$$v = V_0 - ri, \tag{1-30}$$

where V_0 is the voltage for zero external load current, and r is the slope of the response curve.

It is interesting to rearrange this equation. This is done by dividing all terms by r and then rearranging the result in the form,

$$i = \frac{V_0}{r} - \frac{v}{r} = I_0 - \frac{v}{r}, \tag{1-31}$$

where $I_0 = V_0/r$. This equation has the circuit representation in Fig. 1-23. Observe that as far as the terminal conditions are concerned, the two

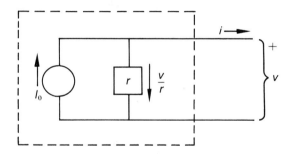

Fig. 1-23. Network representation equivalent to Fig. 1-22.

forms in Figs. 1-22 and 1-23 are equivalent. This establishes the principle of *source transformation*, a principle that is often of considerable value in analysis, since with it and other transformations that will be discussed later, problem solution is often facilitated.

In addition to the independent voltage and current sources, a set of dependent sources also exist. In these, a voltage or current at one pair of terminals of the device results in a voltage or current at a second pair of terminals. There are four possible dependent pairs, and these are of importance in modeling the so-called active elements, e.g., tubes and transistors. (*See* Section 1-24 and Chapter 10.) However, we shall make little reference to active devices until later in the text, and we shall now limit our discussion only to independent voltage and current sources.

1-11 DUALITY

Reference is made to the fact that when discussing the electrical circuit elements we presented the i,v relationships in two forms, with explicit expressions for the cases when the through and the across variable is the dependent or the independent variable. Descriptive forms for which the role of the dependent and independent variables are interchanged are sometimes called dual forms. The concept of duality among the electrical element is best discussed by referring to Fig. 1-24. Each

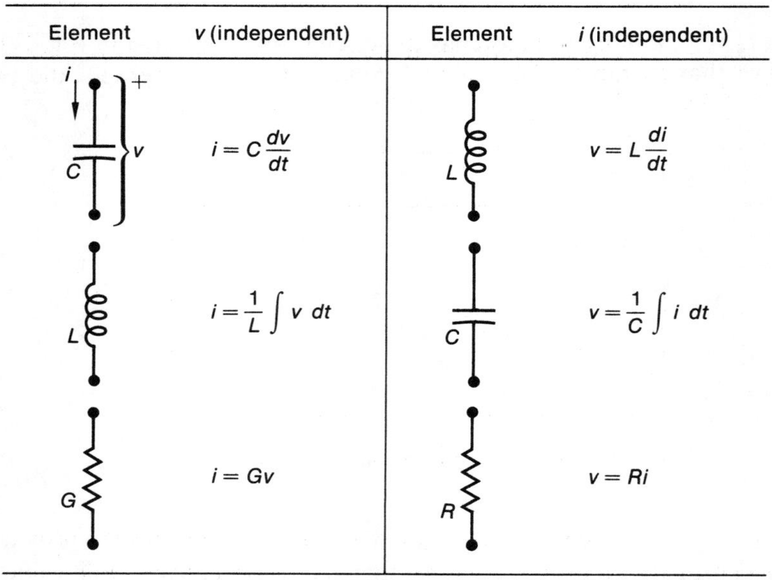

Element	v (independent)	Element	i (independent)
C	$i = C \dfrac{dv}{dt}$	L	$v = L \dfrac{di}{dt}$
L	$i = \dfrac{1}{L} \int v \, dt$	C	$v = \dfrac{1}{C} \int i \, dt$
G	$i = Gv$	R	$v = Ri$

Fig. 1-24. Quantitative aspects of duality in electrical systems.

corresponding pair of equations on a horizontal line of this table are said to be dually related. There is a systematic interchange of the variables and with similar mathematical form. From this it is seen that C and L, G and R are dually related. Note carefully, however, that duality implies only that the mathematical representations are similar in form; no aspect of equivalence is implied.

1-b MECHANICAL ELEMENTS

1-12 THE IDEAL MASS ELEMENT

When considered in terms of Newton's second law of motion, the mass M is defined by the expression,

$$f = M\frac{dv}{dt} = M\frac{d^2x}{dt^2} = Ma \quad \text{(Newtons)} \tag{1-32}$$

which relates the force (a through variable) with the acceleration (an across variable). Actually, this is a special case of the more general formula.

$$f = \frac{d}{dt}(Mv), \tag{1-33}$$

which states that "the force is equal to the time rate of change of momentum and is in the direction of the momentum change." Under ordinary circumstances the velocity of motion is sufficiently small so that the mass M, which from relativistic considerations can be expressed as an explicit function of the speed, remains substantially constant.

Representations of considerable interest are the two equivalent schematic diagrams given in Fig. 1-25. The analytic relations between the

Fig. 1-25. Schematic representation of the mass element.

force and the motional variables, depending on which is selected as the independent variable, are the pair of relations;

$$f = M\frac{dv}{dt}, \quad v = \frac{1}{M}\int f\,dt. \tag{1-34}$$

We call attention to the exact equivalence in form that exists between this set of equations and Eqs. (1-7) and (1-8) for the capacitor. This establishes the *analogy* between the mass M in the mechanical system and the capacitor C in the electrical system. We shall have many occasions to exploit this analogy at later points in the text.

1-13 THE SPRING

A spring element is one which stores energy in the displacement due to the elastic deformation that results from the application of a force. The terminal properties of the spring are shown graphically in Fig. 1-26. Here

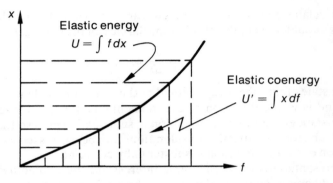

Fig. 1-26. Force-displacement relationship for a spring.

we show the displacement-force relationship to be nonlinear, which is the general case. For the region where the curve is linear, the analytic expression is written,

$$f = Kx \quad \text{(Newtons)} \tag{1-35}$$

where K is the spring constant. Over this linear region, the spring satisfies Hooke's law. Physical springs satisfy this relationship for forces below the elastic limit of the material. Deviations from this linear relation may occur for a number of reasons including, temperature effects, rotation of the spring ends, hysteresis, and others.

The schematic representation of the spring element is given in Fig. 1-27. The analytic relations between the through and the across variables are the pair of equations,

$$f = K(x_1 - x_2) = K \int (v_1 - v_2) \, dt = K \int v \, dt,$$
$$v = \frac{1}{K} \frac{df}{dt}. \tag{1-36}$$

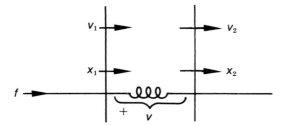

Fig. 1-27. Schematic representation of the spring.

The first of these expressions shows that if $x_1 > x_2$ there is a compressive force and f is positive, for the specified reference conditions. If $x_1 < x_2$ there is a negative or extensive force. Note the analogy between the spring and the inductor, both in the through-across relations and in the schematic representation.

1-14 THE DAMPER

There are three important types of mechanical friction: static, Coulomb, and viscous. Static friction is directly concerned with motion and manifests itself through the greater force required to initiate motion between two surfaces in contact than to maintain the motion. Coulomb friction acts in a direction to oppose motion. The Coulomb friction force between rolling or sliding surfaces is dependent on the normal force, but is substantially independent of velocity. We shall not be particularly concerned with such forces.

The force of friction between moving surfaces that are separated by a viscous fluid, or between a body and a fluid medium (e.g., a rotating fan in air, a rotating propellor in water, a ship moving through water, etc.) depends on the velocity, often in a complicated way. Several representative variations are illustrated in Fig. 1-28. Curve A is reasonably representative of the frictional force relation for a dash-pot, an oil-actuated device, which is illustrated in rough physical design and schematic representation in Fig. 1-29. Curve B in Fig. 1-28 illustrates all three important types of friction. A region exists for which the force is independent of velocity, a region for which the friction is substantially constant and independent of velocity, and a region of nonlinear dependence on velocity.

We shall here limit ourselves to viscous friction of the general class illustrated in Curve A, although a linear dependence is not required. For

32   Modeling of System Elements

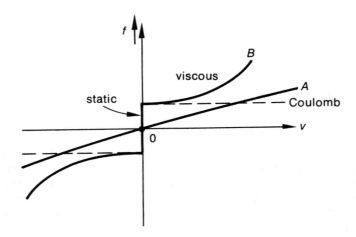

Fig. 1-28.   Representative frictional forces.

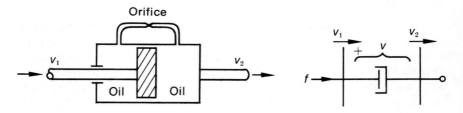

Fig. 1-29.   Physical and diagrammatic representations of a dash pot.

the case of a linear dependence of velocity on force, we write for the v, f dependence the pair of equations,

$$f = Dv, \qquad v = \frac{f}{D}. \tag{1-37}$$

Attention is called to the similarity between this pair of equations and those for the resistor, as given in Eqs. (1-22) and (1-23). Here D is the damping constant.

1-15 RIGID LINKAGE (Mechanical Transformer)

A rigid linkage is a device for transforming force and velocity variables. Consider the simple rigid linkage shown in Fig. 1-30. From the figure we

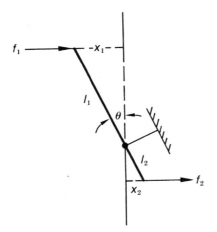

Fig. 1-30. A simple rigid linkage mechanism.

see that,

$$\sin\theta = \frac{x_1}{l_1} = \frac{x_2}{l_2}.$$

We define the quantity n,

$$n = \frac{l_1}{l_2}. \tag{1-38}$$

It follows that,

$$x_1 = nx_2,$$

so that,

$$v_1 = nv_2. \tag{1-39}$$

If there are no losses in the linkage, then the power input to the linkage must equal the power output from the linkage. This requires, for the reference conditions shown, that,

$$f_1 v_1 = f_2 v_2, \tag{1-40}$$

from which it follows that,

$$\frac{f_1}{f_2} = \frac{v_2}{v_1} = \frac{1}{n}. \tag{1-41}$$

The linkage can be given schematic representation, as shown in Fig. 1-31. This representation has been borrowed from electrical symbolism

34 Modeling of System Elements

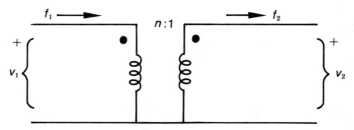

Fig. 1-31. Schematic representation of a rigid linkage (a translational transformer).

(*see* Section 1-8). The dots that appear in Fig. 1-31 are included to specify points which are to be marked with + signs to specify the positive reference for velocity. Also, the arrows specify the positive reference directions for force.

1-16 INDEPENDENT MECHANICAL SOURCES

In the foregoing discussion of M, K, D, which refer to passive elements, it has been assumed that sources or drivers exist which produce forces or velocities. It is convenient to define sources as force or velocity drivers, although velocity sources are less common than force drivers. These sources are represented schematically in Fig. 1-32, and are assumed to maintain their terminal polarities independently of the magnitudes of the power at the terminals; that is, they are regarded as independent sources. This is an idealization, of course, since any practical source would ordinarily be unable to maintain such an independence. In the more practical case, a driver can often be represented by an ideal source with associated lumped elements to account approximately for the terminal properties. This was discussed in connection with electrical drivers in Section 1-10.

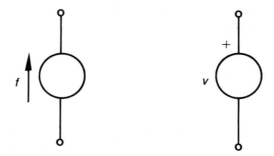

Fig. 1-32. Schematic representation of force and velocity drivers.

1-17 MECHANICAL ELEMENTS—ROTATIONAL

A set of rotational mechanical elements and rotational variables exist which bear a one-to-one correspondence to the translational mechanical elements and the translational variables which have been discussed in the foregoing sections. In the rotational system, torque is the through variable and angular velocity is the motional or across variable. The following are the corresponding fundamental quantities:

J = polar moment of inertia, corresponds to M in translation
K = rotational spring constant, corresponds to K in translation
D = rotational damping, corresponds to D in translation
$\mathcal{T}$ = torque, corresponds to f in translation
ω = $d\theta/dt$, angular velocity, corresponds to $v = dx/dt$ in translation

a. The Inertial Element. In this rotational set, J is the rotational parameters, and is the assumed constant proportionality factor between torque and angular acceleration,

$$\mathcal{T} = J\frac{d\omega}{dt} \quad \text{(Newton-meters)} \tag{1-42}$$

J of a rotational body depends on the mass and the square of a characteristic distance of the body, the radius of gyration, K, and is given by,

$$J = Mk^2. \tag{1-43}$$

For a simple point mass rotating about an axis at a distance r from the center of mass, $k = r$, and

$$J = Mr^2. \tag{1-44}$$

For a simple disk of radius r rotating about its center, $k = r/\sqrt{2}$, and

$$J = \tfrac{1}{2}Mr^2. \tag{1-45}$$

The essential features of the rotational system are given schematically in Fig. 1-33. The pair of equations that apply for this element are,

$$\mathcal{T} = J\frac{d\omega}{dt}, \quad \omega = \frac{1}{J}\int \mathcal{T}\, dt. \tag{1-46}$$

b. The Rotational Spring Element. A rotational spring is one which will twist under the action of a torque. It is depicted in schematic form in Fig. 1-34, and is described by the pair of equations,

$$\mathcal{T} = K\theta = K\int \omega\, dt, \quad \omega = \frac{1}{K}\frac{d\mathcal{T}}{dt} \tag{1-47}$$

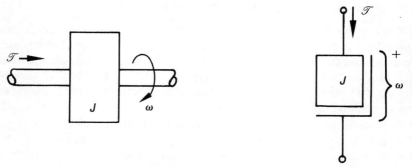

Fig. 1-33. Schematic representation of rotational inertia.

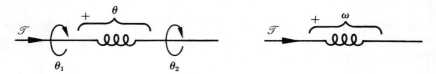

Fig. 1-34. Schematic representation of the rotational spring.

This relationship assumes a linear spring. For a nonlinear θ, $\mathcal{T}$ relation, much of the discussion in Section 1-13 would be applicable here, with the appropriate changes in the basic variables.

c. Rotational Viscous Damper. The rotational damper differs from the translational damper principally in the character of the motion. Otherwise, much of the discussion in Section 1-14 applies, in essence, in the present case also. The schematic diagrams for this element are given in Fig. 1-35. The element is described by the pair of relations, for linear viscous damping,

$$\mathcal{T} = D\omega, \qquad \omega = \frac{\mathcal{T}}{D} \tag{1-48}$$

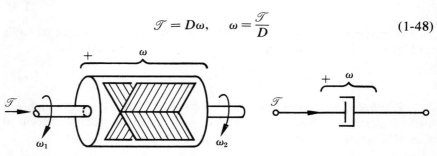

Fig. 1-35. Schematic representation of the rotational viscous damper.

d. Gear Train (Rotational Transformer). A simple gear train provides torque and angle variable transformations in much the same way that the rigid linkage bar provides force and velocity variable transformations in the linear system. The situation is illustrated in Fig. 1-36. If there are n_1

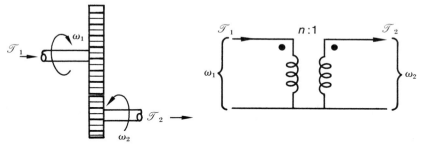

Fig. 1-36. Simple gear train and its schematic representation.

teeth on gear 1 and n_2 teeth on gear 2, then the gear train requires that,

$$\frac{\omega_1}{n_2} = \frac{\omega_2}{n_1},$$

or (1-49)

$$\frac{\omega_1}{\omega_2} = \frac{n_2}{n_1} = n. \quad \text{(gear ratio)}$$

Furthermore, if the gear train is lossless, then the total power input to one gear must equal the total power output from the second gear. This requires, for the specified reference conditions, that,

$$p = \omega_1 \mathcal{T}_1 = \omega_2 \mathcal{T}_2,$$

from which it follows that,

$$\frac{\mathcal{T}_1}{\mathcal{T}_2} = \frac{\omega_2}{\omega_1} = \frac{1}{n}. \quad (1\text{-}50)$$

The parallel to the electrical transformer and to the linear mechanical transformer (the bar linkage) is obvious.

e. Rotational Drivers. Rotational drivers exist which produce torques or which produce angular velocities. Here, as for the translational system, not all drivers maintain constant torque or constant rotational velocity independent of load. Certain devices exist which do meet the idealized conditions, for example, a synchronous motor will maintain constant

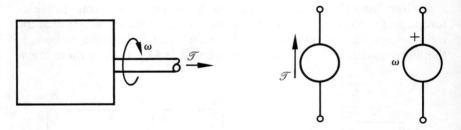

Fig. 1-37. Rotational driver and schematic representations.

angular speed for all values of torque. The schematic representation of rotational drivers is given in Fig. 1-37.

f. Rack and Pinion. The rack and pinion is a device for converting rotational motion into translational motion. The reverse is not always possible, and depends upon the manner of cutting the teeth and the general design of the assembly. In its simplest form, the rack and pinion has the form illustrated in Fig. 1-38.

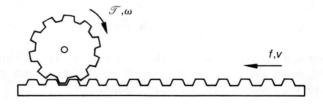

Fig. 1-38. Rack and pinion—rotational to translational transformer.

For N_p teeth on the pinion or rotary gear, and for n_r teeth/meter on the rack, a movement of 2π radians of the pinion, or N_p teeth, results in the passage of N_p teeth on the rack. In general, for a motion of θ radians of the pinion, there will be a linear displacement of the rack such that,

$$\frac{\theta}{2\pi} = \frac{n_r}{N_p} x,$$

or

$$\theta = 2\pi \frac{n_r}{N_p} x. \tag{1-51}$$

From this it follows that,

$$\omega = 2\pi \frac{n_r}{N_p} v. \qquad (1\text{-}52)$$

For a dissipationless system, conservation of energy requires that

$$\mathcal{T}\omega + fv = 0$$

from which we may write

$$\mathcal{T} = \frac{-1}{2\pi n_r/N_p} f. \qquad (1\text{-}53)$$

This expression is given schematic representation in Fig. 1-39. Here, as for the rigid linkage and the gear train, a rigid constraint is imposed by the transformer on the through and the across variables.

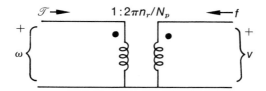

Fig. 1-39. Schematic representation of rack and pinion.

1-c FLUID ELEMENTS

1-18 LIQUID SYSTEMS

Liquid systems are composed of liquid-filled tanks or vessels that are connected by pipes, tubes, orifices and other flow-restricting devices. The analysis of such systems must proceed, of course, by using the fundamental laws that govern the flow of liquids. In our studies, it will be assumed that liquid tanks have a free surface of liquid. Connecting pipes are considered to be full of liquid. In liquid filled systems it is also assumed that fluid accelerations are small; this means that steady flow or nearly steady flow persists.

Two different types of flow are important, laminar flow and turbulent flow. Consider the flow of a liquid through a small pipe under low velocity conditions. Owing to its viscosity, the fluid will be stationary at the wall because of the friction with the wall surface, and the velocity will vary with distance from the wall surface. A "boundary layer" exists at the wall surface, and it is in this region that the fluid motion is affected by the solid boundary. Within the region of constant flow, the mean velocity varies with the distance from the wall. In the case of laminar flow, if one were to think of flow lines as depicting the motion of the fluid particles, these would be parallel to the wall surfaces. This makes it possible to specify the precise flow conditions at all points in the fluid field. Turbulent flow indicates a randomness in the flow, which makes it impossible to specify the precise conditions at all points in the fluid field.

Fluid flow has been studied in considerable detail. The work of Reynolds has yielded a criterion which allows an estimate of whether the flow is laminar or turbulent. From dimensional analysis considerations, an important factor in flow problems is the quantity,

$$Re = \frac{\rho v D}{\eta}, \tag{1-54}$$

where ρ is the density, v is the stream velocity, D is a characteristic dimension (equals the diameter, if the tube or duct is cylindrical) and η is the absolute viscosity. This quantity is known as the Reynolds number. Experiment shows that if the Reynolds number exceeds about 4000 the flow is turbulent. Laminar flow exists when the Reynolds number is less than about 2000. The prediction is not precise between these two values of Reynolds number, the transition depending on the surface roughness, the shape of the surface, dimensions and pressure gradients.

1-19 LIQUID RESISTANCE

For laminar flow in general, the flow in circular tubes or pipes is found from the Pouisuille-Hagen law,

$$h_1 - h_2 = \frac{128\eta l}{\pi D^4} Q \quad \text{(meters)} \tag{1-55}$$

where h = head, m,
  η = absolute viscosity, Newtons-sec/m^2,
  l = length of tube or pipe, m,
  D = inside diameter of pipe, m,
  Q = liquid flow rate, m^3/sec.

Attention is called to the analogy of this laminar flow law with Ohm's law for the electric circuit because the flow (current) is directly proportional to the head (voltage).

The laminar resistance is found from,

$$R = \frac{dh}{dQ} = \frac{128\eta l}{\pi D^4}. \quad (\text{sec}/\text{m}^2) \tag{1-56}$$

Note that laminar resistance is constant, and is directly analogous to electrical resistance. However, laminar flow is not often encountered in industrial practice.

In turbulent flow through pipes, orifices, valves, and other flow restricting devices in general, the flow is found from Bernoulli's law, which may be written in the form,

$$Q = KA\sqrt{2g(h_1 - h_2)} \quad (\text{m}^3/\text{sec}) \tag{1-57}$$

where $K =$ a flow coefficient (usually about 0.6),
$A =$ area of restriction, m^2,
$g =$ acceleration due to gravity, m/sec^2,
$h =$ head of liquid, m.

Unlike Eq. (1-55) for laminar flow, this expression shows a square law flow relation between Q and h. The turbulent resistance is given by the relation,

$$R = \frac{dh}{dQ} = \frac{Q}{gK^2A^2} = \frac{2(h_1 - h_2)}{Q} \quad (\text{sec}/\text{m}^2) \tag{1-58}$$

if the flow coefficient K is considered constant. Clearly, the turbulent resistance is not constant, and depends upon the flow rate and the differential head existing at any time. It is consequently necessary to define a turbulent resistance at a particular value of flow and head, and to employ this value of resistance over a narrow operating range. A new value for R is required for every new operating range.

The schematic representation of fluid resistance is given in Fig. 1-40, with R constant for laminar flow, and nonlinear for turbulent flow.

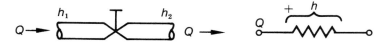

Fig. 1-40. Fluid (liquid) resistance.

1-20 LIQUID CAPACITANCE, INDUCTANCE, AND SOURCES

The concept of liquid capacitance is a simple one to understand; it is simply a measure of the capacity of a tank to store the liquid. Refer to

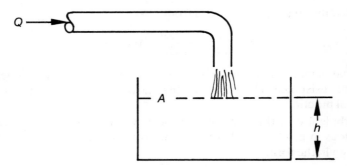

Fig. 1-41. Fluid (liquid) capacitance.

Fig. 1-41 which illustrates the situation. Conservation of mass requires that the change in volume of the liquid being stored in a tank is specified by the fluid flow rate into the tank. Thus,

$$Q = \frac{dV}{dt} = A\frac{dh}{dt} = \frac{dV}{dh}\frac{dh}{dt} \quad (\text{m}^3/\text{sec}) \tag{1-59}$$

where V denotes the volume, and A the sectional area of the tank. Now, in analogy with Eq. (1-7) for electrical capacitance, we define C in terms of flow and time rate of change of head. Thus,

$$C = \frac{dV}{dh}. \quad (\text{m}^2) \tag{1-60}$$

The liquid capacitance of a tank is equal to the cross-sectional area of the tank taken at the liquid surface. If the tank has a constant cross-sectional area, the liquid capacitance is a constant for any head. In the more general case, the liquid capacitance will not be a constant.

To examine the inertance of the flow in a pipe, and which can be denoted as the fluid inductance, consider a pipe of length l containing a fluid. The force necessary to accelerate the mass in the pipe $(\rho A l)$ is given by,

$$f = A(h_2 - h_1) = \rho A l \frac{dv}{dt}.$$

But the flow rate $Q = Av$, so that,

$$h_2 - h_1 = \frac{\rho l}{A}\frac{dQ}{dt} = L\frac{dQ}{dt}, \tag{1-61}$$

where the quantity L is the fluid inductance.

A fluid source must comprise two parts—a reservoir (sump) of the liquid and a pump for causing the liquid to flow under some sort of pressure. A large variety of fluid pumps are available. We refer specifically to gear pumps and centrifugal pumps. The design and operation of the gear pumps is such that liquid is provided at a relatively constant flow rate against a wide range of heads. Essentially, therefore, such a pump is a constant flow device, and is analogous to the current source in electricity. The centrifugal pump is one for which the pressure head remains substantially constant over wide ranges of flow, and is analogous to the voltage source. In essence, both through-type and across-type fluid sources are available.

1-21 GAS SYSTEMS

Gas systems that consist of pressure vessels or chambers and various connecting pipes, valves, etc. may be analyzed by using the basic laws for the flow of compressible gases. However, for systems in which the pressure differentials are less than about 5 percent of the static pressure, the compressibility is not usually important, and the systems may be treated as liquid systems. Ventilating and other air-transport systems where changes of air density are small may be treated as incompressible flow systems. Generally, however, the pressure differentials are substantial, and compressible flow considerations are necessary.

The system elements for a gas system are illustrated in Fig. 1-42. To describe the features of these elements, the flow laws for both laminar and turbulent conditions must be used. For turbulent flow through pipes, orifices and valves, the steady-flow energy equation (the first law of thermodynamics) for adiabatic flow of ideal gases is,

$$w = KAY\sqrt{2g(p_1 - p_2)/\rho} \quad \text{(kg/sec)} \tag{1-62}$$

where $w =$ gas flow rate, kg/sec,
$A =$ area of restriction,
$Y =$ rational expansion factor $= \rho\sqrt{V\dfrac{\gamma}{\gamma - 1}}$,
$\gamma =$ specific heat ratio for gases,
$\rho =$ gas density, kg/m^3,
$K =$ a flow coefficient,
$p =$ pressure, kg/m^2.

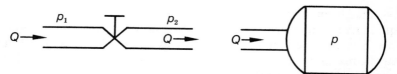

Fig. 1-42. Fluid (gas) resistance and capacitance.

The turbulent gas flow resistance is defined as,

$$R = \frac{dp}{dw}. \quad (\text{sec}/\text{m}^2) \qquad (1\text{-}63)$$

It is not possible to easily calculate R from Eq. (1-63) because the expansion factor Y depends on the pressure. In this case it is easier to determine resistance from a plot of pressure against flow for any particular device. For laminar gas flow through tubes and pipes, Eq. (1-56) can be used.

To discuss the capacitance of a pressure vessel, we first write a flow law for the pressure vessel of Fig. 1-42, thus

$$w = \rho \frac{dV}{dt}. \quad (\text{kg}/\text{sec}) \qquad (1\text{-}64)$$

The gas capacitance is defined as the ratio of change of weight of gas in the vessel to the change in pressure,

$$C = \rho \frac{dV}{dp} = V \frac{d\rho}{dp}. \quad (\text{m}^2) \qquad (1\text{-}65)$$

By combining this expression with Eq. (1-64) we have that

$$w = C \frac{dp}{dt}. \quad (\text{kg}/\text{sec}) \qquad (1\text{-}66)$$

The capacitance expression in Eq. (1-65) must be calculated from thermodynamic relations because the gas expands from a region of high pressure into the vessel at lower pressure, or perhaps expands from the vessel into a region of lower pressure. It is desirable to assume a polytropic expansion process, since the expansion will probably follow a path between an isothermal and an adiabatic process. Thus it is assumed that,

$$\frac{p}{\rho^n} = \text{const.} \qquad (1\text{-}67)$$

where the polytropic exponent n will lie between $n = 1.0$ for isothermal expansion and 1.41, the ratio of specific heats for adiabatic expansion.

Now, use the ideal gas law,

$$pV = n\mathcal{R}\theta \tag{1-68}$$

where $\mathcal{R}$ is the universal gas constant and θ is the temperature, deg K. By combining Eqs. (1-67) and (1-68) with Eq. (1-65) we find that, for $n = 1$

$$C = \frac{V}{n\mathcal{R}\theta}. \quad (\text{m}^2) \tag{1-69}$$

Numerous tests show that the exponent n ranges from 1.0 to 1.2 for uninsulated metal vessels at common pressure and temperature. Thus for constant volume and temperature, the gas capacitance of a vessel is constant; hence it is analogous to electric capacitance.

As in the case of liquid systems, gas will sustain and transmit acoustic vibrations; acoustic pipes and acoustic resonators are well known. Such gas phenomenons are possible, as with liquids, because gas possesses both inertia and it is elastic. Refer to Fig. 1-43, and focus on the mass of

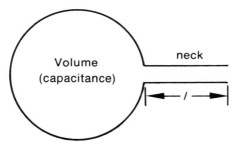

Fig. 1-43. An acoustic resonator.

fluid that is contained in the neck. If the linear dimensions of the neck are small compared with the wavelength of the vibration, it has been found that the inertance is given by,

$$\text{inertance} = \rho l. \tag{1-70}$$

1-d THERMAL ELEMENTS

1-22 THERMAL SYSTEMS

Thermal systems are those that involve the transfer of heat from one point to another or from one substance to another. We shall find that such systems are also characterized by resistance and capacitance. However, there is no thermal element that is analogous to inductance, since such an element would permit actions that would be contrary to the second law of thermodynamics. In our discussion, we shall assume that substances characterized by resistance to heat flow have negligible capacitance or storage of heat, e.g. a thin film of air, wood, cork; also that substances characterized by heat storage have negligible resistance to heat flow, e.g., a block of copper or other heavy metal. The validity of these approximations may often be poor, and we can often endow a material with both properties.

There are three different flow laws for heat, and these are appropriate to three different heat transfer processes. The names of different persons or groups are associated with these heat transfer processes; Fourier with heat conduction, Newton with heat convection, and Stefan-Boltzmann with heat radiation.

For the conduction of heat through a specific conductor, the heat flow is described by the Fourier law. This law relates the heat flow to the gradient of the temperature (*see* Fig. 1-44a),

$$Q = KA \frac{(\theta_1 - \theta_2)}{l} \quad \text{(Joules/sec)} \tag{1-71}$$

where Q = heat flow, Joules/sec,
K = thermal conductivity, Joules/m-sec-deg,
A = area normal to heat flow, m²,
l = length of conductor, m,
θ = temperature, deg K.

The direction of the heat flow is from the region of high temperature to the region of low temperature. The form of this law is identical with Ohm's law, Eq. (1-22), which relates the current in an electric circuit and the difference of potential. The thermal resistance (conduction) is therefore,

$$R = \frac{d\theta}{dQ} = \frac{l}{kA} \quad \text{(deg-sec/Joule)} \tag{1-72}$$

By comparing this expression with Eq. (1-24) for electrical resistance, it is seen that the form is the same, the difference lying in the presence of thermal conductivity in one case and electrical conductivity in the other. It is true, in fact, that pure metals that are good thermal conductors are also good electrical conductors.

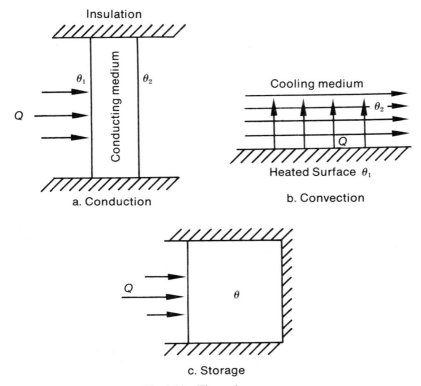

Fig. 1-44. Thermal processes.

A fluid or gas is involved in convection heat transfer, and convection is the process by which heat transfers across a heated surface into (or from) the fluid (*see* Fig. 1-44b). For convection heat transfer, by the Newton law of cooling,

$$Q = hA(\theta_1 - \theta_2) \quad \text{(Joules/sec)} \quad (1\text{-}73)$$

where h is the convection coefficient (Joules/m²-sec-deg). It is expected that h will depend on the fluid and upon the state of the fluid flow, but in a given situation, h is nearly constant. We define a thermal resistance for convection,

$$R = \frac{d\theta}{dQ} = \frac{1}{hA} \quad \text{(deg-sec/Joule)} \quad (1\text{-}74)$$

48 Modeling of System Elements

Since this resistance factor is related to the flow of heat to a fluid boundary layer, no factor of length appears.

Radiant heat transfer is a thermal process that is quite different from conduction and convection, both of which are molecular processes. In radiant heat transfer thermal energy flows from a region of high temperature to one of lower temperature by an electromagnetic radiation process. The heat flow, say in a vacuum, under radiant conditions is given by the Stefan-Boltzmann law for surface radiation,

$$Q = \sigma A \epsilon (\theta_1^4 - \theta_2^4) \quad \text{(Joules/sec)} \quad (1\text{-}75)$$

where σ = Stefan-Boltzmann constant = 5.672×10^{-8} Joules/m²-sec-deg⁴,
ϵ = emissivity of the surface (ranges between 0 and 1).

This is a fourth power flow law between the through variable and the across variable, as against the square-root relation for turbulent flow, and the linear relation for electric current.

The thermal resistance for radiation is written,

$$R = \frac{d\theta}{dQ} = \frac{1}{4\sigma \epsilon A \theta_a^3} \quad \text{(deg-sec/Joule)} \quad (1\text{-}76)$$

where θ_a = average of radiator and receiver temperatures. Owing to the inverse θ_a^3 in the denominator of this formula, a given R can be used only over a fairly small temperature range. The value of R so calculated when the source and receiver temperatures differ by a factor of two is only about 10 percent lower than that using a more elaborate expression based directly on Eq. (1-75).

To discuss thermal capacitance, refer to Fig. 1-44c. By adding heat to a system, its internal energy is increased, according to the relation

$$Q = C \frac{d\theta}{dt} \quad \text{(Joules/sec)} \quad (1\text{-}77)$$

This expresses the heat added (or lost) with the consequent increase (or decrease) of temperature. The thermal capacitance C (Joules/deg) is written

$$C = W C_p \quad \text{(Joules/deg)} \quad (1\text{-}78)$$

where W = weight of the block, kg,
C_p = specific heat at constant pressure, Joules/deg-kg.

Thermal capacitance which relates the heat added with the change in temperature is precisely analogous to electrical capacitance which relates the charge added with the change in voltage.

We have already noted that thermal systems do not always act as isolated thermal resistors or capacitors. Fire brick is a material that has a high resistance to the flow of heat, but because of its appreciable thermal specific heat, it possesses a substantial thermal capacitance. Moreover, because these properties are distributed throughout the material, the foregoing considerations do not apply exactly. But materials with distributed characteristics can be approximated by considering the material to be made up of slabs, each of which can be described by both resistance and capacitance, and considering a series collection of such quantities. More will be said about this approximating process later. But suffice it to say that there are many physical problems of importance in which thermal resistance and thermal capacitance are sufficiently isolated to allow some calculations of significance to be made.

Thermal sources ordinarily produce heat at a constant rate, and depend on the rate at which the excitation (fuel, electrical power, friction, etc.) is applied. Hence thermal sources are analogous to the electrical current source, or the constant flow liquid pump.

1-e N-PORT DEVICES

1-23 TRANSDUCERS

In addition to the class of two-port devices which have been classified as transformers, we wish to consider another very important class of two-port devices. These are known as transducers or energy converters, and usually have the through and across variables at the two-ports in different physical forms. An electromechanical transducer such as a radio loudspeaker will have one electrical port and one mechanical port (plus, in fact, an acoustic port); a thermoelectric transducer will have one thermal port and one electrical port; an accelerometer will produce an electrical output corresponding to some function of the mechanical motion.

As one example, we consider an accelerometer, somewhat in the form illustrated in Fig. 1-45. A coupling field is established between the electrical and the mechanical circuits. When the case of the device, and so also the electrical coil which is rigidly attached thereto, moves, owing to its attachment to the mechanical system whose acceleration is to be

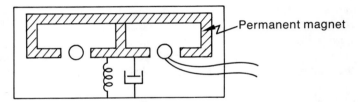

Fig. 1-45. The basic features of a mechano-electric accelerometer.

measured, there is a relative motion of the magnet with respect to the winding. As a result, a voltage is induced in the coil which depends on the relative motion of the magnet with respect to the coil. By the proper adjustment of the spring constant, the voltage is proportional to the acceleration, over wide ranges of acceleration.

A second example is a telephone transmitter, an electromechanical device which receives acoustic energy which is immediately converted into the mechanical motion of a metallic diaphragm. In one type of telephone microphone the mechanical motion of the diaphragm alters the magnetic properties of the electromagnetic circuit, thereby producing an electrical output which is proportional to the acoustic-mechanical input.

Transducers that operate over a linear range can be represented in two-port form as shown in Fig. 1-46 (1)*. Such a two-port device can be represented analytically by equations of the form,

$$\tau_1 = a_{11}\alpha_1 + a_{12}\alpha_2$$
$$\tau_2 = a_{21}\alpha_1 + a_{22}\alpha_2 \qquad (1\text{-}79)$$

which may be written in matrix form,

$$\begin{bmatrix} \tau_1 \\ \tau_2 \end{bmatrix} = \begin{bmatrix} a_{11} & a_{12} \\ a_{21} & a_{22} \end{bmatrix} \begin{bmatrix} \alpha_1 \\ \alpha_2 \end{bmatrix}. \qquad (1\text{-}80)$$

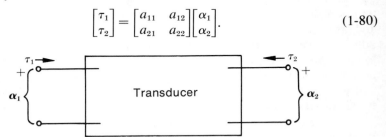

Fig. 1-46. Two-port representation of a transducer.

*References will be denoted in parentheses. These are listed on pp. 509 to 511.

To determine analytically the a-factors in these equations is a complicated problem in most cases, owing to the energy conversion principles that are involved in such devices. Moreover, the theoretical demands are beyond those expected of the present readers, and so this aspect of the transducer will not be pursued here. However, certain features of the transducer are possible from a study of Eqs. (1-79). Thus, it follows directly from Eqs. (1-79) that the a-factors may be specified in terms of measurements made at the ports. Specifically, it follows that,

$$a_{11} = \left.\frac{\tau_1}{\alpha_1}\right|_{\alpha_2=0} = \text{Input terminal driving point admittance when the output variable } \alpha_2 = 0 \text{ (this requires shorting or blocking the output port).}$$

$$a_{12} = \left.\frac{\tau_1}{\alpha_2}\right|_{\alpha_1=0} = \text{Transfer admittance between output } \alpha_2 \text{ and the input } \tau_1 \text{ when } \alpha_1 = 0.$$

$$a_{21} = \left.\frac{\tau_2}{\alpha_1}\right|_{\alpha_2=0} = \text{Transfer admittance between output } \alpha_1 \text{ and input } \tau_2 \text{ when } \alpha_2 = 0. \quad (1\text{-}81)$$

$$a_{22} = \left.\frac{\tau_2}{\alpha_2}\right|_{\alpha_1=0} = \text{Output terminal driving point admittance when the input variable } \alpha_1 = 0.$$

Because α_1 or α_2 are set at zero in these definitions, they are called short-circuit parameters. A physical understanding of these expressions must be deferred until later in our study.

It is here noted without proof that certain classes of transducers exist for which $a_{21} = a_{12}$; these are known as *reciprocal* types. Other classes of transducers exist, for which $a_{21} \neq a_{12}$; these are *nonreciprocal* types. For our present purposes, it is sufficient to know that transducers can be represented by the set of linear equations, Eq. (1-79), at least to a first approximation, over a reasonable range of operation. For other than incremental motion, most transducers are nonlinear devices.

1-24 ACTIVE NETWORKS

Electronic amplifiers constitute a very important class of two-port networks. These are often referred to as *active* networks because they involve electronic devices which serve to convert power from the d-c power sources into a-c power in the output, in accordance with signals that are applied to the input circuit. This conversion process is modeled by means of equivalent sources, the strength of which depends on the input signals. In so far as a simple representation is concerned, Fig. 1-46 will apply for the active network, where, in general, both the input and the output variables are electrical.

For transistor driven circuits, the convenient input-output variables are i_1 and v_2. Thus in place of Eq. (1-79) one uses a description of the following form,

$$v_1 = h_{11}i_1 + h_{12}v_2,$$
$$i_2 = h_{21}i_1 + h_{22}v_2. \tag{1-82}$$

These h-coefficients are called hybrid parameters because, in their description, both input and output variables are involved. Specifically,

$$h_{11} = \left.\frac{v_1}{i_1}\right|_{v_2=0} = \text{Driving point impedance when the output is short-circuited.}$$

$$h_{12} = \left.\frac{v_1}{v_2}\right|_{i_1=0} = \text{Inverse of voltage gain ratio when input is open-circuited.}$$

$$h_{21} = \left.\frac{i_2}{i_1}\right|_{v_2=0} = \text{Current gain ratio when output is short-circuited.}$$

$$h_{22} = \left.\frac{i_2}{v_2}\right|_{i_1=0} = \text{Output admittance when the input is open-circuited.}$$

(1-83)

Equations (1-82) can be given network representation, as shown in Fig. 1-47. The transistor is an active device and reciprocity does not exist, thus $h_{12} \neq h_{21}$. Note in Fig. 1-47 that the sources that are included in the model depend on certain input and output variables, and are known, therefore, as dependent sources. This matter will be discussed in greater detail in Chapter 10.

A study of electronic circuits is a specialized one that involves special configuration of driver and network elements. Further, modeling of the

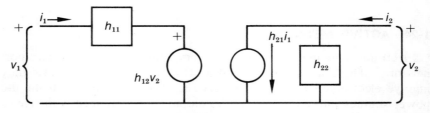

Fig. 1-47. An equivalent active network for Eq. (1-57).

active device (tube, transistor, field effect transistor, etc.) is particularly important since the operation of an electronic circuit is not necessarily in the linear mode. When the linear mode of operation is employed, Eqs. (1-82) will apply, but not otherwise. We note, therefore, that much of our study is fundamental to a study of electronic circuits, but it does not provide all of the techniques of analysis suitable to electronic circuit study.

1-25 MODELING OF COMPLICATED SITUATIONS

We have already noted that if we wish to study a problem analytically, we must mathematically model the important features of the problem. For the case of simple physical elements, basic physical laws provide the necessary effect-cause interrelationships, as already discussed above in some detail. In many physical problems the modeling process is a less accurate science, owing to the complexity of the devices in the process. For example, the model of a chemical distillation column is quite complicated, being both a multiport and nonlinear device. Sometimes complicated systems can be modeled from physical principles; often important port relationships are obtained experimentally. However the important port relationships are determined, there will be a general representation of the form shown in Fig. 1-48. The mathematical description will be of the general form

$$\begin{bmatrix} \text{output} \\ \text{variables} \\ 1,\ldots,m \end{bmatrix} = \begin{bmatrix} \text{system} \\ \text{description} \end{bmatrix} \cdot \begin{bmatrix} \text{input} \\ \text{variables} \\ 1,\ldots,n \end{bmatrix}. \quad (1\text{-}84)$$

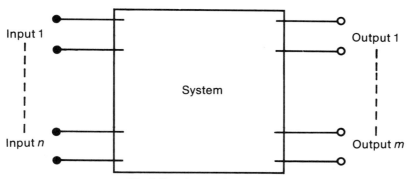

Fig. 1-48. A general N-port system.

54   Modeling of System Elements

When the variables are stochastic ones, experimental methods may be used to provide an applicable mathematical model. For example, suppose that a mathematical model of a simple traffic interchange is required, say the junction of two corridors in a school. The situation can be represented as in Fig. 1-49, where each arrow denotes the flow direction at each port (point of entry to the junction).

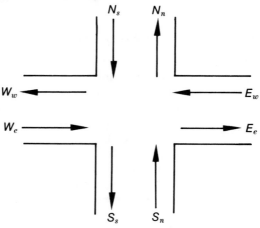

Fig. 1-49.   The junction of two corridors.

The general form of the interrelationship to be expected among the flow rates (number of students per unit time) and the several directions of travel can be written driectly. It would be of the form

$$\begin{bmatrix} F_{Nn} \\ F_{Ns} \\ F_{Ee} \\ F_{Ew} \\ F_{Ss} \\ F_{Sn} \\ F_{Ws} \\ F_{We} \end{bmatrix} = \begin{bmatrix} 0 & a_{12} & a_{13} & \cdots & a_{18} \\ a_{21} & 0 & a_{23} & \cdots & a_{28} \\ \cdot & \cdot & \cdot & \cdots & \cdot \\ \cdot & \cdot & \cdot & \cdots & \cdot \\ \cdot & \cdot & \cdot & \cdots & \cdot \\ \cdot & \cdot & \cdot & \cdots & \cdot \\ a_{71} & a_{72} & \cdot & \cdots & a_{78} \\ a_{81} & a_{82} & \cdot & \cdots & 0 \end{bmatrix} \begin{bmatrix} N_n \\ N_s \\ E_e \\ E_w \\ S_s \\ S_n \\ W_w \\ W_e \end{bmatrix} \quad (1\text{-}85)$$

where the coefficients a_{ij} would be rather wildly fluctuating quantities in time. There are no fixed physical laws concerning such flow rates, but an experimental procedure may be adopted to determine these. This could be done by having a number of persons stand at each junction keeping a count of the number of students per unit time in each direction during the course of a typical day. The measured flow-time chart could be approximated by mathematical functions to yield the coefficients $a_{ij}(t)$. Of course, not all coefficients in the coefficient matrix are independent; they can be determined from the 8 measurements that must be made.

1-26 SUMMARY

Many results of the discussion of this chapter are summarized in Table 1-1. Not only does this table give the through-across relations of the system elements, but also the analogies that exist among the elements are placed in clear evidence in this table.

TABLE 1-1 Linear System Elements.

System element	Graphical representation	Independent variable Through	Across
Electrical capacitor C		$v = \dfrac{1}{C}\int i\,dt$	$i = C\dfrac{dv}{dt}$
inductor L		$v = L\dfrac{di}{dt}$	$i = \dfrac{1}{L}\int v\,dt$
resistor R		$v = Ri$	$i = \dfrac{v}{R}$
Mechanical (translational) mass M		$v = \dfrac{1}{M}\int f\,dt$	$f = M\dfrac{dv}{dt}$
spring K		$v = \dfrac{1}{K}\dfrac{df}{dt}$	$f = K\int v\,dt$
damper D		$v = \dfrac{f}{D}$	$f = Dv$
Mechanical (rotational) inertia J		$\omega = \dfrac{1}{J}\int \mathcal{T}\,dt$	$\mathcal{T} = J\dfrac{d\omega}{dt}$
spring K		$\omega = \dfrac{1}{K}\dfrac{d\mathcal{T}}{dt}$	$\mathcal{T} = K\int \omega\,dt$
damper D		$\omega = \dfrac{\mathcal{T}}{D}$	$\mathcal{T} = D\omega$

TABLE 1-1 *continued*

System element	Graphical representation	Independent variable Through	Across
Fluid (liquid) capacitance C		$h = \dfrac{1}{C}\int Q\,dt$	$Q = C\dfrac{dh}{dt}$
inertance L		$h = L\dfrac{dQ}{dt}$	$Q = \dfrac{1}{L}\int h\,dt$
resistance		$h = RQ$	$Q = \dfrac{h}{R}$
Fluid (gas) capacitance C		$p = \dfrac{1}{C}\int w\,dt$	$w = C\dfrac{dp}{dt}$
inertance L		$p = L\dfrac{dw}{dt}$	$w = \dfrac{1}{L}\int p\,dt$
resistance		$p = Rw$	$w = \dfrac{p}{R}$
Thermal capacitance C			$q = C\dfrac{d\theta}{dt}$
resistance R			$q = \dfrac{\theta}{R}$

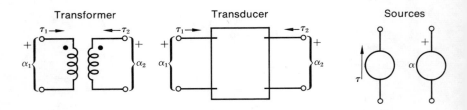

Transformer   Transducer   Sources

PROBLEMS

1-1. The voltage-charge characteristic of a capacitor is given graphically in the accompanying sketch.

Problems   57

a. Determine the capacitance of the capacitor.
b. How much energy is stored in the electric field of the capacitor, if it is charged to 10^3 volts?

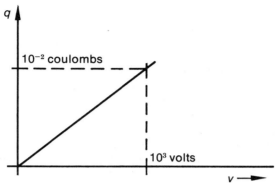

1-2. A capacitor is made up of 34 sheets of aluminum foil separated by a 0.003 inch dielectric film having a relative permittivity of 3.8. Alternate sheets of foil are connected together to form two intermeshed electrically insulated stacks. The area per sheet is 1.6 sq. in. Determine the capacitance of this capacitor.

1-3. The B-H curve of a given ferromagnetic material, neglecting hysteresis, is illustrated. If this curve is to be represented by piecewise linear approximations between O-H_1 and H_1-H_2, write the appropriate description for B as a function of H over the full range of operation a-a'.

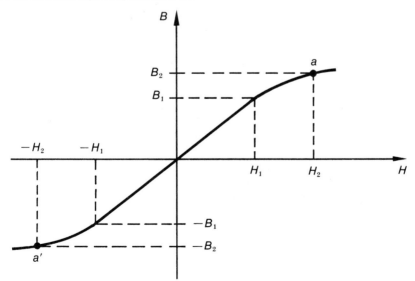

58 Modeling of System Elements

1-4. Two windings are closely wound on a toroid that has a mean circumference of 10 inches with a coil radius of 1 inch. There are 1,000 turns per layer.

a. Calculate the inductance of each winding when the core is of nonmagnetic material.

b. If the relative permeability of the core is $\mu_r = 1,400$, what is the inductance?

c. Find the mutual inductance between the two windings under the conditions in b.

d. What is the total inductance of the toroidal coil, if the windings are connected series-aiding; series opposing?

1-5. The field structure of a 15 horsepower d-c motor has an inductance of 2.2 henrys and carries a current of 6.4 amp.

a. Determine the flux linkages of this field.

b. How much energy is stored in the magnetic field?

c. What must be the waveshape of the applied voltage if the current is to increase linearly with time from 0 to 6.4 amp. in 0.2 sec?

1-6. The $\psi - i$ characteristic of an inductor is given. If the steady state current is $i = 1.5 \sin \omega t$, determine graphically the waveform for ψ. From this, determine the approximate waveform of the applied voltage.

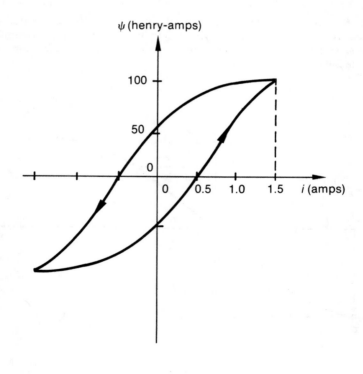

1-7. Repeat Problem 1-6 when the $\psi - i$ characteristic is approximately that shown in the accompanying sketch.

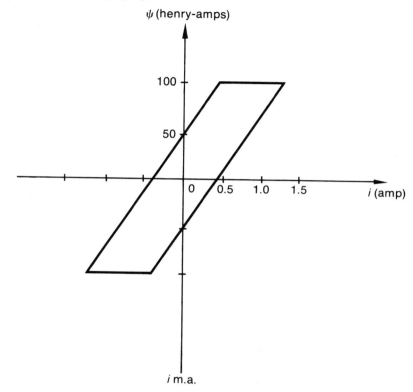

1-8. Consider two mutually coupled coils, each with an inductance of 1 mH and with $k = 0.2$. One of the coils is mounted on pivots so that it can move inside of the other coil. Sketch the variation of M with angle of rotation, as the angle varies from 0 to 360 deg.

1-9. The approximate characteristic of a diode is shown in the accompanying sketch. Write expressions for the terminal characteristics of this device.

1-10. A piecewise linear approximation to a tunnel diode $i - v$ characteristic is shown. Specify this approximate characteristic over the range from 0 to 0.35 volts.

1-11. a. A triangular voltage wave is applied to a diode which has the static characteristic shown. Determine graphically the waveform of the current.
  b. Repeat a. for the case when the $i - v$ characteristic is approximated by the two section piecewise approximation shown.

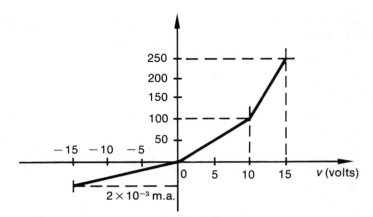

Problem 1-9

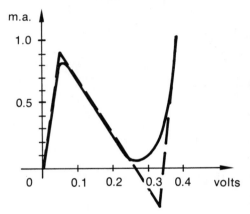

Problem 1-10

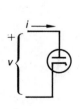 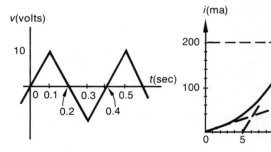

Problem 1-11

1-12. A nonlinear spring is described by the f, x relation,

$$f = 2x - \tfrac{1}{2}x^2.$$

If a constant $f(t) = 1.8$ is applied, determine approximately by graphical means, the displacement $x(t)$.

1-13. The damping factor of a linear bilateral damper is $D = 20$ Newton-sec/m. A sinusoidal velocity with a peak amplitude of 20 m/sec and with a period of 1 sec is applied. Calculate the average energy dissipated during one complete period of the applied velocity wave.

1-14. Calculate the laminar resistance of 100 feet of $\tfrac{3}{16}$ ID tubing for air and water at 68 degF. η, cp (poise gm/cm-sec $\times$ 6.72 $\times$ 10^{-2} lb/ft-sec) for air: 1.813×10^{-2}; for water, 1.002.

1-15. Water flow through a triangular weir is given by the equation,

$$Q = C_v \sqrt{2gh^5}.$$

Find an expression for the effective resistance of the weir.

1-16. The flow versus differential head for a 1-in. orifice is given in the figure. The orifice coefficient is assumed to be constant at 0.6 except for very low flow where the Reynolds number is small.
   a. Show that the resistance at any head is given by $R = 2\sqrt{h/2g}/KA \,(\text{sec/ft}^2)$.
   b. Determine R at the following values of h: 16, 20, 24.
   c. Relative to R at $h = 20$, what is the percentage variation in R at the 16 and 24 ft. values.

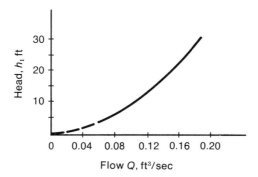

1-17. Calculate the capacitance as a function of head of a spherical liquid storage tank of radius a.

1-18. The output flow from a vessel of unknown shape is specified by Eq. (1-57). Determine the shape of the vessel in order that;
   a. The change in head should be constant in time.
   b. The rate of change of head is proportional to the head.

62   Modeling of System Elements

1-19. Calculate the gas capacitance of a 20-gal pressure vessel containing air at 200 degF.

1-20. One side of a steel plate is maintained at 350 degF, the other side being at 175 deg. The important data are: thickness $\frac{3}{8}''$, area $A = 1$ ft^2 thermal conductivity $= 29$ Btu/ft-deg-hr.
  a. Calculate the thermal resistance for conduction.
  b. Calculate the heat flow through A.

1-21. The walls of an electrically heated furnace are maintained at 1100°C. Contained in the furnace is a large steel casting at 875°C. The surface area is 0.1 m^2. Calculate the thermal resistance for radiation. Assume an emissivity of 0.9.

1-22. Determine the thermal capacitance of the following:
  a. a 50 gal. tank of water. $C_p = 4182$ Joules/kg-deg.
  b. a block of copper 2″ on each side. $C_p = 385$ Joules/kg-deg; specific gravity = 8.9.
  c. 1 kg. of fire brick. $C_p = 190$ Joules/kg-deg.

Part II   Interconnected Systems

2
Interconnected Systems: Equilibrium Formulations

The foregoing chapter was concerned with the modeling process of elements that comprise the systems to be examined in our subsequent studies. For linear one-port elements, it was found that a simple terminal relationship exists between the through and the across variables that analytically describe each element. This chapter will be concerned with certain features of the interconnection of a number of elements into a system.

2-1 INTERCONNECTED ELEMENTS

We begin by considering two connected systems, one mechanical and one electrical, which are illustrated in Fig. 2-1. Four separate aspects of such systems can be studied, (a) the terminal properties of the elements in the systems (these were discussed at some length in Chapter 1), (b) the geometrical or topological constraints that are imposed by the interconnection, (c) a description of the dynamics of the system, and (d) the response of the system when specified excitations are applied at one or more places in the system. This chapter will be devoted to (c). Later chapters will be concerned with (b) as it relates to (c), and with (d).

In establishing the equilibrium or dynamical equations of an interconnected system, and these include a mathematical description of the system elements and the constraints imposed by their interconnection, a number of rather different, though fundamentally equivalent forms, may be used. The system may be described in terms of one or a set of integro-differential equations; the equilibrium description may be given in state

66   Interconnected Systems: Equilibrium Formulations

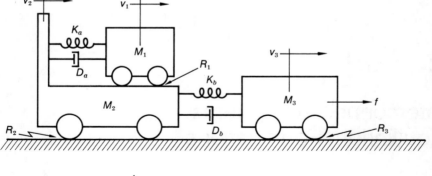

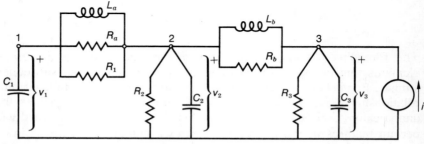

Fig. 2-1.   A mechanical and an electrical system for study.

variable form; the description may be given graphically by means of a signal flow graph or computer diagram. Other representations also exist, but they all lead, in the final analysis, to the same equilibrium representation of the system. We shall not discuss all available methods for interconnected system description, but we shall devote considerable attention to the three methods mentioned above, namely, a Kirchhoff description that leads to integro differential equation formulations, a state variable formulation, and the use of the signal flow graph in analysis.

2-a   KIRCHHOFF FORMULATION

We shall find that a characteristic feature of a Kirchhoff formulation of the dynamic equations of an interconnected system is that, for linear systems, the description appears in terms of sets of linear integrodifferential equations that relate the input and the output variables.

2-2 OPERATIONAL NOTATION

We shall find it convenient in our subsequent studies to introduce the symbol p to denote the time derivative operator d/dt. An equivalent notation that is often used is the so-called dot notation, with one dot over a variable to denote the first time derivative, two dots to denote the second time derivative, etc. With the p notation we would write,

$$u = \frac{dv}{dt} \triangleq pv = \dot{v}, \tag{2-1}$$

where the symbol $\triangleq$ means equal, by definition. From purely formal considerations, it follows that,

$$v = \int u \, dt.$$

But by comparison with Eq. (2-1) we see that this expression may be written,

$$v = \frac{u}{p}. \tag{2-2}$$

Thus, in summary, by definition, we write the functional pair involving p,

$$\text{differentiation} \quad p = \frac{d}{dt},$$
$$\text{integration} \quad \frac{1}{p} = \int dt. \tag{2-3}$$

Using these definitions, the electrical elements can be described by the following expressions, depending on whether i or v is the dependent variable.

TABLE 2-1. Properties of Electrical Elements

Element	independent variable	
	voltage v	current i
Capacitor	$i = Cpv$	$v = \frac{1}{Cp}i$
Resistor	$i = \frac{v}{R}$	$v = Ri$
Inductor	$i = \frac{1}{Lp}v$	$v = Lpi$

A similar description is possible for all the elements that are contained in Table 1-1.

It is general practice, especially in the literature of electrical engineering, to introduce the symbol $z(p)$ to describe the functional v, i relationship, and $y(p)$ to describe the functional i, v relationship. In particular, the one-port or two-terminal element, as illustrated in Fig. 2-2 is specified in general by the relation,

$$v = z(p)i, \qquad (2\text{-}4)$$

or by,

$$i = y(p)v. \qquad (2\text{-}5)$$

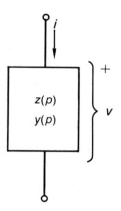

Fig. 2-2.   Block representation of the one-port device.

Clearly, of course, for a given element,

$$y(p) = \frac{1}{z(p)}. \qquad (2\text{-}6)$$

The operator $z(p)$ is called an *impedance* operator, and $y(p)$ is called an *admittance* operator.

We shall extend the meanings of the functions $z(p)$ and $y(p)$ when we consider more complicated arrays of elements. In fact, the general forms for $z(p)$ and $y(p)$ play very important roles in the study of interconnected systems, both for analysis, with which we shall be largely interested, and in the general study of system synthesis.

2-3 THROUGH-ACROSS EQUILIBRIUM LAWS

Two equilibrium laws exist which provide the basis for writing the input-output equilibrium equations. One of these is expressed in terms of the through variables, the second being expressed in terms of the across variables. The first of these, when applied to systems of interconnected electrical elements, is the Kirchhoff current law, KCL. When applied to systems of interconnected mechanical elements, this is D'Alembert's principle. The second relationship is the Kirchhoff voltage law, KVL.

The KCL is essentially a statement of conservation of charge, and states that at any point in the circuit, and usually the point is chosen at the junction of several branches, the rate at which charge reaches the junction or node must equal the rate at which it leaves the node. More often the KCL is written, "the sum of the currents toward a node must be equal to the sum of the currents away from the node," or equivalently, "the algebraic sum of the currents toward a node must be zero." Symbolically, this law is written,

$$\sum_{\text{node}} i_b = 0, \qquad (2\text{-}7)$$

where i_b denotes the branch currents.

The KVL is essentially a statement of conservation of energy when applied to a closed loop in the circuit. This law recognizes that the voltage across an element is the work done in carrying unit charge from the point of one potential at one terminal of the element to the second point of the element. This law states that, the algebraic sum of the voltages around any closed loop of a network must be zero," or alternatively "the sum of the voltage rises must equal the sum of the voltage drops in traversing a closed path around a network." Symbolically, this law is written,

$$\sum_{\text{loop}} v_b = 0. \qquad (2\text{-}8)$$

We shall examine several examples to show how one employs these two Kirchhoff laws.

2-4 NODE EQUILIBRIUM EQUATIONS

A study of the node-pair equilibrium law is best accomplished by means of a specific example. We consider for study the mechanical system of Fig. 2-1, or equivalently, the electric circuit of Fig. 2-1b. Thus we examine

Fig. 2-3 which is Fig. 2-1a redrawn and appropriately marked for our present needs. The fact that a force driver is used strongly suggests that we choose the through variables as the independent ones. Thus the equilibrium law that is appropriate to the present problem is that given in Eq. (2-7). As already mentioned, when applied to mechanical systems, the point law is known as D'Alembert's principle, which is usually written,

$$\sum f - M\frac{dv}{dt} = 0. \qquad (2-9)$$

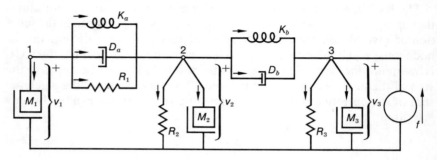

Fig. 2-3. The physical system under study.

The term $M(dv/dt)$ is often referred to as the *kinetic reaction*. This law essentially requires that each node be isolated in the analysis and that one distinguish the $M(dv/dt)$ force from all of the other forces. Actually, it is very convenient to use Eq. (2-7) directly rather than write it in the form given in Eq. (2-9).

From an examination of Fig. 2-3 we write the following set of equilibrium equations by an application of the point law at the three nodes:

Node 1 $\quad f_{M_1} + f_{K_a} + f_{D_a} + f_{R_1} = 0$

Node 2 $\quad -f_{K_a} - f_{D_a} - f_{R_1} + f_{R_2} + f_{M_2} + f_{K_b} + f_{D_b} = 0 \qquad (2\text{-}10)$

Node 3 $\quad -f_{K_b} - f_{D_b} + f_{R_3} + f_{M_3} - f = 0$

This is the starting point for a set of equations which may be used to determine the subsequent motion of the system. To deduce this set of equations requires that we employ the known through-across relationships for the various elements, as given in Table 1-1. The set of equations in Eq. (2-10) is now written as follows, with care in establishing the variables and in applying the reference conditions.

Node Equilibrium Equations

Node 1 $M_1 \dfrac{dv_1}{dt} + K_a \displaystyle\int (v_1 - v_2)\,dt + D_a(v_1 - v_2) + R_1(v_1 - v_2) = 0$

Node 2 $-K_a \displaystyle\int (v_1 - v_2)\,dt - D_a(v_1 - v_2) - R_1(v_1 - v_2) + R_2 v_2 + M_2 \dfrac{dv_2}{dt}$

$\qquad + K_b \displaystyle\int (v_2 - v_3)\,dt + D_b(v_2 - v_3) = 0 \qquad (2\text{-}11)$

Node 3 $-K_b \displaystyle\int (v_2 - v_3)\,dt - D_b(v_2 - v_3) + R_3 v_3 + M_3 \dfrac{dv_3}{dt} - f = 0$

This set of integrodifferential equations completely describes the dynamics of the system.

It is convenient to write this set of equations in terms of the operational variable $p = d/dt$ which was introduced in Section 2-1. We now have,

Node 1 $M_1 p v_1 + \dfrac{K_a}{p}(v_1 - v_2) + D_a(v_1 - v_2) + R_1(v_1 - v_2) = 0$

Node 2 $-\dfrac{K_a}{p}(v_1 - v_2) - D_a(v_1 - v_2) - R_1(v_1 - v_2) + R_2 v_2 + M_2 p v_2$

$\qquad + \dfrac{K_b}{p}(v_2 - v_3) + D_b(v_2 - v_3) = 0 \qquad (2\text{-}12)$

Node 3 $-\dfrac{K_b}{p}(v_2 - v_3) - D_b(v_2 - v_3) + R_3 v_3 + M_3 p v_3 - f = 0$

This set of equations is the *branch* set, since each term is written in a form which explicitly indicates the through-across relationship on an element by element (or branch by branch) basis.

It often proves to be much more convenient to rearrange this set of equations in the following manner:

$\left[\dfrac{K_a}{p} + (D_a + R_1) + M_1 p\right] v_1 - \left[\dfrac{K_a}{p} + (D_a + R_1)\right] v_2 = 0$

$-\left[\dfrac{K_a}{p} + (D_a + R_1)\right] v_1 + \left[\left(\dfrac{K_a + K_b}{p}\right) + (D_a + R_1 + R_2 + D_b) + M_2 p\right] v_2$

$\qquad -\left[\dfrac{K_b}{p} + D_b\right] v_3 = 0 \qquad (2\text{-}13)$

$-\left[\dfrac{K_b}{p} + D_b\right] v_2 + \left[\dfrac{K_b}{p} + (D_b + R_3) + M_3 p\right] v_3 = f$

These equations may be written in the following functional form,

$$y_{11}(p)v_1 + y_{12}(p)v_2$$
$$y_{21}(p)v_1 + y_{22}(p)v_2 + y_{23}(p)v_3 = 0 \qquad (2\text{-}14)$$
$$y_{32}(p)v_2 + y_{33}(p)v_3 = f$$

where the various operational factors $y(p)$ are seen to be,

$$y_{11}(p) = \frac{K_a}{p} + (D_a + R_1) + M_1 p$$

$$y_{12}(p) = y_{21}(p) = -\left[\frac{K_a}{p} + (D_a + R_1)\right]$$

$$y_{22}(p) = \frac{K_a + K_b}{p} + (D_a + R_a + R_2 + D_b) + M_2 p \qquad (2\text{-}15)$$

$$y_{23}(p) = y_{32}(p) = -\left[\frac{K_b}{p} + D_b\right]$$

$$y_{33}(p) = \frac{K_b}{p} + (D_b + R_3) + M_3 p$$

It is observed that the operators $y(p)$ have the following significance:

y_{11} = sum of all through-across admittance functions connected to node 1
y_{22} = sum of all admittance functions connected to node 2
y_{33} = sum of all admittance functions connected to node 3
$y_{12} = y_{21}$ = negative of sum of all admittance functions common to nodes 1 and 2
$y_{23} = y_{32}$ = negative of sum of all admittance functions common to nodes 2 and 3
f = algebraic sum of all force drivers to node 3.

It is often convenient to express this set of equations in matrix form, as follows

$$\begin{bmatrix} y_{11} & y_{12} & 0 \\ y_{21} & y_{22} & y_{23} \\ 0 & y_{32} & y_{33} \end{bmatrix} \begin{bmatrix} v_1 \\ v_2 \\ v_3 \end{bmatrix} = \begin{bmatrix} 0 \\ 0 \\ f \end{bmatrix} \qquad (2\text{-}16)$$

In more compact form, this equation set may be written,

$$[YV] = [F], \qquad (2\text{-}17)$$

where each matrix is defined in Eq. (2-14), with the matrix elements of the Y-matrix being specified in Eq. (2-15). Observe that this is just a generalization of Eq. (2-5) that expresses the entire network.

Example 2-4.1

Two fluid flow problems are illustrated. Write the equilibrium equations for each, and draw network models.

Solution: These are flow problems, each with a single node appropriate to the head h. By an application of an equilibrium equation for fluid flow that corresponds to Eq. (2-7) we write,

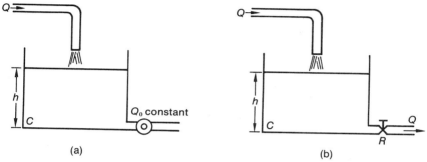

Fig. 2-4.1-1

For Fig. 2-4.1-1a
$$Cph = Q - Q_0$$

For Fig. 2-4.1-1b
$$Cph = Q - Q_1$$
$$Q_1 = \frac{h}{R}$$

which can be combined to yield,

$$\left(Cp + \frac{1}{R}\right)h = Q.$$

The appropriate network models are given in Figs. 2-4.1-1.

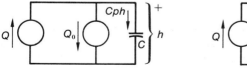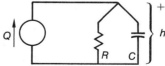

Fig. 2-4.1-2

Example 2-4.2

Write the equilibrium equations and draw a network model for the fluid system shown

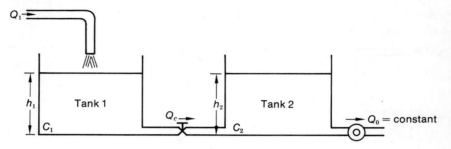

Fig. 2-4.2-1

Solution: We write directly that,

$$Q_1 = C_1 p h_1 + \left(\frac{h_1 - h_2}{R_1}\right),$$

$$0 = -\left(\frac{h_1 - h_2}{R_1}\right) + C_2 p h_2 + Q_0.$$

The network model is given in Fig. 2-4.2-2.

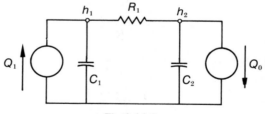

Fig. 2-4.2-2

Example 2-4.3

In the figure shown heat at a constant rate is supplied by an immersion heater to a fluid. The agitators are provided to insure that perfect stirring occurs and thereby to insure uniform heat distribution throughout the fluid in each container. Write the equilibrium equations and draw a network model.

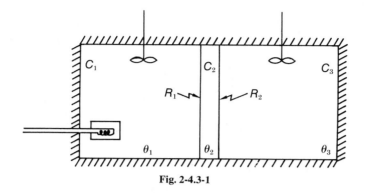

Fig. 2-4.3-1

Solution: We shall assume that a convection layer exists at each surface of the metallic divider. These are described respectively by R_1 and R_2. Thus we have

$$Q = C_1 \frac{d\theta_1}{dt} + \left(\frac{\theta_1 - \theta_2}{R_1}\right)$$

$$0 = -\left(\frac{\theta_1 - \theta_2}{R_1}\right) + C_2 \frac{d\theta_2}{dt} + \left(\frac{\theta_2 - \theta_3}{R_2}\right)$$

$$0 = -\left(\frac{\theta_2 - \theta_3}{R_2}\right) + C_3 \frac{d\theta_3}{dt}$$

The network model is given in Fig. 2-4.3-2.

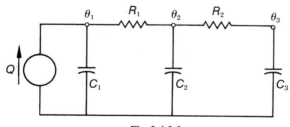

Fig. 2-4.3-2

Example 2-4.4

The fluid system discussed in Example 2-4.2 is modified as shown in Fig. 2-4.4-1, where *LC* is to denote a level controller. It is the function of this device to maintain automatically the input flow rate at such a value that the level in tank 2 is maintained at a specified reference level h_0. It

76   Interconnected Systems: Equilibrium Formulations

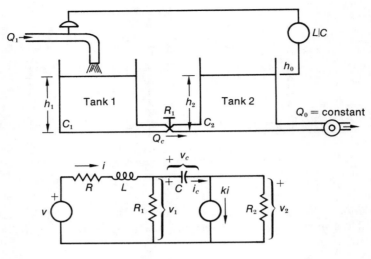

Fig. 2-4.4-1

will be assumed that the flow rate Q_1 that results is proportional to the difference between the reference level h_0 and the tank level h_2. Note that a level controller with other than a proportional relation should be quite possible. Write the controlling equations for this modified system.

Solution: We write directly that;

a. For tank 1 and connecting line

$$Q_1 = C_1 p h_1 + \left(\frac{h_1 - h_2}{R_1}\right).$$

b. For level control, for proportional output,

$$Q_1 = K(h_0 - h_2).$$

c. For tank 2,

$$0 = -\left(\frac{h_1 - h_2}{R_1}\right) + C_2 p h_2 + Q_0.$$

Combining Eqs. a. and b. gives,

$$\left(C_1 p + \frac{1}{R_1}\right)h_1 - \left(\frac{1}{R_2} - K\right)h_2 = K h_0,$$

$$-\frac{h_1}{R_1} + \left(C_2 p + \frac{1}{R_1}\right)h_2 = -Q_0.$$

It is too early to comment specifically about these equations, but it is noted that the presence of the level control essentially converts the input into a dependent source and in consequence the final set of equations has lost some of the symmetry that exists in the system without the level control (or the feedback line).

2-5 LOOP EQUILIBRIUM EQUATIONS

We shall here proceed by considering a specific example for detailed study. Now, however, we shall choose an electrical network for study. Refer to the circuit of Fig. 2-4. In this example, a voltage driver is applied

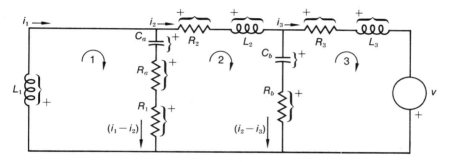

Fig. 2-4. The electric circuit under study.

to the circuit; hence we shall choose the voltage variables as the independent ones. To proceed, we employ the KVL law as the basis for our study. We use the closed loops shown to denote the paths that we shall employ when we write the KVL expressions. We have,

$$\text{loop 1} \quad v_{L_1} + v_{C_a} + v_{R_a} + v_{R_1} = 0$$

$$\text{loop 2} \quad -v_{R_1} - v_{R_a} - v_{C_a} + v_{R_2} + v_{L_2} + v_{C_b} + v_{R_b} = 0 \quad (2\text{-}18)$$

$$\text{loop 3} \quad -v_{R_b} - v_{C_b} + v_{R_b} + v_{L_3} - v = 0$$

Now employ the across-through relationships of the electrical elements from Table 1-1. These are often referred to as the generalized Ohm's law relationships for the electrical elements. Equations (2-18) now assume the form,

loop 1 $L_1 di_1 + \dfrac{1}{C_a}\int (i_1 - i_2)\,dt + R_a(i_1 - i_2) + R_1(i_1 - i_2) = 0$

loop 2 $-R_1(i_1 - i_2) - R_a(i_1 - i_2) - \dfrac{1}{C_a}\int (i_1 - i_2)\,dt + R_2 i_2 + L_2 \dfrac{di_2}{dt}$

$\qquad + \dfrac{1}{C_b}\int (i_2 - i_3)\,dt + R_b(i_2 - i_3) = 0$ (2-19)

loop 3 $-R_b(i_2 - i_3) - \dfrac{1}{C_b}\int (i_2 - i_3)\,dt + R_3 i_3 + L_3 \dfrac{di_3}{dt} - v = 0$

This set of integrodifferential equations describes the dynamics of this electrical system.

We again use the operational form $p = (d/dt)$ and write this set of equations in the form

loop 1 $L_1 p i_1 + \dfrac{1}{C_a p}(i_1 - i_2) + R_a(i_1 - i_2) + R_1(i_1 - i_2) = 0$

loop 2 $-R_1(i_1 - i_2) - R_a(i_1 - i_2) - \dfrac{1}{C_a p}(i_1 - i_2) + R_2 i_2 + L_2 p i_2$

$\qquad + \dfrac{1}{C_b p}(i_2 - i_3) - R_b(i_2 - i_3) = 0$ (2-20)

loop 3 $-R_b(i_2 - i_3) - \dfrac{1}{C_b p}(i_2 - i_3) + R_3 i_3 + L_3 p i_3 - v = 0$

This is a set of branch equations since each term explicitly relates to an element of the circuit. Attention is called to the fact that the current variables i_1, i_2 and i_3 as here selected can be interpreted to be currents associated with the separate loops, and if they are assumed to circulate in the loops, then the current through the branch common to loops 1 and 2 will be $i_1 - i_2$, and that through the branches common to loops 2 and 3 will be $i_2 - i_3$, as chosen in writing Eqs. (2-19). Note carefully, however, that while the concept of the loop current is a convenient topological scheme for writing equations, these do not have physical reality, since they cannot be associated directly with the actual conduction of charged carriers, as discussed in Chapter 1.

Now we rearrange Eqs. (2-20) by separating the various terms on the basis of the dependent variables. Thus,

$$\left[\dfrac{1}{C_a p} + (R_a + R_1) + L_1 p\right] i_1 - \left[\dfrac{1}{C_a p} + (R_a + R_1)\right] i_2 = 0$$

$$-\left[\frac{1}{C_a p}+(R_a+R_1)\right]i_1 + \left[\left(\frac{1}{C_a}+\frac{1}{C_b}\right)\frac{1}{p}+(R_1+R_a+R_2+R_b)+L_2 p\right]i_2$$
$$-\left[\frac{1}{C_b p}+R_b\right]i_3 = 0 \qquad (2\text{-}21)$$
$$-\left[\frac{1}{C_b p}+R_b\right]i_2 + \left[\frac{1}{C_b p}+(R_b+R_3)+L_3 p\right]i_3 = v$$

These equations are now written in functional form,

$$\begin{aligned}
z_{11}(p)i_1 + z_{12}(p)i_2 &= 0 \\
z_{21}(p)i_1 + z_{22}(p)i_2 + z_{23}(p)i_3 &= 0 \\
z_{32}(p)i_2 + z_{33}(p)i_3 &= v
\end{aligned} \qquad (2\text{-}22)$$

where the operational factors are seen to be,

$$\begin{aligned}
z_{11}(p) &= \frac{1}{C_a p}+(R_a+R_1)+L_1 p \\
z_{12}(p) &= z_{21}(p) = -\left[\frac{1}{C_a p}+(R_a+R_1)\right] \\
z_{22}(p) &= \left(\frac{1}{C_a}+\frac{1}{C_b}\right)\frac{1}{p}+(R_1+R_a+R_2+R_b)+L_2 p \qquad (2\text{-}23) \\
z_{23}(p) &= z_{32}(p) = -\left[\frac{1}{C_b p}+R_b\right] \\
z_{33}(p) &= \frac{1}{C_b p}+(R_b+R_3)+L_3 p
\end{aligned}$$

It is observed that the various z-factors in these equations have the following meanings:

z_{11} = sum of all impedances on the contour of loop 1
z_{22} = sum of all impedances on the contour of loop 2
z_{33} = sum of all impedances on the contour of loop 3
$z_{12} = z_{21}$ = negative of the sum of all impedance common to loops 1 and 2
$z_{23} = z_{32}$ = negative of the sum of all impedances common to loops 2 and 3
v = algebraic sum of all voltage sources in loop 3

This set of equations may be expressed conveniently in matrix form,

$$\begin{bmatrix} z_{11}(p) & z_{12}(p) & 0 \\ z_{21}(p) & z_{22}(p) & z_{23}(p) \\ 0 & z_{32}(p) & z_{33}(p) \end{bmatrix} \begin{bmatrix} i_1 \\ i_2 \\ i_3 \end{bmatrix} = \begin{bmatrix} 0 \\ 0 \\ v \end{bmatrix} \qquad (2\text{-}24)$$

where the matrix elements are defined in Eqs. (2-22) and (2-23). In general matrix notation, this may be written,

$$[ZI] = [V]. \tag{2-25}$$

This has the general form of Ohm's law, but applies to the entire system.

Example 2-5.1

The accompanying diagram shows a simplified model of the lungs. R represents the fluid resistance (air viscance) in the bronchi; K represents the fluid spring due to the compressibility of air and the elasticity of the chest and lung tissue; p_c is the pressure within the chest cavity and is

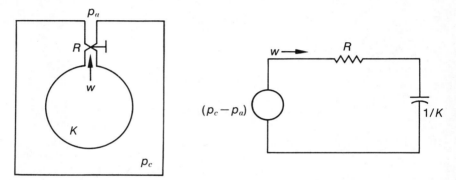

Fig. 2-5.1-1   Fig. 2-5.1-2

assumed identical with the pressure within the alveoli; p_a is the ambient air pressure measured at the buccal cavity, and w is the flow rate of air. Write the controlling dynamic equation for this system, and draw the network-model.

Solution: This is a simple system that is described by the equation,

$$p_c - p_a = Rw + K \int w \, dt.$$

Example 2-5.2

The system illustrated in the accompanying figure is driven by a synchronous motor, whence the input is a known velocity function. Assume that the frictional forces are proportional to velocity.
a. Draw a network model of the system.
b. Write a set of equations from which v_2 can be found.

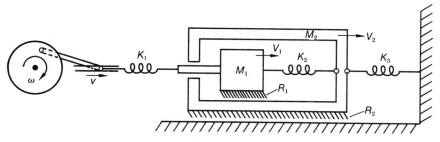

Fig. 2-5.2-1

Solution: Observe that the procedure called for in this example is the reverse of that in our prior discussion, namely, to draw the network model first and then to write the equations from this model.

By a careful application of Table 1-1, the network model is easily found to be that given in Fig. 2-5.2-2. From a loop point of view, this involves

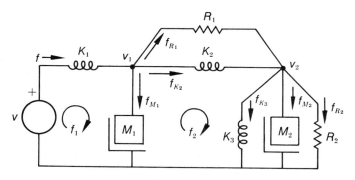

Fig. 2-5.2-2

elements in parallel: R_1 and K_2 in parallel (and this can be written conveniently as $R_1 \| K_2$), and K_3, M_2 and R_2 in parallel (which can be designated as $K_3 \| M_2 \| R_2$). If it were possible to write equivalent $z(p)$-s for each parallel group, and the methods for doing so will be discussed later, two equations would suffice to describe the system, in terms of f_1 and f_2 as independent variables, where,

$$f_1 = f = f_{R_1} + f_{K_2} + f_{M_1},$$

$$f_2 = f_{R_1} + f_{K_2} = f_{K_3} + f_{M_2} + f_{R_2}.$$

In the absence of such techniques for combining the effects of elements in parallel, we must write,

$$v_1 = f_{R_1} R_1 + v_2$$

$$v = \frac{1}{K_1} \frac{df}{dt} + v_1$$

where

$$v_1 = \frac{1}{M_1} \int f_{M_1} dt$$

$$v_2 = f_{R_2} R_2$$

with R_2 the inverse of the mechanical D_2. By the proper combination of the above equations, an explicit equation for v_2 can be deduced in terms of the system parameters and the applied v. The details of such a reduction will not be undertaken at this point.

2-b STATE VARIABLES AND STATE EQUATIONS

A feature of the node and loop formulation of the dynamic equations of an interconnected system is that, for linear systems, these are specified by sets of linear integrodifferential equations that related the input and output variables. We shall find, when we consider these integrodifferential equations for solution, that each equation of the set will be replaced by an equivalent linear differential equation. That is, in Eqs. (2-13) and (2-21) we would multiply each $y(p)$ and $z(p)$ respectively by the factor p, which is the equivalent of differentiating all terms of the equation. When this is done, each functional form for $y(p)$ and $z(p)$, the contribution associated with each node or loop variable, can contain terms involving p to orders p^2, p^1 and p^0. In general, therefore, the system is described by sets of differential equations, each term of which can be of second order in p. Another equally important system formulation is now to be considered which involves sets of first order differential equations only.

2-6 INTRODUCTION TO THE STATE FORMULATION

An alternative description of a connected system is given by what has become known as a *state* formulation. This description differs in several important respects from the Kirchhoff description. Now in addition to the input and the output variables, there is a set of variables known as the state variables. In the state formulation, account is taken of the fact explicitly that the behavior of a dynamic system depends not only on the presently applied forces, but also upon those that had been applied in the past. That is, inherent in a dynamic situation is an essential "memory" which is spread over the many elements in which the effects of past applied forces have been stored as reflected in charges on capacitors and fluxes in inductors. Moreover, it is a feature of a state formulation that a knowledge of the state of the system $x(t_0)$ at any time t_0, and there is no need for the history of the system in reaching the state $x(t_0)$ at this time, plus a knowledge of the forces subsequently applied, will allow a determination of the output and the state at any time $t > t_0$.

The input, state, and output consist, in general, of sets of variables. It is convenient to represent them by vector quantities. For example, an input set of m variables would be written as an m-vector.

$$u = \begin{bmatrix} u_1 \\ u_2 \\ \cdot \\ \cdot \\ \cdot \\ u_m \end{bmatrix} \qquad (2\text{-}26)$$

The state is denoted by the vector x, say of order n, and the output is denoted by the vector y, say of order r. We talk of the input space U which represents the set of all possible inputs u to the system. Similarly, the state space X represents the set of all possible states of the system, and the output space Y represents the set of all possible outputs. Also, the set of all time values for which u, x, y, are defined is the time space. If the time space is continuous, the system is known as a continuous time system. If the input and state vectors are defined only for discrete instants of time, the time space is discrete, and the system is called a discrete-time system. We shall have occasion to distinguish between functions that are functions of a discrete variable, and continuous functions which are sampled at discrete instants of time (sampled data systems).

A feature of the state formulation which is now to be discussed in some detail, is that it is accomplished in terms of a set of n first order differential equations instead of a set of $n/2$ second order differential equations by a Kirchhoff formulation. Thus the equilibrium of the dynamic system is described by a set of first order differential equations in an n-dimensional "state space". This description is given by an equation of the form,

$$\dot{x} = f(x,u,t),$$

or more specifically

$$\dot{x}_i(t) = X_i(x_1, x_2, \ldots, x_n, t) + u_i(t) \qquad (t > 0 \text{ and } i = 1, 2, \ldots, n) \qquad (2\text{-}27)$$

The quantities $x_1, x_2, \ldots, x_n$ are the dynamic state variables, and $u_i(t)$ are the inputs. As already noted, the components $x_1, x_2, \ldots$, usually denote charges on capacitors and fluxes of inductors, although special attention is required if capacitors form closed loops and for junctions of inductors. Note that here we are using the dot notation for time differentiation instead of the p notation that was used in Part 2-a. It is convenient for our subsequent discussion to write this system of simultaneous first order differential equations in the more compact vector (matrix) form,

$$\dot{x}(t) = X(x,t) + u(t) \qquad (t > 0) \qquad (2\text{-}28)$$

where x, u, X are $n \times 1$ column vectors. We shall find that linear systems can be described by a matrix equation of the form,

$$\dot{x} = Ax + Bu + G\dot{u}, \qquad (2\text{-}29)$$

where $x(t)$ is the state vector which describes the system at any time t, A is an $n \times n$ square matrix that is specified by the system, (in general, A may be time dependent, although for the applications to be considered in this book, the elements will be constants), u is the source or driver matrix, and y is the output vector. If y depends only on u, the system is said to be *memoryless*.

In the special case when the G coefficient is zero, Eq. (2-29) reduces to the form,

$$\dot{x} = Ax + Bu. \qquad (2\text{-}30)$$

Actually, however, Eq. (2-29) can always be reduced to the form in Eq. (2-30) by a simple change of state variable. That is, if we let $\hat{x}$ represent a new state vector, where,

$$\hat{x} = x - Gu, \qquad (2\text{-}31)$$

then Eq. (2-29) can be written,

$$\dot{\hat{x}} = A\hat{x} + (B + AG)u, \qquad (2\text{-}32)$$

which is of the same general form as Eq. (2-30). Because of this, we shall assume in subsequent developments that the state equations for the general system will be given by Eq. (2-30). Of course, if the system is force-free, then the second term is absent in Eq. (2-30).

In the general form given in Eq. (2-28), whether $X(x,t)$ are linear or nonlinear, the system is called *nonautonomous* when the set of equations are such that $X(x)$ is independent of t. Otherwise the system is called *autonomous*.

2-7 STATE EQUATIONS FOR LINEAR SYSTEMS

The state vector $x(t)$ is defined as the minimal set of state variables which uniquely determines the future state of a dynamic system, if the present values are known. This means that given $x(t_0)$, the state of the system at time $t = t_0$, then the state vector at any future time $x(t)$ for $t > t_0$ is uniquely determined by Eq. (2-29) or some appropriate form thereof. But selecting state variables is not a unique process; many acceptable state variables exist for any given system. Any minimum set of through and across variables that will completely describe the system would be a suitable state vector. These must be chosen so as to allow a formulation in terms of a set of first order differential equations.

While the selection of the state variables is not unique, a reasonably systematic selection is possible. Consider a simple *RLC* electrical network. This will later be discussed in some detail, but we here note the fact that if the voltage across the capacitor and the current through the inductor are known at some initial time t_0, then the circuit equations will allow a description of the system behavior for all subsequent times. This suggests the following guide for the selection of acceptable state variables:

1. Through variables associated with inductance type elements (L, K).
2. Across variables associated with capacitance type elements (C, M).
3. Dissipative type elements do not specify independent state variables.
4. Special consideration is required when closed loops of capacitors or junctions of inductors exist. In these cases not all state variables are independent, if chosen according to rules 1 and 2.

We shall examine the situation somewhat more systematically below using topological aspects of the network in the discussion. However, the rules given do allow us to formulate the equilibrium equations of many systems in state variable form.

Example 2-7.1

Write the system equation in state form for the circuit shown.

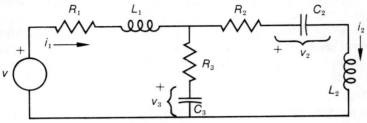

Fig. 2-7.1-1

Solution: In accordance with the rules given, the state variables are selected to be i_1, i_2, v_2, v_3. We write the following equations by inspection.

$$v = R_1 i_1 + L_1 \frac{di_1}{dt} + R_3(i_1 - i_2) + v_3,$$

$$0 = R_2 i_2 + v_2 + L_2 \frac{di_2}{dt} - v_3 - R_3(i_1 - i_2),$$

$$C_3 \frac{dv_3}{dt} = i_1 - i_2,$$

$$C_2 \frac{dv_2}{dt} = i_2.$$

From these we write,

$$L_1 \frac{di_1}{dt} = v - (R_1 + R_3) i_1 + R_3 i_2 - v_3,$$

$$L_2 \frac{di_2}{dt} = v_3 + R_3 i_1 - (R_2 + R_3) i_2 - v_2,$$

$$C_2 \frac{dv_2}{dt} = i_2,$$

$$C_3 \frac{dv_3}{dt} = i_1 - i_2.$$

These results are written in matrix form,

$$\frac{d}{dt} \begin{bmatrix} v_2 \\ v_3 \\ i_1 \\ i_2 \end{bmatrix} = \begin{bmatrix} 0 & 0 & 0 & 0 \\ 0 & 0 & \frac{1}{C_3} & -\frac{1}{C_3} \\ 0 & -\frac{1}{L_1} & -\frac{R_1 + R_3}{L_1} & \frac{R_3}{L_1} \\ -\frac{1}{L_2} & \frac{1}{L_2} & \frac{R_3}{L_2} & -\frac{R_2 + R_3}{L_2} \end{bmatrix} \cdot \begin{bmatrix} v_2 \\ v_3 \\ i_1 \\ i_2 \end{bmatrix} + \begin{bmatrix} 0 \\ 0 \\ v \\ 0 \end{bmatrix}$$

Example 2-7.2

It is required to write a system description for the vertical motion of an automobile that is moving on a rough road of known description.

a. Draw physical and network models of the system taking the shock absorbers into account, but neglecting damping in the tires and in the springs.

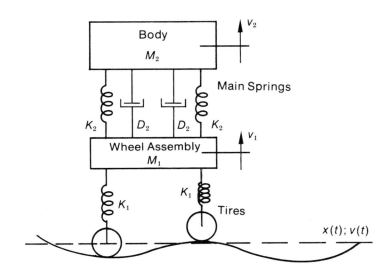

Fig. 2-7.2-1a

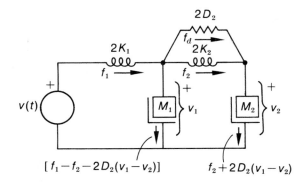

Fig. 2-7.2-1b

88 Interconnected Systems: Equilibrium Formulations

b. Write the dynamic equations in state form.

Solution: The physical and network models are given in the figures. Using the guides for selecting state variables, appropriately interpreted for mechanical variables, we select f_1, f_2, v_1, v_2 as state variables. The following equations are written by inspection,

$$f_1 = 2K_1 \int (v - v_1) dt = \frac{2K_1}{p}(v - v_1),$$

$$f_2 = \frac{2K_2}{p}(v_1 - v_2),$$

$$v_1 = \frac{1}{M_1 p}[f_1 - f_2 - 2D_2(v_1 - v_2)],$$

$$v_2 = \frac{1}{M_2 p}[f_2 + 2D_2(v_1 - v_2)].$$

These equations are now rewritten in the form,

$$pf_1 = 2K_1(v - v_1),$$

$$pf_2 = 2K_2(v_1 - v_2),$$

$$pv_1 = \frac{1}{M_1}[f_1 - f_2 - 2D_2(v_1 - v_2)],$$

$$pv_2 = \frac{1}{M_2}[f_2 + 2D_2(v_1 - v_2)].$$

In matrix form, these are,

$$\frac{d}{dt}\begin{bmatrix} f_1 \\ f_2 \\ v_1 \\ v_2 \end{bmatrix} = \begin{bmatrix} 0 & 0 & -2K_1 & 0 \\ 0 & 0 & 2K_2 & -2K_2 \\ \frac{1}{M_1} & -\frac{1}{M_1} & -\frac{2D_2}{M_1} & \frac{2D_2}{M_1} \\ 0 & \frac{1}{M_2} & \frac{2D_2}{M_2} & -\frac{2D_2}{M_2} \end{bmatrix} \begin{bmatrix} f_1 \\ f_2 \\ v_1 \\ v_2 \end{bmatrix} + \begin{bmatrix} 2K_1 v \\ 0 \\ 0 \\ 0 \end{bmatrix}$$

Example 2-7.3

Refer to the network shown which includes a dependent source. Write the dynamic equations in state form.

State Equations for Linear Systems

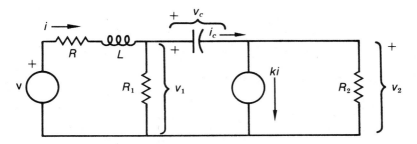

Fig. 2-7.3-1

Solution: Using our normal guide, we select i and v_c as state variables. We write the following equations by inspection,

$$v = (R + Lp)i + v_c + v_2,$$

$$i_c = i - \frac{v_1}{R_1} = ki + \frac{v_2}{R_2},$$

$$v_2 = v_1 - v_c,$$

$$v_1 = R_1 i - R_1 \frac{C dv_c}{dt}.$$

Combining the second and third of these equations gives,

$$i_c = \frac{C dv_c}{dt} = ki + \frac{v_1}{R_2} - \frac{v_c}{R_2} = ki - \frac{v_c}{R_2} + \frac{1}{R_2}\left(R_1 i - RC\frac{dv_c}{dt}\right).$$

Thus

$$\frac{dv_c}{dt} C\left(1 + \frac{R}{R_2}\right) = -\frac{v_c}{R_2} + \left(k + \frac{R_1}{R_2}\right)i. \tag{1}$$

Also,

$$(1-k)i - \frac{v_1}{R_1} = \frac{v_1 - v_c}{R_2},$$

$$(1-k)i + \frac{v_c}{R_2} = v_1\left(\frac{1}{R_1} + \frac{1}{R_2}\right) = \left(\frac{1}{R_1} + \frac{1}{R_2}\right)[v - (R+Lp)i].$$

From these,

$$\frac{v_c}{R_2} + \left[(1-k) + R\left(\frac{1}{R_1} + \frac{1}{R_2}\right)\right]i = \left(\frac{1}{R_1} + \frac{1}{R_2}\right)v - L\left(\frac{1}{R_1} + \frac{1}{R_2}\right)pi.$$

Hence,

$$\frac{di}{dt} = \frac{v}{L} - \frac{v_c}{LR_2\left(\frac{1}{R_1}+\frac{1}{R_2}\right)} + \frac{1}{L}\left[R + \frac{1-k}{\left(\frac{1}{R_1}+\frac{1}{R_2}\right)}\right]i. \qquad (2)$$

These equations are written in matrix form,

$$\frac{d}{dt}\begin{bmatrix} v_c \\ i \end{bmatrix} = \begin{bmatrix} -\frac{1}{C(R_2+R)} & \frac{kR_2+R_1}{C(R_2+R)} \\ -\frac{R_1}{L(R_1+R_2)} & \frac{R}{L}+\frac{(1-k)R_1R_2}{L(R_1+R_2)} \end{bmatrix}\begin{bmatrix} v_c \\ i \end{bmatrix} + \begin{bmatrix} 0 \\ \frac{v}{L} \end{bmatrix}$$

2-8 DIFFERENTIAL EQUATIONS IN NORMAL FORM

If we had the description of a system in terms of an nth order differential equation, or in terms of a set of second order differential equations, as deduced by an application of the KCL or the KVL, it is readily possible to convert these higher order differential equations into sets of first order differential equations. In this case we are transforming the higher order differential equations into *normal form*. Such a transformation is of interest not only here in providing a set of state equations, but we shall also be interested in this transformation when we consider the numerical solution of differential equations. Further, we shall find the normal form very useful in setting up diagrams for analog computer simulation.

Example 2-8.1

Replace the van der Pol oscillator equation by a pair of first order differential equations. The van der Pol equation is,

$$\ddot{x} + k(x^2-1)\dot{x} + x = 0.$$

Solution: Define the variable,

$$\dot{x} = \xi,$$

and combine with the original differential equation to get,

$$\dot{\xi} = -k(x^2-1)\xi - x.$$

Thus the required solution is,

$$\frac{d}{dt}\begin{bmatrix} x \\ \xi \end{bmatrix} = \begin{bmatrix} 0 & 1 \\ -1 & -k(x^2-1) \end{bmatrix}\begin{bmatrix} x \\ \xi \end{bmatrix}.$$

It is observed that this equation is an autonomous nonlinear function of x.

Example 2-8.2

Convert the following set of differential equations into a set of first order differential equations, and write the result in vector form.

$$\frac{d^2y}{dt^2} + 5\frac{dy}{dt} + 2(y-x) = f_1(t),$$

$$\frac{d^2x}{dt^2} + 2\frac{dx}{dt} - 4\frac{dy}{dt} + x = f_2(t).$$

Solution: Define the quantities,

$$\dot{x} = \xi, \qquad \dot{y} = \eta.$$

Combine these with the given set of differential equations. This gives,

$$\frac{d\eta}{dt} + 5\eta + 2(y-x) = f_1(t),$$

$$\frac{d\xi}{dt} + 2\xi - 4\eta + x = f_2(t).$$

In matrix form, this set of equations is given by,

$$\frac{d}{dt}\begin{bmatrix} x \\ y \\ \xi \\ \eta \end{bmatrix} = \begin{bmatrix} 0 & 0 & 1 & 0 \\ 0 & 0 & 0 & 1 \\ -1 & 0 & -2 & 4 \\ -2 & 2 & 0 & -5 \end{bmatrix}\begin{bmatrix} x \\ y \\ \xi \\ \eta \end{bmatrix} + \begin{bmatrix} 0 \\ 0 \\ f_2 \\ f_1 \end{bmatrix}$$

Let us now consider the reduction to normal form of a general linear differential equation of nth order of the form,

$$a_n\frac{d^n y}{dt^n} + a_{n-1}\frac{d^{n-1}y}{dt^{n-1}} + \cdots + a_0 y = b_m\frac{d^m u}{dt^m} + b_{m-1}\frac{d^{m-1}u}{dt^{m-1}} + \cdots + b_0 u. \quad (2\text{-}33)$$

We defer our attention of this particular differential equation, and first consider the reduced equation,

$$a_n\frac{d^n \eta}{dt^n} + a_{n-1}\frac{d^{n-1}\eta}{dt^{n-1}} + \cdots + a_0\eta = u, \quad (2\text{-}34)$$

which is seen to have the same form on the left, but has the simple function u on the right. We also observe the very important feature of the linear differential equation, that the *r*th order response of a system is the solution of a system to the *r*th order of the initial excitation function. That is, if $\eta(t)$ is the response to the excitation $u(t)$ of a system that is specified by Eq. (2-34), then the response to an excitation $d^r u/dt^r$ will be $d^r \eta/dt^r$. That this is so follows by a straightforward differentiation of Eq. (2-34), with,

$$a_n \frac{d^n}{dt^n}\left(\frac{d^r \eta}{dt^r}\right) + a_{n-1}\frac{d^{n-1}}{dt^{n-1}}\left(\frac{d^r \eta}{dt^r}\right) + \cdots + a_0\left(\frac{d^r \eta}{dt^r}\right) = \frac{d^r u}{dt^r}. \quad (2\text{-}35)$$

This result is displayed graphically in Fig. 2-5, which shows a sequence of situations for designated inputs and the corresponding outputs. In the

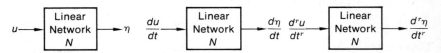

Fig. 2-5. The input-output relations for a linear network.

light of this result, we can now find a solution to Eq. (2-33) in terms of the solution of Eq. (2-34). This is given by the expression,

$$y = b_m \frac{d^m \eta}{dt^m} + b_{m-1}\frac{d^{m-1}\eta}{dt^{m-1}} + \cdots + b_0 \eta. \quad (2\text{-}36)$$

This development shows that the differential equation Eq. (2-33) may be replaced by the pair of equations given by Eqs. (2-35) and (2-36).

Based on the foregoing, we may write Eq. (2-33) in normal form quite readily. We define the quantities,

$$x_1 = x$$
$$\dot{x}_1 = x_2 = \frac{dx}{dt}$$
$$\dot{x}_2 = x_3 = \frac{d^2 x}{dt^2}$$
$$\cdot \quad \cdot \quad \cdot$$
$$\cdot \quad \cdot \quad \cdot$$
$$\cdot \quad \cdot \quad \cdot$$
$$\dot{x}_{n-1} = x_n = \frac{d^{n-1}x}{dt^{n-1}}$$

$$(2\text{-}37)$$

The final equation is obtained by combining these forms with Eq. (2-33),

$$a_n \dot{x}_n = -a_{n-1}x_n - a_{n-2}x_{n-1} - \cdots - a_0 x_1 + u. \tag{2-38}$$

In addition, the expression for $y(t)$ of Eq. (2-36) is then,

$$y = b_m x_{m+1} + b_{m-1} x_m + \cdots + b_0 x_1. \tag{2-39}$$

In matrix form these results have the form,

$$\frac{dx}{dt} = Ax + Bu$$
$$y = Px \tag{2-40}$$

where,

$$A = \begin{bmatrix} 0 & 1 & 0 & \cdots & 0 \\ 0 & 0 & 1 & & 0 \\ \cdot & \cdot & \cdot & \cdots & \cdot \\ \cdot & \cdot & \cdot & \cdots & \cdot \\ \cdot & \cdot & \cdot & \cdots & \cdot \\ 0 & 0 & 0 & & 1 \\ -\frac{a_0}{a_n} & -\frac{a_1}{a_n} & -\frac{a_2}{a_n} & & -\frac{a_{n-1}}{a_n} \end{bmatrix} \quad B = \begin{bmatrix} 0 \\ 0 \\ \cdot \\ \cdot \\ \cdot \\ 0 \\ \frac{1}{a_n} \end{bmatrix}$$

$$P = [b_0 \quad b_1 \cdots b_m \quad 0 \cdots 0]$$

The notational simplicity of Eq. (2-40) is evident, since a system of order n has the appearance of a first order system. In addition to the compact notation, we shall find that advantages exist when we undertake the solution of these equations.

If $m > n$, the system is said to be improper. When the system is improper, it would be found that derivatives of the input appear in the output state equation.

2-9 STATE VARIABLE TRANSFORMATION

Once the normal form state equation representation has been deduced, it is quite possible to use this set to obtain a new set of state variables that is a linear combination of the original state variables. Suppose that we begin with a system of equations in normal form,

$$\dot{x} = Ax + Bu,$$
$$y = Px + Qu. \tag{2-41}$$

We introduce a new set of state variables, $x_1, x_2, x_3, \ldots, x_n$, each of which is a linear combination of the state variables x_i; that is,

$$\hat{x}_i = \sum_{j=1}^{n} m_{ij} x_i \quad (i = 1, 2, \ldots, n) \tag{2-42}$$

or, in vector notation,

$$\hat{x} = Mx, \tag{2-43}$$

where M is an $n \times n$ matrix with matrix elements m_{ij}. Now assuming that the components of $\hat{x}$ are linearly independent, then M is nonsingular and its inverse M^{-1} exists. Thus,

$$x = M^{-1}\hat{x}, \tag{2-44}$$

$$\dot{x} = M^{-1}\dot{\hat{x}}.$$

These are combined with Eq. (2-41) to give,

$$M^{-1}\dot{\hat{x}} = AM^{-1}\hat{x} + Bu,$$
$$y = PM^{-1}\hat{x} + Qu. \tag{2-45}$$

The first equation is premultiplied by M, and yields,

$$\dot{\hat{x}} = MAM^{-1}\hat{x} + MBu.$$

Thus the normal form in terms of the new state variables is,

$$\dot{\hat{x}} = \bar{A}\hat{x} + \bar{B}u,$$
$$y = \bar{P}\hat{x} + \bar{Q}u, \tag{2-46}$$

where,

$$\bar{A} = MAM^{-1} \qquad \bar{B} = MB \qquad \bar{P} = PM^{-1} \qquad \bar{Q} = Q.$$

For example, the transformation,

$$\begin{bmatrix} \hat{x}_1 \\ \hat{x}_2 \\ \cdot \\ \hat{x}_n \end{bmatrix} = \begin{bmatrix} b_0 & b_1 & \cdots & b_{n-1} \\ & b_0 & \cdots & b_{n-2} \\ & & \cdots & \\ & & & b_0 \end{bmatrix} \begin{bmatrix} x_1 \\ x_2 \\ \\ x_n \end{bmatrix}$$

when applied to Eq. (2-40) produces

$$\begin{bmatrix} \dot{\hat{x}}_1 \\ \cdot \\ \cdot \\ \cdot \\ \cdot \\ \dot{\hat{x}}_n \end{bmatrix} = \begin{bmatrix} -\frac{a_0}{a_n} & 1 & 0 & \cdots & \\ \cdot & 0 & 1 & \cdots & \\ \cdot & 0 & 0 & 1 & \cdots \\ \cdot & & & & 1 \\ -\frac{a_{n-1}}{a_n} & & & & 0 \end{bmatrix} \begin{bmatrix} \hat{x}_1 \\ \cdot \\ \cdot \\ \cdot \\ \hat{x}_n \end{bmatrix} + \begin{bmatrix} 0 \\ 0 \\ b_m \\ \cdot \\ b_0 \end{bmatrix} u$$

$$y(t) = [1 \quad 0 \cdots 0]\hat{x}(t)$$

2-10 DISCRETE AND SAMPLED TIME SYSTEMS

In the previous work of this chapter, it has been assumed that the system components can be represented by ordinary differential (continuous time) equations and/or algebraic equations. When the system is assumed to undergo instantaneous changes at regular intervals of time, the system is called a discrete time system. The digital computer is perhaps the most important example of a discrete time system; it is assumed to remain quiescent during the basic time intervals established by an electronic clock, with instantaneous state transitions at the end of these intervals. Systems which sample the driving function on an instantaneous basis at regular time intervals, or on a quantized basis (staircase or sample and hold) between sampling periods, are common and very important sampled time systems.

If the digital computer is to be used to compute the solution of a differential equation, it is first necessary to approximate the differential equation by a difference equation, a form that the computer can handle directly. To obtain the finite difference approximation to a differential equation, derivatives are replaced by finite-differences. It is here assumed that time t has been sliced into equal increments T sec. long, and we concern ourselves only with the values of x at the times given by the sequence $t = \{0, T, 2T, \ldots, nT, \ldots\}$. For values of x between any pair of values of T, we may interpolate or repeat the study for smaller time increments T. We consider a first order differential equation for study,

$$\frac{dx}{dt} = f[x(t), u(t), t]. \tag{2-47}$$

The difference equation approximation to this is obtained by noting that the derivative dx/dt is written approximately,

$$\frac{dx}{dt} = \frac{x(t_1+T)-x(t_1)}{T} = \frac{x_{n+1}-x_n}{T}. \quad (2\text{-}48)$$

Thus appropriate to Eq. (2-47) are the two equivalent forms

$$x(t+T) = Tf[x(t),u(t),t]+x(t),$$

and (2-49)

$$x(n+1) = Tf[x(n),u(n),n]+x(n).$$

These are recursive relations which allow the calculation of $x[t_0+(n+1)T]$ from a knowledge of all previous states $x(t_0+nT)$ including the initial state $x(t_0)$. Correspondingly, the output, which in the continuous time case is given by a functional expression,

$$y = g[x(t),u(t),t], \quad (2\text{-}50)$$

becomes, when determined from the finite difference calculation,

$$y(n) = g[x(n),u(n),n]. \quad (2\text{-}51)$$

We define a discrete time system as one which can be represented by a set of equations of the form given by Eqs. (2-49) and (2-51). Further, the system is linear if the functions f and g are linear in x and u. For the linear discrete time system, the governing set of equations are of the form,

$$\begin{aligned} x(n+1) &= A(n)x(n)+B(n)u(n), \\ y(n) &= P(n)x(n)+Q(n)u(n), \end{aligned} \quad (2\text{-}52)$$

where the matrixes A, B, P, Q may depend on the discrete time variable n. If these matrixes are constant, the system is time invariant.

From the manner of their development, it is clear that these equations apply for a linear time-invariant system that is excited by piecewise constant (quantized) inputs in which the transitions occur at multiples of a fundamental period T. When the component models are discrete state forms, a system of nonhomogeneous discrete state equations of the general form given in Eq. (2-52) would be developed directly from the structure of the system.

If we were to consider a discrete time approximation to a differential equation of higher order than the first, appropriate finite difference approximations to the higher order derivatives would be used. These follow from a logical extension of Eq. (2-46). Thus for the second order derivative we can write,

$$\frac{d^2x}{dt^2} = \frac{d}{dt}\left(\frac{dx}{dt}\right) \doteq \frac{1}{T}\left[\frac{x_{n+2}-x_{n+1}}{T} - \frac{x_{n+1}-x_n}{T}\right],$$

$$= \frac{x_{n+2}-2x_{n+1}+x_n}{T^2}.$$
(2-53)

Similarly, for the third order derivative, we can show that,

$$\frac{d^3x}{dt^3} \doteq \frac{x_{n+3}-3x_{n+2}+3x_{n+1}-x_n}{T^3}.$$
(2-54)

Using these general results, a kth order differential equation (we here write the order as k to avoid confusion with the nth term in the sequence) of the form in Eq. (2-33) in difference equation form will be,

$$\alpha_k y(n+k) + \alpha_{k-1} y(n+k-1) + \cdots + \alpha_0 y(n)$$
$$= \beta_m u(n+m) + \beta_{m-1} u(n+m-1) + \cdots + \beta_0 u(n). \quad (2\text{-}55)$$

It is of interest and often of convenience to introduce the symbol z to denote a unit time advance operator. This means that z^r denotes the time advance of r time intervals and correspondingly z^{-r} denotes the delay of r time intervals. With this notation, the difference equation, Eq. (2-55) can be written,

$$(\alpha_k z^k + \alpha_{k-1} z^{k-1} + \cdots + \alpha_0) y(n) = (\beta_m z^m + \beta_{m-1} z^{m-1} + \cdots + \beta_0) u(n). \quad (2\text{-}56)$$

By comparing this equation with Eq. (2-33) it is seen that the discrete time equation has the same form as the continuous time equation but with the delay operator replacing the derivative operator.

Equation (2-55) is a recursive relation that is suitable for numerical calculation. It permits the calculation of $y(n+k)$ from a knowledge of all previous values of y. Thus one would begin with the initial state, say $y(0)$, and determine the successive values of y by taking into account the appropriate past values of y and u, as required by the equation. To discuss this process further, a change in notation is helpful. Let us denote the latest value of y as $y(n)$; hence the previous values will be $y(n-1)$, $y(n-2)$, etc. Correspondingly, the latest value of u will be $u(n)$ with prior values $u(n-1)$, $u(n-2)$, etc. Using these changes, Eq. (2-55) is written,

$$\xi_0 y(n) + \xi_1 y(n-1) + \cdots + \xi_k y(n-k)$$
$$= \zeta_0 u(n) + \zeta_1 u(n-1) + \cdots + \zeta_m u(n-m), \quad (2\text{-}57)$$

98   Interconnected Systems: Equilibrium Formulations

where

$$\xi_i = \alpha_{k-i} \qquad \zeta_i = \beta_{m-i}.$$

This equation can be written simply as,

$$\sum_{j=0}^{k} \xi_j y(n-j) = \sum_{j=0}^{m} \zeta_j u(n-j). \tag{2-58}$$

The latest value $y(n)$ can be written explicitly as,

$$y(n) = \sum_{j=0}^{m} \frac{\zeta_j}{\xi_0} u(n-j) - \sum_{j=1}^{k} \frac{\xi_j}{\xi_0} y(n-j),$$

which is written,

$$y(n) = \sum_{j=0}^{m} b_j u(n-j) - \sum_{j=1}^{k} a_j y(n-j), \tag{2-59}$$

where,

$$b_j = \frac{\zeta_j}{\xi_0} \qquad a_j = \frac{\xi_j}{\xi_0}.$$

Clearly, if all the a_j-s are zero, the output $y(n)$ is a linear weighting of the previous $(n-1)$ samples of the input. This is called a *nonrecursive* process.

To carry out the details of Eq. (2-59) for the present value of $y(n)$ requires that the prior m values of u and k values of y be available. Such a *recursive* process imposes certain memory demands on the computer. It is possible to reduce the memory requirement by replacing this one recursion relation by two expressions, in a manner that parallels replacing the differential equation of Eq. (2-33) by the pair of equations, Eqs. (2-35) and (2-36). We begin with the reduced equation,

$$\eta(n) + a_1 \eta(n-1) + \cdots + a_k \eta(n-k) = u(n), \tag{2-60}$$

or equivalently,

$$\eta(n) = u(n) - \sum_{j=1}^{k} a_j \eta(n-j). \tag{2-61}$$

Furthermore, we can write a similar relation that is appropriate to each value of $u(n-j)$ and then sum the results. This gives,

$$y(n) = \sum_{j=0}^{m} b_j \eta(n-j). \tag{2-62}$$

Clearly, therefore, Eqs. (2-61) and (2-62) together replace the single equation Eq. (2-59). Observe that now it is necessary to save only the m or k previous values of η depending on which is greater.

We can employ a procedure for writing the difference equation in state form that parallels that used in connection with the reduction to normal form of the differential equation given in Eq. (2-33). We begin with Eq. (2-55), and establish the reduced equation,

$$\alpha_k \lambda(n+k) + \alpha_{k-1} \lambda(n+k-1) + \cdots + \alpha_0 \lambda(n) = u(n). \quad (2\text{-}63)$$

We define the quantities,

$$\begin{aligned} x_1(n) &= \lambda(n) \\ x_2(n) &= \lambda(n+1) \\ &\vdots \\ x_k(n) &= \lambda(n+k+1) \end{aligned} \quad (2\text{-}64)$$

From these it follows that,

$$\begin{aligned} x_1(n+1) &= x_2(n) \\ x_2(n+1) &= x_3(n) \\ &\vdots \\ x_{k-1}(n+1) &= x_k(n) \end{aligned} \quad (2\text{-}65)$$

By combining these expressions with Eq. (2-63) we find the final equation,

$$x_k(n+1) = \frac{1}{\alpha_k}[u(n) - \alpha_{n-1} x_k(n) - \alpha_{k-2} x_{k-1}(n) - \cdots - \alpha_0 x_1(n)].$$

Consequently Eq. (2-64) has now been replaced by the set of equations,

$$x(n+1) = Ax(n) + Bu(n),$$

where,

$$A = \begin{bmatrix} 0 & 1 & 0 & \cdots & 0 & 0 \\ 0 & 0 & 1 & \cdots & 0 & 0 \\ \vdots & \vdots & \vdots & \cdots & \vdots & \vdots \\ 0 & 0 & 0 & \cdots & 0 & 1 \\ -\dfrac{\alpha_0}{\alpha_k} & -\dfrac{\alpha_1}{\alpha_k} & -\dfrac{\alpha_2}{\alpha_k} & \cdots & -\dfrac{\alpha_{k-2}}{\alpha_k} & -\dfrac{\alpha_{k-1}}{\alpha_k} \end{bmatrix} \quad B = \begin{bmatrix} 0 \\ 0 \\ \vdots \\ 0 \\ \dfrac{1}{\alpha_k} \end{bmatrix}$$

100   Interconnected Systems: Equilibrium Formulations

$$x(n) = \begin{bmatrix} x_1(n) \\ x_2(n) \\ \cdot \\ \cdot \\ \cdot \\ x_k(n) \end{bmatrix} \quad (2\text{-}66)$$

We now use these results in connection with Eq. (2-55). Thus we write,

$$y(k) = \beta_m x_{m+1}(k) + \beta_{m-1} x_m(k) + \cdots + \beta_0 x_1(k), \quad (2\text{-}67)$$

which is,

$$y(k) = Px(k),$$

where P is the row matrix,

$$P = [\beta_0 \quad \beta_1 \cdots \beta_m, \quad 0 \cdots 0]. \quad (2\text{-}68)$$

That is, the state equations for the nth order difference equation are,

$$\begin{aligned} x(k+1) &= Ax(k) + Bu(k), \\ y(k) &= Px(k), \end{aligned} \quad (2\text{-}69)$$

where A and B are given in Eq. (2-66) and P is given in Eq. (2-68). Techniques for the solution of these equations will be discussed later.

PROBLEMS

2-1. The equivalent models of a number of different mechanical systems are shown. Draw network models for each, label each carefully, and write the equilibrium equations for each.

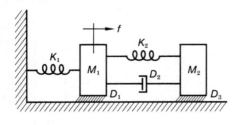
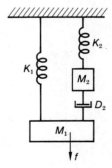

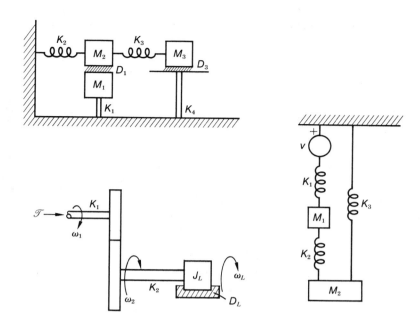

2-2. Write the equilibrium equations on a loop basis for each of the given circuits.

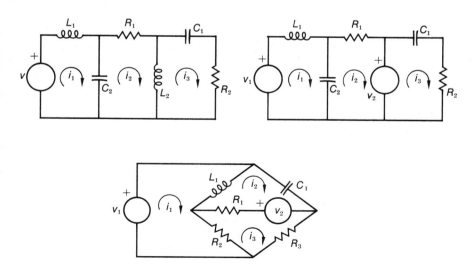

2-3. Write a set of equilibrium equations on a loop basis for each of the following coupled systems.

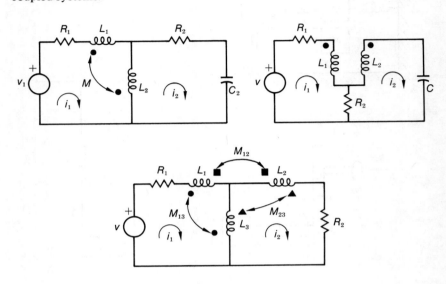

2-4. Write a set of equilibrium equations on a nodal basis for each of the following systems.

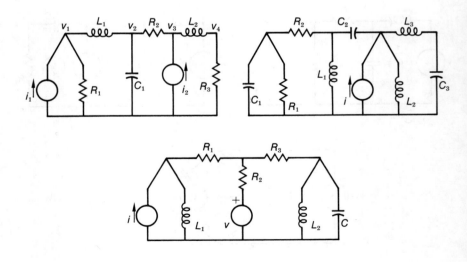

2-5. Draw network models and write the system equations for the given fluid systems. From these write differential equations for determining h_2 in terms of Q.

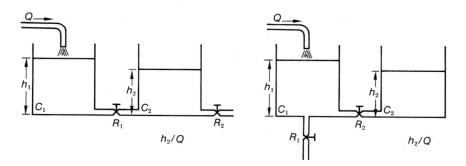

2-6. Derive differential equations for relating the variables noted.

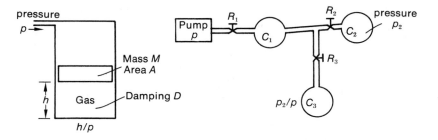

2-7. Derive equations for relating the variables noted.

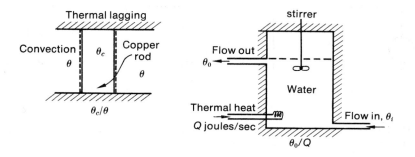

2-8. Express the equilibrium equations of the network shown in state variable form.

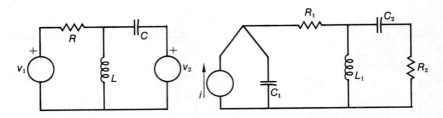

2-9. Write state equations for each of the following:
   a. An armature controlled d-c motor driving a mechanical system, as illustrated:

$$\text{electrical circuit} \qquad v_a = R_a i_a + L_a p i_a + K_a \omega$$
$$\text{mechanical circuit} \quad K_a i_a = J p \omega + D \omega + T_e$$

where R_a, L_a are motor armature electrical parameters, $K_a =$ motional emf constant; $J =$ inertia of rotating system, $T_e =$ electrical torque by motor.
   b. The electrical circuits shown.

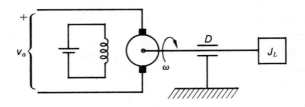

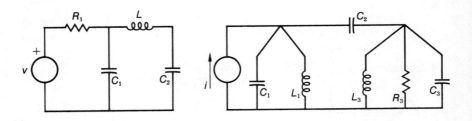

2-10. Consider the mutually coupled circuit illustrated. Write the differential equations of the system in normal form.

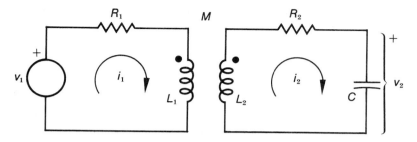

2-11. Deduce state models for each of the systems shown.

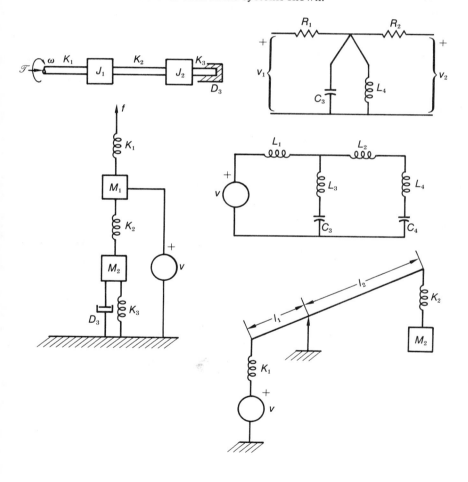

2-12. Write the following differential equations in normal form, and specify the A and B matrixes for each,

$$p^2y + 2py + y = u,$$
$$(p^3 + 3p^2 - 2p + 4)y = pu.$$

2-13. Deduce the matrixes, A,B,P,Q of the normal form for the differential equation,

$$(p^3 + p^2 + 2p + 5)y = u_1 + 3u_2,$$

where u_1 and u_2 are inputs, and y is the output.

2-14. Deduce the state and matrixes A,B,P,Q for the equation,

$$(p^3 + p^2 + 2p + 5)y = (1 + 3p)x.$$

2-15. Refer to the accompanying figure.
  a. Write differential equations for i_1 and i_2.
  b. Write these in normal form, giving the appropriate A and B matrixes.
  c. Deduce a state model for this network. Relate this to b.

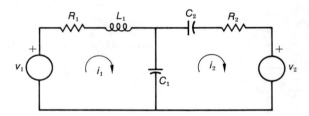

2-16. Refer to the accompanying figure. Select the voltages and inputs and the loop currents as outputs. Write the transfer matrix for the system, (the transfer matrix gives the operational relation between output and input).
  a. Using the Kirchhoff formulation.
  b. By writing the differential equations of the network in normal form.

2-17. Consider the simple RC ladder network shown.
  a. Write the equilibrium description of the system in normal form.
  b. Formulate the description in difference equation form.

2-18. A continuous system is described by the differential equation

$$(5p^2 + 3p + 2)y = (3p + 1)x$$

  a. Write the equation in normal form.
  b. Write the difference equation approximation to this differential equation.
  c. Write the difference equation in b. in normal form.

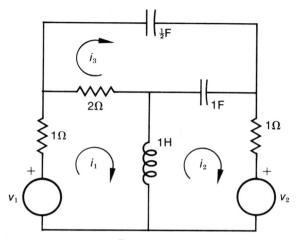

Problem 2-16

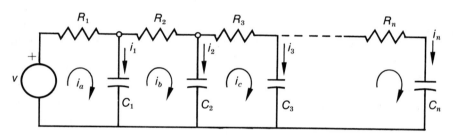

Problem 2-17

2-19. A system is characterized by the difference equation,
$$5y(n+2) + 3y(n+1) + 2y(n) = 3x(n+1) + x(n).$$
Express this system in state variable form.

2-20. A system is characterized by the simultaneous difference equations,
$$3y_1(n+1) + 5y_1(n) - 2y_2(n) = x_1(n),$$
$$y_1(n) + y_2(n+2) + y_2(n) = x_2(n).$$
Write these equations in normal form.

3

Signal Flow Graphs

The signal flow graph (SFG) (2) is a graphical portrayal of the signal interconnections of a system. It may represent the complicated interconnection of all or a portion of the system. It has the feature that the signal paths from input to output may be placed in sharp focus without the complications of the hardware of the system.

From our point of view, the SFG possesses three features that will be of importance in our subsequent work, (1) it provides a tool of analysis which often possesses advantages over other methods, (2) it provides a means for easily establishing the program for an analog computer study of a system, (3) it provides a convenient means for relating the Kirchhoff network formulation with the state variable formulation. These features will be discussed, in part, in this chapter, and the SFG will be used elsewhere as conditions require.

3-1 PROPERTIES OF SFG

The variables in a signal flow graph are represented by points called *nodes*. Connections between nodes are by directed lines which are unilateral (signals can flow only in the direction indicated by the arrows), and are called *transmittances*. Generally the independent variables (real or assumed) are viewed as system inputs or driving variables. The dependent variables, representing the unknown quantities are viewed as system responses. The paths of interaction and the value of the transmittances along them exhibit the manner and the degree of effect of one quantity on another.

The SFG allows a ready visualization of the system interactions. Further, rules exist which allow certain algebraic transformations to be effected readily, thus allowing the rapid transformation of graphs into forms which may have some desired attribute for subsequent study. Often, in fact, the SFG makes apparent those combinations or transformations which are likely to lead to some desired form. However, it must be recognized that the SFG technique is of importance principally in representing and handling the set of equations which form the mathematical model of the interconnected physical system. It is less useful in studying the physical system itself.

The rules for drawing a signal flow graph are the following:

1. Signals travel along branches only in the direction of the arrows.
2. A signal traveling along a branch from one node to another is multiplied by the transmittance of that branch.
3. The value of the variable represented by any node is the sum of all signals entering that node.
4. The value of the variable represented by any node is transmitted to all branches leaving that node.

These rules are illustrated by several examples.

Example 3-1.1

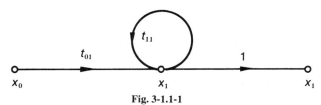

Fig. 3-1.1-1

Draw the SFG of the equation,

$$x_1 = t_{01}x_0 + t_{11}x_1.$$

Solution: The graph is drawn according to the rules above.

Example 3-1.2

Draw the SFG of the related equations,

$$x_1 = t_{01}x_0 + t_{11}x_1 + t_{12}x_2,$$
$$x_2 = t_{02}x_2 + t_{12}x_1 + t_{22}x_2.$$

110   Signal Flow Graphs

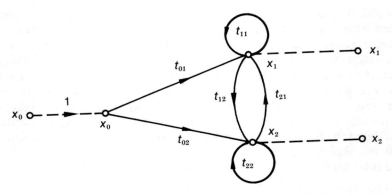

Fig. 3-1.2-1

Solution: In accordance with the rules above, the graph follows.

For convenience, the SFG is usually drawn in such a way that no branch enters an input node or leaves an output node. These conditions can be designated readily on the SFG by introducing additional nodes connected by appropriately directed unity transmittances to effect the desired isolation. For example, if in Example 3-1.2 x_0 denotes the input node, and both x_1 and x_2 denote output nodes, we would add the broken lines shown.

3-2 GRAPHING DIFFERENTIAL EQUATIONS

To graph a differential equation we proceed in a direct manner, remembering that the symbol $1/p$ defines integration, but otherwise using the rules prescribed above. To examine the principles involved, refer to the following differential equation,

$$\frac{d^2y}{dt^2} + 5\frac{dy}{dt} + 2y = 3x. \tag{3-1}$$

We wish to develop a SFG for this equation. In this equation x is the independent variable, and we wish to determine the SFG for y. To accomplish our ends, we write the differential equation in the form,

$$y = \frac{3}{2}x - \frac{5}{2}\frac{dy}{dt} - \frac{1}{2}\frac{d^2y}{dt^2}. \tag{3-2}$$

We shall make use of the following implicit relations, even though they are not specified explicitly,

$$\frac{dy}{dt} = py,$$

$$\frac{d^2y}{dt^2} = p^2y. \tag{3-3}$$

A direct SFG representation for Eq. (3-1) is given in Fig. 3-1. Now we modify this graph to include the implicit relation between a variable and its derivatives, as given in Eq. (3-3). The resulting graph is given in Fig. 3-2.

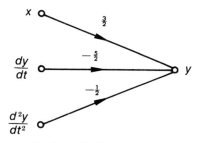

Fig. 3-1. SFG of Eq. (3-2).

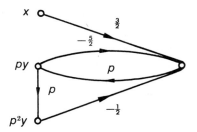

Fig. 3-2. The complete SFG of Eq. (3-1).

A more interesting form of the graph is that which arranges the nodes in such a way that a clear signal path extends from the source node to the output node, the dependent node at which the response is observed. For the present case this is most conveniently obtained by writing the differential equation in the form,

$$p^2y = 3x - 5py - 2y. \tag{3-4}$$

112   Signal Flow Graphs

The SFG now attains the form shown in Fig. 3-3. We shall later find that in this form the SFG provides the scheme for programming an analog computer for studying Eq. (3-1).

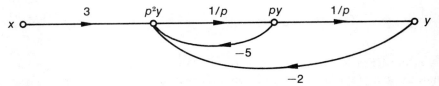

Fig. 3-3. The SFG for Eq. (3-1) illustrating the signal transmission features.

3-3 SIMULTANEOUS DIFFERENTIAL EQUATIONS

The procedure in graphing a set of differential equations is a logical extension of the procedure discussed above. In the manner of Eq. (3-4) each equation is solved for the highest order derivative of the different variables existing in the set. For example, if two differential equations in x and y exist, then one of these equations will be written for the highest order derivative in y, and the second will be written for the highest order derivative in x. The particular choice will be dictated by the highest order that appears in either of the set.

The following steps formalize the procedure in graphing a set of simultaneous differential equations:

1. Each equation of the set is associated with a different variable, the selection being such as to permit the highest order derivative of the chosen variable to be selected for each equation.
2. Each equation is solved for the highest order derivative of the chosen variable.
3. Draw a partial SFG for each equation, proceeding as though the highest order derivative specified in 2 is independent, with all other terms being dependent.
4. Include auxiliary nodes and branches in each partial SFG to relate successive order derivatives, to provide a signal flow path to all nodes representing derivatives of lower order than those written under rule 2.
5. Combine the partial SFG-s, interconnecting identically labeled nodes.
6. Redraw the graph in a form which clearly displays the transmission paths from the source node to the response node.

Simultaneous Differential Equations 113

Example 3-3.1

Graph the following set of differential equations,

$$\frac{d^3y}{dt^3} + 5\frac{d^2y}{dt^2} + 3y - \frac{dx}{dt} = f(t),$$

$$\frac{d^2y}{dt^2} + y - 2\frac{d^2x}{dt^2} + 4\frac{dx}{dt} = 0.$$

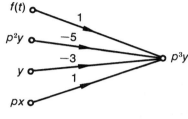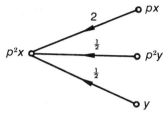

Fig. 3-3.1-1

Solution: We follow the procedure outlined above.
1. The highest order derivative is that in y in the first equation.
2. The equations are now solved to display the highest order derivatives for the different variables on the left. Thus,

$$p^3y = f(t) - 5p^2y - 3y + px,$$
$$2p^2x = 4px + p^2y + y.$$

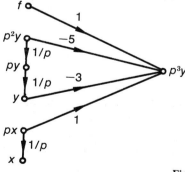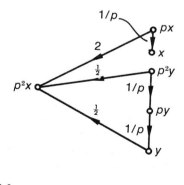

Fig. 3-3.1-2

3. Partial graphs appropriate to each equation are drawn. These are shown in the accompanying figures.

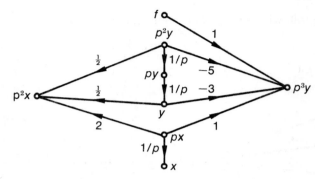

Fig. 3-3.1-3

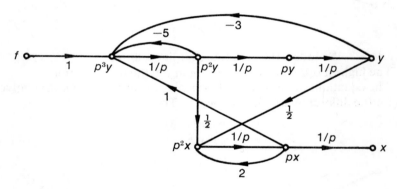

Fig. 3-3.1-4

4. Auxiliary nodes and branches are added to the partial graphs to relate derivatives of successive order. The partial graphs are modified appropriately, as shown.
5. The partial graphs are combined to give the single SFG.
6. The combined graph is redrawn to show direct paths for signal flow. The result is given.

Example 3-3.2

Illustrated is the circuit of a low pass filter. Draw a SFG for the network in a form that shows the signal flow characteristics through the network from input to output.

Simultaneous Differential Equations

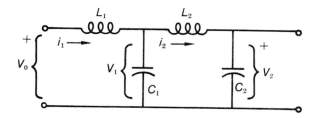

Fig. 3-3.2-1

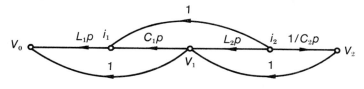

Fig. 3-3.2-2

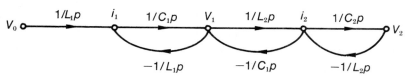

Fig. 3-3.2-3

Solution: From an inspection of the circuit, we write the following equations.

$$v_0 = L_1 p i_1 + v_1,$$
$$v_1 = L_2 p i_2 + v_2,$$
$$v_1 = \frac{1}{C_1 p}(i_1 - i_2),$$
$$v_2 = \frac{1}{C_2 p} i_2.$$

A SFG is drawn directly from these equations, as shown.

This is a valid SFG but it is not in a form that shows the signal flow through the network from input to output, but rather from output to input. Suppose, therefore, that we rewrite the equations above as follows,

$$i_1 = \frac{1}{L_1 p}(v_0 - v_1),$$

$$i_2 = \frac{1}{L_2 p}(v_1 - v_2),$$

$$v_1 = \frac{1}{C_1 p}(i_1 - i_2),$$

$$v_2 = \frac{1}{C_2 p} i_2,$$

These expressions are now graphed, the result being that shown.

It is observed in the second rewriting of the equations in Example 3-3.2 that the independent variable has been chosen, and then each equation has been written for a different one of the dependent variables. Clearly, since some latitude exists in the choice of the dependent variable in any given equation, the resulting form of the SFG will be different, and will depend on the selection of the variables. This lack of a single unique SFG for a given set of equations is both the strength and the weakness of the SFG representation. The strength lies in the fact that a portrayal of the system interconnections might be selected from a number to best meet the needs of a given situation. The weakness lies in the fact that one must examine a number of alternatives. We shall find that the versatility of the SFG is a distinct advantage in many cases.

3-4 THE ALGEBRA OF SFG-s

In the several examples above, it has been found that sets of simultaneous differential equations, when graphed, lead to a complicated interconnection pattern among the nodes, but that between a specified input and a specified output node there is a set of paths involving the entire array. It is anticipated therefore that this interconnection pattern should be reducible to a form that involves the variable associated with the output with that associated with the input. In essence, of course, what this means is that the set of differential equations must be solved simultaneously to eliminate all variables except those associated with the input and output. A formal set of reduction rules for SFG is possible which allows writing the total transmittance of a configuration from the transmittances of the branches making up the SFG. We proceed to develop the important reduction rules.

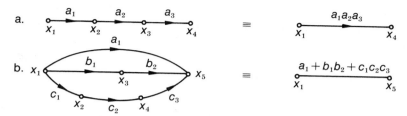

Fig. 3-4. (a) Cascaded branches; (b) Parallel branches.

The reduction rules for cascaded and parallel branches follow without difficulty. These are given in Fig. 3-4.

Another type of configuration that has been encountered in our work involves closed signal paths or loops. Many such configurations exist, and we shall examine a number of the important types. First refer to Fig. 3-5 which shows a graph containing a closed loop. The equations

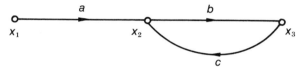

Fig. 3-5. A graph involving a feedback loop.

appropriate to this graph are written by inspection. These are,

$$x_2 = ax_1 + cx_3,$$
$$x_3 = bx_2. \tag{3-5}$$

These are combined to yield,

$$\frac{x_3}{b} = ax_1 + cx_3.$$

The total transmittance, which is the ratio of x_3/x_1, is,

$$\frac{x_3}{x_1} = \frac{ab}{1-bc}. \tag{3-6}$$

This result may be interpreted in two ways.

1. If the feedback path c is absent, the signal path from x_1 to x_3 has a transmittance ab.
2. The feedback loop introduces the factor $1/(1-bc)$, which is 1 minus

the product of the branches of the loop (called the loop transmittance). Further, we may describe the situation in terms of a single loop that shares nodes with a path, thereby modifying this transmittance of that path. In terms of a notation that will later be used, we may write the results of Eq. (3-6) in the form,

$$T = \frac{P}{1-L}, \tag{3-7}$$

where P denotes the path transmittance, and L denotes the loop transmittance.

Refer to Fig. 3-6. The transmittances of the two different graphs may

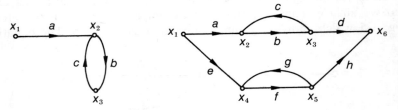

Fig. 3-6. (a) Single isolated loop; (b) Two isolated loops.

be written, in accordance with the foregoing: for the single isolated loop of Fig. 3-6a,

$$T = \frac{x_2}{x_1} = \frac{a}{1-bc}, \tag{3-8}$$

for the two isolated loops of Fig. 3-6b,

$$T = \frac{x_6}{x_1} = \frac{P_1}{1-L_1} + \frac{P_2}{1-L_2},$$

which is,

$$T = \frac{abd}{1-bc} + \frac{efh}{1-fg}. \tag{3-9}$$

Now let us consider the SFG of Fig. 3-7. This graph is conveniently divided into two cascaded parts, as indicated. The total transmittance is

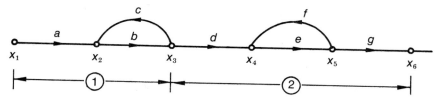

Fig. 3-7. Two isolated loops on a single transmission path.

the product of the transmittance of each subgraph, or,

$$T = \frac{x_6}{x_1} = \frac{P_1}{1-L_1} \frac{P_2}{1-L_2} \qquad (3\text{-}10)$$
$$= \frac{P_1 P_2}{1-L_1-L_2+L_1L_2} = \frac{P}{1-L_1-L_2+L_1L_2},$$

which is,

$$T = \frac{x_6}{x_1} = \frac{abdeg}{1-bc-ef+bcef}. \qquad (3\text{-}11)$$

Now refer to Fig. 3-8, which is similar to Fig. 3-7, but which involves what are called interacting loops, since two loops share a single node.

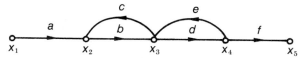

Fig. 3-8. A SFG with interacting loops.

Because of this interaction, the isolated loop procedure used in connection with Fig. 3-7 is not permitted. Instead, we must give detailed attention to the SFG. The set of equations appropriate to this graph are the following,

$$\begin{aligned} x_2 &= ax_1 + cx_3, \\ x_3 &= bx_2 + ex_4, \\ x_4 &= dx_3, \\ x_5 &= fx_4. \end{aligned} \qquad (3\text{-}12)$$

120   Signal Flow Graphs

By a systematic elimination of x_2, x_3, and x_4, the transmittance from x_1 to x_5 is found to be,

$$T = \frac{x_5}{x_1} = \frac{abdf}{1 - bc - de}. \tag{3-13}$$

In terms of our general notation, this is,

$$T = \frac{P}{1 - L_1 - L_2}. \tag{3-14}$$

By comparing this expression with Eq. (3-10) we see that interaction of the loops has resulted in the product term $L_1 L_2$ being missing from the equation.

To generalize the foregoing results, refer to Fig. 3-9, which illustrates a rather general type of graph. The total transmittance of this graph may

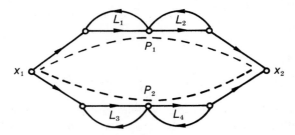

Fig. 3-9. A SFG with two sets of interacting loops.

be written on the basis of the foregoing discussion. The result is,

$$T = \frac{x_2}{x_1} = \frac{P_1}{1 - L_1 - L_2} + \frac{P_2}{1 - L_3 - L_4}. \tag{3-15}$$

Now suppose that this expression is expanded. This yields,

$$T = \frac{P_1(1 - L_1 - L_2) + P_2(1 - L_3 - L_4)}{1 - (L_1 + L_2 + L_3 + L_4) + (L_1 L_3 + L_1 L_4 + L_2 L_3 + L_2 L_4)}. \tag{3-16}$$

This expression is of the form

$$T = \frac{P_1 \Delta_1 + P_2 \Delta_2}{\Delta} \tag{3-17}$$

which shows it to be the sum of the individual path transmittances P_k, each of which is weighted by a path factor Δ_k, the sum being divided by a quantity Δ that involves all of the loops, but which reflects the manner of the interconnection.

In its most general form, the transmittance of a SFG may be written as

$$T = \sum_k \frac{P_k \Delta_k}{\Delta} \qquad (3\text{-}18a)$$

where P_k denotes the path transmittance from input to output for every possible direct path through the network. In writing the P_k, no node should be encountered more than once along path k. The graph determinant Δ involves, as is seen from Eq. (3-15), only the closed loops of the graph and their interconnections, if any. The rule for evaluating Δ in terms of the loop transmittances $L_1, L_2, \ldots$, is,

$\Delta = 1 - $ (sum of all different loop transmittances)
$\quad +$ (sum of transmittance products of all possible pairs of nontouching loops)
$\quad -$ (sum of transmittance products of all possible triples of nontouching loops)
$\quad + \cdots \qquad (3\text{-}18b)$

The path factor Δ_k, which is essentially a weighting factor for each path transmittance, involves all of the loops in the graph which are isolated from path k. When a path touches all of the loops of a graph, the path factor is unity. The path factor is unity when the path contains no loops. In general, the path factor Δ_k is the value of the graph determinant with path k removed from the network. A general proof of this theorem is rather involved, and will not be given (*see* Reference 2).

Example 3-4.1

Specify the path factors: P_k, Δ_k, and Δ for the graph shown.

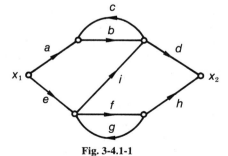

Fig. 3-4.1-1

Solution: The required quantities are obtained from an inspection of the graph, in accordance with the Mason rules given above. These are,

$$P_1 = abd \qquad P_2 = efh \qquad P_3 = eid$$
$$\Delta_1 = 1 - fg \qquad \Delta_2 = 1 - bc \qquad \Delta_3 = 1$$
$$\Delta = 1 - (bc + fg) + bcfg$$

Example 3-4.2

Specify the path factors: P_k, Δ_k, Δ for the graph shown.

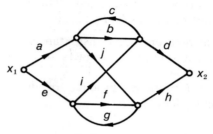

Fig. 3-4.2-1

Solution: The required quantities are the following:

$$P_1 = abd \quad P_2 = efh \quad P_3 = eid \quad P_4 = ajh \quad P_5 = ajgid \quad P_6 = eicjh$$
$$\Delta_1 = 1 - fg \quad \Delta_2 = 1 - bc \quad \Delta_3 = 1 \quad \Delta_4 = 1 \quad \Delta_5 = 1 \quad \Delta_6 = 1$$
$$\Delta = 1 - (bc + fg + cjgi) + bcfg$$

3-5 STATE EQUATIONS AND THE SFG

The state equations of a network lend themselves to very ready graphing, since they are in a form for easy partial graph description. The problem is little more than that of combining the partial graphs, as discussed in Section 3-3. Suppose that we consider the state equations that result from the normal form decomposition of the differential equation,

$$\frac{d^3y}{dt^3} + 3\frac{d^2y}{dt^2} + 4\frac{dy}{dt} + 7y = 2\frac{du}{dt} + 5u. \qquad (3\text{-}19)$$

In accordance with the discussion in Section 2-9, a state description is given by a normal form decomposition of this differential equation, which is,

$$\begin{aligned}\dot{x}_1 &= x_2 \\ \dot{x}_2 &= x_3 \\ \dot{x}_3 &= -3x_3 - 4x_2 - 7x_1 + u \\ y &= 2x_2 + 5x_1\end{aligned} \quad (3\text{-}20)$$

which is the matrix form,

$$\frac{d}{dt}\begin{bmatrix} x_1 \\ x_2 \\ x_3 \end{bmatrix} = \begin{bmatrix} 0 & 1 & 0 \\ 0 & 0 & 1 \\ -7 & -4 & -3 \end{bmatrix}\begin{bmatrix} x_1 \\ x_2 \\ x_3 \end{bmatrix} + \begin{bmatrix} 0 \\ 0 \\ u \end{bmatrix} \quad y = [5 \ 2 \ 0]\begin{bmatrix} x_1 \\ x_2 \\ x_3 \end{bmatrix} \quad (3\text{-}21)$$

Clearly, the nodes are evident from the state equations, and the SFG is drawn directly, as indicated in Fig. 3-10.

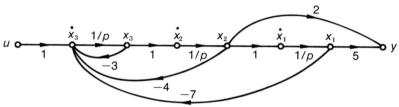

Fig. 3-10. The SFG for the state equations, Eq. (3-20).

We note that an application of the rules of Section 3-4 for the transmittance of the graph will return the differential equation given in Eq. (3-19). Conversely, if we were to graph the differential equation, Eq. (3-19) in a form that involves only $1/p$ or integrating functions, then the state equations can be read from the SFG. Further, since it is possible to find more than one SFG that would meet the description of the given differential equation, then more than one state description of a given system is possible, as already discussed.

Example 3-5.1

Graph the state equations,

$$\begin{aligned}\dot{x}_1 &= -3x_1 + x_2 & \dot{x}_3 &= -7x_1 + 5u \\ \dot{x}_2 &= -4x_1 + x_3 + 2u & y &= x_1\end{aligned}$$

and from the SFG determine the system transmittance.

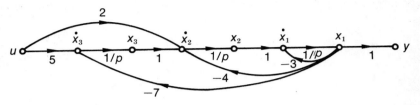

Fig. 3-5.1-1

Solution: The SFG is deduced directly from the equations, and has the form shown. We now employ Eq. (3-18) for the graph transmittance. The important graph factors are,

$$P_1 = \frac{5}{p^3} \quad P_2 = \frac{2}{p^2}$$

$$\Delta_1 = 1 \quad \Delta_2 = 1$$

$$\Delta = 1 + \frac{3}{p} + \frac{4}{p^2} + \frac{7}{p^3}$$

Hence

$$T = \frac{y}{u} = \frac{(5/p^3) + (2/p^2)}{1 + (3/p) + (4/p^2) + (7/p^3)} = \frac{2p+5}{p^3 + 3p + 4p + 7}$$

Observe that this is exactly Eq. (3-19). Thus the given state equations and those in Eq. (3-20) are different representations of the same system. This implies that given the system transmittances, one can use the SFG as an intermediate point in finding a state representation of the system, and vice versa, given the state representation of the system, the SFG may be used as an intermediate point in finding the total transmittance of the system. Actually, of course, purely analytical methods exist which permit the transformation from the transmittance function to a set of state variables. Such a transformation is an important step in a synthesis procedure for finding a network configuration appropriate to a given transmittance function, using state equations as a starting point for the synthesis procedure.

Discrete time systems can also be simulated by an analog type SFG which has the same general topology as that for the continuous time system except that the integrator is replaced by a unit time delay operator

z^{-1}. The discrete time SFG for a system described by the equations,

$$x(n+1) = zx(n) = Ax(n) + Bu(n),$$
$$y(n) = Px(n) + Qu(n).$$

has the form shown in Fig. 3-11.

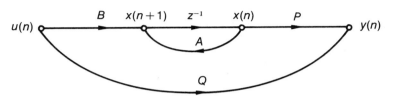

Fig. 3-11. The flow graph of a discrete time system.

PROBLEMS

3-1. Draw a SFG for the following set of equations

$$x_1 + 3x_2 - 4x_3 = 0$$
$$2x_2 + x_3 - 5x_4 = 0$$
$$x_3 + 2x_4 = 0$$

when
 a. x_1 denotes the input and x_4 denotes the output.
 b. x_4 denotes the input and x_1 denotes the output.

3-2. Deduce the controlling equations for the following systems, and draw a SFG for each.

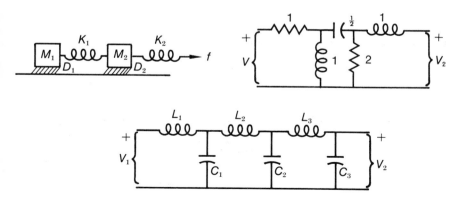

3-3. Draw SFG-s for the differential equations,

$$(p^2+3p+5)y = 7$$

$$\ddot{y}+3y = 2x+5$$

$$\frac{d^3y}{dt^3}+3\frac{d^2y}{dt^2}+7y = 2x$$

3-4. Draw SFG-s for the following sets of differential equations, for zero initial conditions:

$$(p^2+3p+2)y = x+7 \qquad (p^2+5)y = 3x$$
$$(p+2)x = (3-2p)y \qquad (p^2+7)x = 5y+3$$

3-5. Write the equations that are portrayed in the following SFG-s.

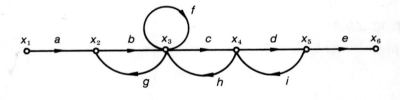

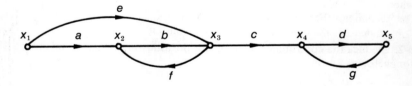

3-6. Find the transmittance of the SFG illustrated.

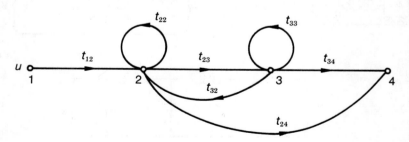

3-7. Calculate the transmittances of the following SFG-s.

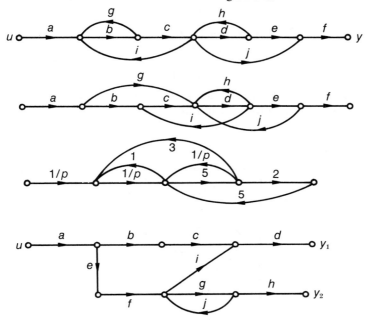

3-8. For the following SFG-s, write:
a. The transmittances.
b. The state equations for each system.

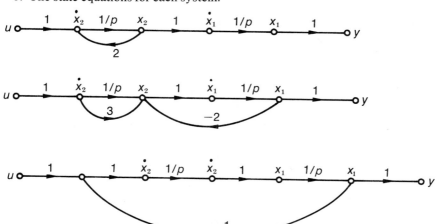

3-9. The controlling equations of a motor that developes a torque T_m and that drives an inertia load coupled to the motor through a flexible shaft, are,

$$T_m = (J_m p^2 + D_m p)\theta_m + K(\theta_m - \theta_L)$$
$$0 = -K(\theta_m - \theta_L) + J_L p^2 \theta_L + T_L$$

a. Draw a SFG for this system.
b. Write a set of state equations for this system.

3-10. a. Draw a SFG for the general system differential equation given in Eq. (2-33), with the normal form equations given in Eq. (2-40).

b. It can be shown that the system SFG can be given in the ladder form shown. Prove that this satisfies Eq. (2-33).

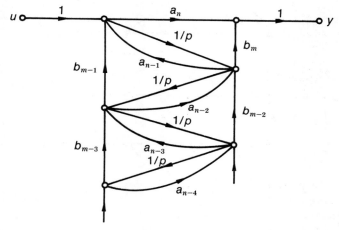

3-11. Consider the simultaneous difference equations,

$$y_1(n+1) + 2y_1(n) - 3y_2(n) = x_1(n),$$
$$y_1(n) + 5y_2(n+1) = x_2(n).$$

a. Give a SFG representation of this system.
b. Write the equations in normal form.

3-12. Specify the difference equation of the system that is defined by the SFG shown.

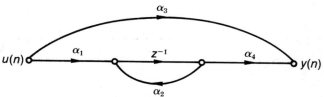

3-13. The SFG of a circuit that produces a discrete-time signal $y(n)$ is shown. Specify the expression for $y(n)$.

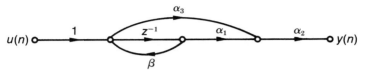

3-14. The SFG of a system is given.
  a. Write the differential equation of the system.
  b. Write the state equations from the SFG.
  c. If the integrators are replaced by unit delays, what is the difference equation of the system?
  d. Write the state equations of the system specified in c.

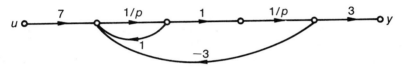

4

System Geometry and Constraint Equations

It was noted in the Introduction to Chapter 2 that among the features of a system that are important to an understanding of systems are the topological or geometrical constraints that are imposed by the interconnection of the elements into the connected network. We wish to examine these properties and the consequences thereof.

4-1 INTERCONNECTED ELEMENTS

We know from our studies that a given set of elements can be connected in a variety of ways; hence the particular connection pattern imposes constraints on the physical variables that describe the system. To keep account of these constraints in any problem is little more than a bookkeeping chore. It has been found that the organizational and bookkeeping needs can be systematized in terms of matrixes that describe the connectivity properties of the system. These connectivity properties are deduced from a study of the topology of the network, a study that involves only rather elementary aspects of graph theory, a branch of algebraic topology.

We note that in addition to the connectivity properties that are described by the topological structure, an algebraic structure that consists of the interrelations between the variables associated with the nodes, branches, and meshes of the graph, is superimposed on the topological structure. It is the physical variables of the network that correspond to the quantities in the algebraic structure. The matrix approach permits a direct correlation between these two structures. Moreover, it has the additional merit of identifying precisely the separate roles of the element characteristics and of the interconnections in describing the dynamics of the system. These system features will also be studied.

4-2 GRAPH OF A NETWORK

Suppose first that the mechanical system shown in Fig. 2-1a is redrawn in its network model form using the graphical symbols that were discussed in Chapter 1 and which are given in Table 1-1. This schematic diagram is given in Fig. 4-1. Suppose now that we represent each element in Fig. 4-1

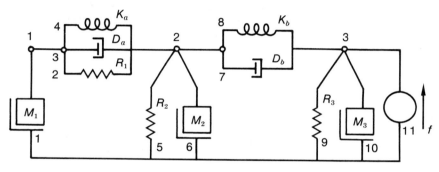

Fig. 4-1. The schematic representation of the mechanical system of Fig. 2-1a.

by a line segment, the ends of the line segment denoting the terminals. However, common points will be drawn as single points. The result is the *graph* which is given in Fig. 4-2. The arrows which are arbitrarily added, orient the graph. In this diagram only the vertexes and the connect-

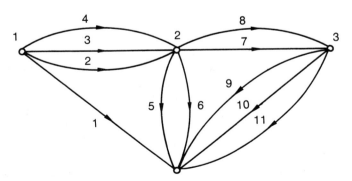

Fig. 4.2. The graph of the network of Fig. 4-1.

132   System Geometry and Constraint Equations

ing branches are important; the curvature, length, or location in space of the branches are of no significance in our subsequent studies.

To avoid misunderstanding later, it is desirable to define the terms which are used to discuss network graphs. For this purpose, suppose that we consider a somewhat more elaborate network than that in Fig. 4-1. This is given in Fig. 4-3. This figure shows a general electrical network in

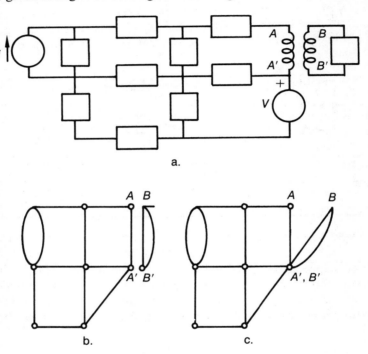

Fig. 4-3. A general network and its corresponding graph.

Fig. 4-3a and its graph in Fig. 4-3b. The magnetically coupled portion of the network makes up what is called a *separate part* of the network. In Fig. 4-3c the nodes marked A' and B' on the network graph have been made to coalesce. This procedure is permissible since no added loops have been formed in the process; otherwise the procedure would not be permissible. Special attention is required when handling multiterminal devices such as transformers, transducers, gyrators, dependent sources, etc., since the terminal graphs of such devices include rigid constraints that preclude the independence that is assumed in what follows.

The important topological terms, with explanations of the meaning of many of these, are given in the following:

Branch: a line segment replacing one network element.

Node: the intersection of two or more branches.

Degree of a node: the number of branches connected to the node.

Edge: an oriented branch, together with its two end points.

Tree: any set of branches which includes all of the nodes of a given graph, but which contains no loops.

Tree branch: a branch of a tree.

Chord (tree link, link): A branch of the graph which does not belong to the tree under consideration.

Tree complement: the totality of chords.

Loop: any closed contour selected in a graph such that every node on the contour is of degree two.

Fundamental circuit (basic loop, tie set loop): a loop selected in a graph such that it contains only one chord and some or all branches of a tree of the graph.

Tie set (fundamental loop set): the set of branches of a graph which form a fundamental circuit with respect to a tree.

Cutset: Imagine that two closed surfaces (say, plastic bags) are used to enclose groups of tree branches in such a way as to divide the tree into two parts, and such that the two surfaces are connected by a single tree branch. The cutset consists of the connecting tree branch plus all chords that are common to the two surfaces. Stated another way, the cutset is the sum of all branches (tree branch plus chords) common to the two closed surfaces.

The meanings of several of these terms are made clear by reference to Fig. 4-4 which shows two of many possible trees of the graph of Fig. 4-3c. The basis for the name tree is evident from this figure, since the structure has no closed paths, a characteristic of a physical tree. Refer now to Fig. 4-5a, which is to illustrate the fundamental circuit, and Fig.

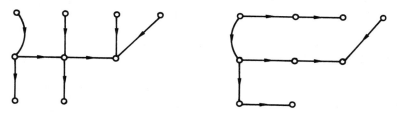

Fig. 4-4. Two trees of the graph of Fig. 4-3c.

4-5b which is used to illustrate the cut set. In Fig. 4-5a, those tree branches which, together with the chord *CD*, form a closed loop, constitutes a fundamental circuit. In Fig. 4-5b, the tree branch plus chords *CD* and *EF* which bridge the two parts into which the tree has been segregated, constitute the cutset.

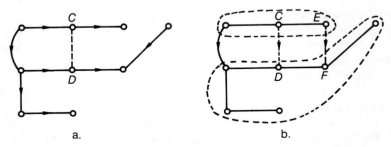

Fig. 4-5. To illustrate: (a) the fundamental circuit, (b) the cutset.

4-3 THE CONNECTION MATRIX

From the manner of drawing the graph of a network, it is clear that the connection properties of the network are contained in the graph. Furthermore, this connection information can be presented in matrix form, thereby translating the geometrical connection information into a convenient algebraic form. The basis for such a connection matrix was first introduced into the network literature by Kirchhoff in 1847.

The method suggested by Kirchhoff relates a rectangular array of numbers with a given directed linear graph. The entries in the rectangular array will be $0, +1, -1$. The rows of the matrix correspond to the nodes of the graph and the columns correspond to the branches. A direction is arbitrarily assigned to each branch of the network. Now, if a branch of the network is connected between nodes j and k, with the branch direction from j toward k, then $+1$ will appear as the jk entry of the matrix. If the branch direction is away from node k, then -1 will appear as the jk entry. If no branch connects j to k, a 0 will appear as the jk entry in the matrix.

As a specific example, refer to Fig. 4-6, which characterizes a particular network. We proceed in the manner specified above to write the connection matrix of this network. This is given by the following table.

TABLE 4-1. Connection matrix for the network of Fig. 4-6.

Nodes	Connection matrix	Branches
$\begin{bmatrix} a \\ b \\ c \\ d \\ e \end{bmatrix}$	$\begin{bmatrix} -1 & 0 & 0 & 0 & -1 & -1 & -1 & 0 \\ +1 & -1 & 0 & 0 & 0 & 0 & 0 & +1 \\ 0 & +1 & -1 & 0 & 0 & +1 & 0 & 0 \\ 0 & 0 & +1 & -1 & 0 & 0 & +1 & 0 \\ 0 & 0 & 0 & +1 & +1 & 0 & 0 & -1 \end{bmatrix}$	$\begin{bmatrix} 1 \\ 2 \\ 3 \\ 4 \\ 5 \\ 6 \\ 7 \\ 8 \end{bmatrix}$

It is observed that each column of the connection matrix contains both a +1 and a −1 as its nonzero entries. This shows that the rows are linearly dependent; that is, their sum is zero. Hence, any row is equal to the negative sum of all the other rows. This means that the matrix contains redun-

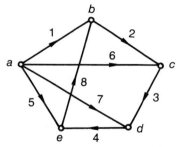

Fig. 4-6. A typical network graph.

dent information, since any one row may be deleted without affecting the information content. In particular, if we were to delete the row corresponding to node e in Fig. 4-6, we would regard this node as the datum or ground node.

We have only limited interest in this connection matrix, except that it does show that it is possible to translate graphical data into algebraic form, a concept that is of utmost importance in our subsequent work.

4-4 GENERAL FORM OF TOPOLOGICAL CONSTRAINTS

We wish to examine the two important matrixes for the network of Fig. 4-6. This is done by first selecting a tree from the network graph of Fig. 4-6. This is given in Fig. 4-7. For definiteness, we show in Fig. 4-8 the

136 System Geometry and Constraint Equations

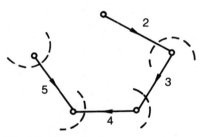

Fig. 4-7. One tree from the graph of Fig. 4-6.

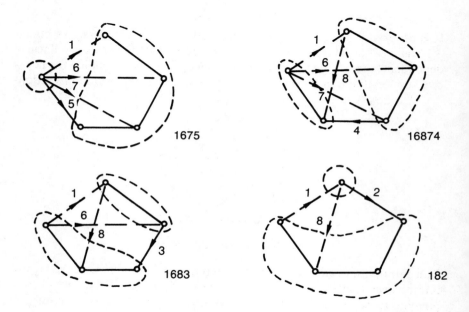

Fig. 4-8. Cutsets for the tree of Fig. 4-7.

details of marking the tree for writing the outsets appropriate to this tree. In general it is not necessary to show these in detail, since they can be written directly from Fig. 4-7. We observe from Fig. 4-8 that the cutsets appropriate to the tree of Fig. 4-7 include the graph edges: 1675, 16874, 1683, 182, where 2,3,4,5 are tree branches, and 1,6,7,8 are chords (links). In writing the cutset matrix, we use the orientation of the tree branch to specify the orientation of the cutset. Thus from Fig. 4-8 we write the cut-

General Form of Topological Constraints

set matrix (observe the rearrangement of the order of the cutsets to yield the diagonal form in the partitioned form of the cutset matrix.

$$\begin{matrix} \text{Cutset} \\ 182 \\ 1683 \\ 16874 \\ 1675 \end{matrix} \begin{bmatrix} 1 & 0 & 0 & 0 & | & -1 & 0 & 0 & 1 \\ 0 & 1 & 0 & 0 & | & -1 & -1 & 0 & 1 \\ 0 & 0 & 1 & 0 & | & -1 & -1 & -1 & 1 \\ 0 & 0 & 0 & 1 & | & 1 & 1 & 1 & 0 \end{bmatrix} \begin{bmatrix} 2 \\ 3 \\ 4 \\ 5 \\ --- \\ 1 \\ 6 \\ 7 \\ 8 \end{bmatrix} \begin{matrix} \\ \text{Branches} \\ \\ \\ \\ \text{Chords} \\ \\ \end{matrix} \quad (4\text{-}1)$$

Clearly, each row of the matrix shows which branches are included in each cutset. Further, the submatrix referring to the branches is a unit matrix.

We note that a row by row reading of this matrix contains significant information regarding the network, when interpreted in the light of the node equilibrium law. That is, since, under conditions of equilibrium, $\Sigma \tau = 0$ at each node, then we see that each row actually contains such point-law information. We can write the cutset equations in the form

$$[Q][\mathcal{T}] = [\mathbf{I} \mid S] \begin{bmatrix} \mathcal{T}_b \\ \mathcal{T}_c \end{bmatrix} = 0 \quad (4\text{-}2)$$

where the $\mathcal{T}$-s represent the through variables, and where the subscripts b refer to branches and c refer to chords, $\mathbf{I}$ is the unit matrix to the left of the partitioning line in Eq. (4-1), and S is the cutset matrix to the right of the partitioning line. It follows from Eq. (4-2) that,

$$[\mathcal{T}_b] = -[S\mathcal{T}_c]. \quad (4\text{-}3)$$

These equations constitute a set of across variable constraints, and show that the branch through-variables are uniquely determined by the chord through-variables.

Also from Fig. 4-7 we can write the fundamental circuit matrix, which specifies the circuits formed by chords plus tree branches. These circuits are formed by adding chords one at a time, and noting those graph edges that make up the circuits that are thus formed. The actual constructions are contained in Fig. 4-9. Clearly, by adding chords one at a time, the circuits thus formed are: 12345, 6345, 745, 8234. In writing the fundamental circuit matrix, we use the orientation of the added chord to specify

138   System Geometry and Constraint Equations

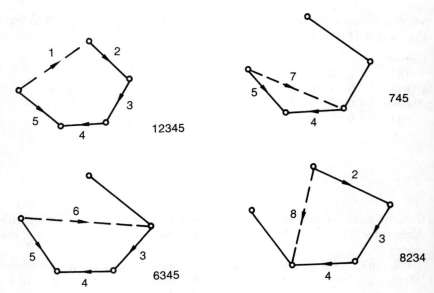

Fig. 4-9. The fundamental circuits for the tree of Fig. 4-7.

the direction of traversing the loop thus formed, with branches having the chord orientation carrying the + sign, and with branches of opposite orientation carrying the − sign. We thus have

$$
\begin{array}{c}
\text{Circuits} \\
\begin{array}{c} 12345 \\ 6345 \\ 745 \\ 8234 \end{array}
\end{array}
\begin{bmatrix}
1 & 1 & 1 & -1 & 1 & 0 & 0 & 0 \\
0 & 1 & 1 & -1 & 0 & 1 & 0 & 0 \\
0 & 0 & 1 & -1 & 0 & 0 & 1 & 0 \\
-1 & -1 & -1 & 0 & 0 & 0 & 0 & 1
\end{bmatrix}
\begin{bmatrix} 2 \\ 3 \\ 4 \\ 5 \\ --- \\ 1 \\ 6 \\ 7 \\ 8 \end{bmatrix}
\begin{array}{l} \text{Branches} \\ \\ \\ \text{Chords} \end{array}
\qquad (4\text{-}4)
$$

Observe that the rows specify which branches are included in each circuit. Furthermore, the submatrix referring to the links is a unit matrix.

A row by row reading of the fundamental circuit matrix specifies, when interpreted in the light of the loop equilibrium law, that $\Sigma\alpha = 0$ around any closed path. Thus we may write the fundamental circuit equations in

the form,
$$[M][\mathscr{A}] = [F \ \ I]\begin{bmatrix}\mathscr{A}_b\\ \mathscr{A}_c\end{bmatrix} = 0 \qquad (4\text{-}5)$$

where the $\mathscr{A}$-s represent the across variables identified with the branches and chords, where the subscripts b and c again refer to branches and chords, respectively, I is the unit matrix, and F is the circuit matrix. It follows from Eq. (4-5) that,

$$\mathscr{A}_c = -F\mathscr{A}_b. \qquad (4\text{-}6)$$

These equations constitute a set of through-variable constraints, and show that the chord across-variables are uniquely determined by the branch across-variables.

4-5 NODE PAIR AND LOOP VARIABLES

The cutsets of Fig. 4-8 and the fundamental circuits of Fig. 4-9 can be used to specify two additional network properties of some interest. Refer to Fig. 4-9 and suppose that $\mathscr{A}_n(\alpha_2, \alpha_3, \alpha_4, \alpha_5)$ denotes the across variable associated with the cutsets (i.e., with the tree branches 2,3,4,5). Suppose that in Fig. 4-9 we associate an across variable $\alpha_1, \alpha_2, \ldots, \alpha_8$ with each edge 1,2,...,8. From Fig. 4-9 and from the defining equations in Eq. (4-4) we write equations for the four fundamental circuits. These are,

$$\begin{aligned}\alpha_2 + \alpha_3 + \alpha_4 - \alpha_5 + \alpha_1 &= 0\\ \alpha_3 + \alpha_4 - \alpha_5 + \alpha_6 &= 0\\ \alpha_4 - \alpha_5 + \alpha_7 &= 0\\ -\alpha_2 - \alpha_3 - \alpha_4 + \alpha_8 &= 0\end{aligned} \qquad (4\text{-}7)$$

Observe that this set of equations may be written analytically in the form,

$$\mathscr{A} = Q^T \mathscr{A}_n, \qquad (4\text{-}8)$$

where Q^T is the transpose of the cutset matrix given in Eq. (4-1).

Now refer to Fig. 4-8 and suppose that we let $\mathscr{T}_L(\tau_1, \tau_6, \tau_7, \tau_8)$ denote the through variables associated with each fundamental circuit, where the sense of the through variable is defined by the chord that establishes the circuit. From Fig. 4-8 and Eq. (4-1) which defines the cutset matrix we write the following equations,

$$\begin{aligned}\tau_2 - \tau_1 + \tau_8 &= 0\\ \tau_3 - \tau_1 - \tau_6 + \tau_8 &= 0\\ \tau_4 - \tau_1 - \tau_6 - \tau_7 + \tau_8 &= 0\\ \tau_5 + \tau_1 + \tau_6 + \tau_7 &= 0\end{aligned} \qquad (4\text{-}9)$$

We observe that this set of equations can be written more compactly as,

$$\mathcal{T} = M^T \mathcal{T}_L, \qquad (4\text{-}10)$$

where M^T is the transpose of the fundamental circuit matrix in Eq. (4-1).

It was noted that in going from Eq. (4-1) to Eq. (4-2) and from Eq. (4-4) to Eq. (4-5) that we proceeded by a row by row reading of the appropriate matrix set. A study of Eqs. (4-7) and (4-9) will show that these denote a column by column reading of these matrix sets, when appropriately interpreted.

A general feature of the cutset matrix [I S] and the fundamental circuit matrix [F|I] for the same tree of a connected graph, when the edges are taken in the same order in each matrix is, from Eq. (4-8),

$$MQ^T = [F \quad I][I \quad S]^T = [F \quad I]\begin{bmatrix} I \\ S^T \end{bmatrix} = 0, \qquad (4\text{-}11)$$

and from Eq. (4-10),

$$QM^T = [I \quad S][F \quad I]^T = [I \quad S]\begin{bmatrix} F^T \\ I \end{bmatrix} = 0. \qquad (4\text{-}12)$$

It follows from these that,

$$F = -S^T, \quad \text{and} \quad S = -F^T. \qquad (4\text{-}13)$$

These relationships will be found of importance in our subsequent studies.

A consequence of this result is that the scalar product of through and across variables vanishes identically. That is,

$$\mathcal{A}^T \mathcal{T} = \mathcal{T}^T \mathcal{A} = 0. \qquad (4\text{-}14)$$

This result follows by examining,

$$\mathcal{A}^T \mathcal{T} = [\mathcal{A}_b^T \mathcal{A}_c^T] \begin{bmatrix} \mathcal{T}_b \\ \mathcal{T}_c \end{bmatrix}.$$

By Eqs. (4-3) and (4-6),

$$= \mathcal{A}_b^T [I \quad -F^T] \begin{bmatrix} -S \\ I \end{bmatrix} \mathcal{T},$$

which vanishes identically, using Eq. (4-13).

4-6 BRANCH PARAMETER MATRIXES

The presentations of the loop and the node-pair analysis in Sections 2-4 and 2-5 can be formalized, using the foregoing topological considerations. That is, a knowledge of the cutset matrix or the fundamental circuit matrix to establish the topological structure, together with a knowledge of the detailed features of the elements of the network to establish the algebraic structure, is sufficient to describe the connected system. Now, however, we shall separate those edges which result from sources, and shall take separate account of these. Hence the cutset and fundamental circuit matrixes will be those for the network alone. Furthermore, we shall suppose that each branch of the network can consist of the general configuration shown in Fig. 4-10. Thus each branch may be described analytically by the expressions,

$$\tau + \tau_s = y(p)(\alpha + \alpha_s),$$
$$\alpha + \alpha_s = z(p)(\tau + \tau_s). \tag{4-15}$$

These equations are dually related.

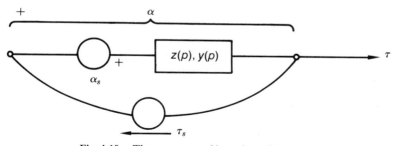

Fig. 4-10. The most general branch configuration.

To establish expressions for the algebraic structure, suppose that the network consists of b branches. Then the branch impedance matrix Z_b and the corresponding branch admittance matrix Y_b will each be $b \times b$ square matrixes. Now, for convenience in the material to follow, we shall suppose that the branch impedance matrix may be written in terms of element submatrixes; an inductor, a resistor, and a capacitor submatrix. The general form of these for electrical elements are the following, where we assume that each branch consists of a single parameter, a convenient but not an essential condition.

142 System Geometry and Constraint Equations

$$Z_b = \begin{bmatrix} \begin{bmatrix} l_{11}p & l_{12}p & l_{13}p \\ l_{21}p & l_{22}p & l_{23}p \\ l_{31}p & l_{32}p & l_{33}p \end{bmatrix} & & \\ & \begin{bmatrix} r_4 & 0 \\ 0 & r_5 \end{bmatrix} & \\ & & \begin{bmatrix} \dfrac{1}{C_6 p} & & \\ & \dfrac{1}{C_7 p} & \\ & & \dfrac{1}{C_8 p} \end{bmatrix} \end{bmatrix} \qquad (4\text{-}16)$$

In this expression it is assumed that each inductor branch may be mutually coupled to every other inductor branch in the group.

This branch impedance matrix is now written in the more general form,

$$Z_b = \begin{bmatrix} lp & 0 & 0 \\ 0 & r & 0 \\ 0 & 0 & \dfrac{1}{cp} \end{bmatrix} \qquad (4\text{-}17)$$

In this partitioned form, Z_b is a diagonal matrix. This allows ready inversion, and so we may write,

$$Y_b = Z_b^{-1} = \begin{bmatrix} lp & 0 & 0 \\ 0 & r & 0 \\ 0 & 0 & \dfrac{1}{cp} \end{bmatrix}^{-1} = \begin{bmatrix} (lp)^{-1} & 0 & 0 \\ 0 & r^{-1} & 0 \\ 0 & 0 & \left(\dfrac{1}{cp}\right)^{-1} \end{bmatrix} \qquad (4\text{-}18)$$

Further, for the system of b branches there will be b branch variables τ and α, so that we have,

$$\mathcal{T} = \begin{bmatrix} \tau_1 \\ \tau_2 \\ \cdot \\ \cdot \\ \cdot \\ \tau_b \end{bmatrix} \qquad \mathcal{A} = \begin{bmatrix} \alpha_1 \\ \alpha_2 \\ \cdot \\ \cdot \\ \cdot \\ \alpha_b \end{bmatrix} \qquad (4\text{-}19)$$

Also, the source vectors are specified explicitly as,

$$T_s = \begin{bmatrix} T_{1s} \\ T_{2s} \\ \cdot \\ \cdot \\ \cdot \\ T_{bs} \end{bmatrix} \quad A_s = \begin{bmatrix} \alpha_{1s} \\ \alpha_{2s} \\ \cdot \\ \cdot \\ \cdot \\ \alpha_{bs} \end{bmatrix} \quad (4\text{-}20)$$

for the through and across sources associated with the passive branches. Now, since Eq. (4-15) specifies the relations for each branch element, and since the foregoing matrixes are merely generalizations to represent the totality of all passive branches, then the generalization of Eq. (4-15) for the entire system can be written,

$$\begin{aligned} \mathcal{T} + T_s &= Y_b(\mathcal{A} + A_s), \\ \mathcal{A} + A_s &= Z_b(\mathcal{T} + T_s). \end{aligned} \quad (4\text{-}21)$$

These compact-looking expressions relate to the branches of the network, but do not involve any interconnection information.

Example 4-6.1

Write the branch parameter matrixes for the following network.

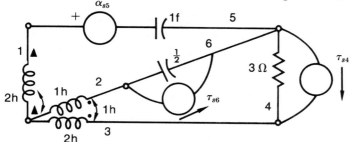

Fig. 4-6.1-1

Solution: The required matrixes are written down by inspection of the network.

$$A_s = \begin{bmatrix} 0 \\ 0 \\ 0 \\ 0 \\ -\alpha_{s5} \\ 0 \end{bmatrix} \quad T_s = \begin{bmatrix} 0 \\ 0 \\ 0 \\ -T_{s4} \\ 0 \\ -T_{s6} \end{bmatrix} \quad Z_b = \left[\begin{array}{ccc|c|cc} 2p & -p & 0 & & & \\ -p & 2p & -p & & & \\ 0 & -p & 2p & & & \\ \hline & & & 3 & & \\ \hline & & & & \dfrac{1}{p} & 0 \\ & & & & 0 & \dfrac{2}{p} \end{array} \right]$$

4-7 EQUILIBRIUM EQUATIONS ON A NODE-PAIR BASIS

To examine the connected network, we shall combine Eqs. (4-21), which specify the properties of the unconnected network (the algebraic structure) and the Q and M matrixes, which specify the cutset and the fundamental circuit constraints imposed by the connection properties of the system (the topological structure). But it must be remembered that now the Q and M matrixes apply for the passive network alone. This requires, when drawing the network graph for the passive system alone, that all across sources are to be replaced by short circuits, and all through sources are to be replaced by open circuits. We initially use the cutset matrix Q, and thereby deduce a set of equations which express the connected system in node-pair variable form.

We begin this discussion by multiplying the first of Eq. (4-21) on the left by Q. Thus,

$$Q(\mathcal{T} + T_s) = QY_b(\mathcal{A} + A_s), \tag{4-22}$$

which is now rewritten in the form,

$$Q\mathcal{T} = Q(Y_b\mathcal{A} + Y_b A_s - T_s). \tag{4-23}$$

But it is noted that $Q\mathcal{T}$ specifies the net current in those edges that bridge the two parts of the cutset, as discussed in connection with Eq. (4-2). This in electricity is the KCL and hence will equal zero. Upon rearranging the other terms in Eq. (4-23) we have,

$$QY_b\mathcal{A} = Q(T_s - Y_b A_s). \tag{4-24}$$

This expression is combined with Eq. (4-10), which relates the across variable of the chords with the across variables associated with the cutset pairs. This transforms Eq. (4-24) to a node-pair description. There results,

$$QY_b Q^T \mathcal{A}_n = Q(T_s - Y_b A_s). \tag{4-25}$$

The right hand side of this matrix equation expresses the net equivalent through sources feeding the node-pairs. The term T_s expresses the through sources associated with the passive branches. The term $Y_b A_s$ expresses the transformation of the across sources associated with the passive branches into equivalent through sources. The minus sign arises from the reference conditions that have been specified. Clearly, since Q depends on the selected tree, then the subsequent through and across variables constitute a complete, but not a unique set. That is, many acceptable node-pair variables are possible. The triple matrix product QY_bQ^T, which specifies a congruence transformation, yields a square matrix which expresses the net admittance at each of the nodes. The element y_{ik} in this matrix specifies the net admittance common to nodes i and k. Clearly, Eq. (4-25) is the more general form of Eq. (2-17).

Equation (4-25) can be translated into more recognizable form. First we recall from Eq. (4-18) that Y_b is a diagonal matrix, comprising the submatrixes for l, r, and c. Now we partition the matrix Q into three submatrixes,

$$Q = [Q_l \mid Q_r \mid Q_c], \qquad (4\text{-}26)$$

where the submatrix Q_l is appropriate to the submatrix $(l_p)^{-1}$; Q_r is appropriate to the submatrix r^{-1}; and Q_c refers to the submatrix $(1/cp)^{-1}$. Under these circumstances, the triple matrix product QY_bQ^T becomes,

$$QY_bQ^T = [Q_l \mid Q_r \mid Q_c] \begin{bmatrix} (lp)^{-1} & & \\ & r^{-1} & \\ & & \left(\dfrac{1}{cp}\right)^{-1} \end{bmatrix} \begin{bmatrix} Q_l^T \\ Q_r^T \\ Q_c^T \end{bmatrix} \qquad (4\text{-}27)$$

In expanded form, this yields,

$$QY_bQ^T = Q_l l^{-1} Q_l^T p^{-1} + Q_r r^{-1} Q_r^T + Q_c c Q_c^T p. \qquad (4\text{-}28)$$

We define the node-pair admittance matrixes,

$$\Gamma = Q_l l^{-1} Q_l^T, \qquad G = Q_r r^{-1} Q_r^T, \qquad C = Q_c c Q_c^T, \qquad (4\text{-}29)$$

which are, respectively, the inverse inductance, the conductance and the capacitance parameter matrixes on the node basis. Eq. (4-25) assumes the form,

$$(\Gamma p^{-1} + G + Cp)\mathscr{A}_n = Q(T_s - Y_b A_s), \qquad (4\text{-}30)$$

which, more simply is,

$$(\Gamma p^{-1} + G + Cp)\mathscr{A}_n = T_n, \qquad (4\text{-}31)$$

146 System Geometry and Constraint Equations

where T_n represents the net equivalent current sources feeding the node-pairs. In expanded form this equation defines the set,

$$\begin{bmatrix} (\Gamma_{11}p^{-1}+G_{11}+C_{11}p) & \cdots & (\Gamma_{1n}p^{-1}+G_{1n}+C_{1n}p) \\ \vdots & & \vdots \\ (\Gamma_{n1}p^{-1}+G_{n1}+C_{n1}p) & \cdots & (\Gamma_{nn}p^{-1}+G_{nn}+C_{nn}p) \end{bmatrix} \begin{bmatrix} \alpha_1 \\ \vdots \\ \alpha_n \end{bmatrix} = \begin{bmatrix} \tau_1 \\ \vdots \\ \tau_n \end{bmatrix} \quad (4\text{-}32)$$

This set of equations is now written in the usual form already discussed in Eq. (2-16)

$$\begin{bmatrix} y_{11} & y_{12} & \cdots & y_{1n} \\ y_{21} & y_{22} & \cdots & y_{2n} \\ \vdots & \vdots & \cdots & \vdots \\ \vdots & \vdots & \cdots & \vdots \\ y_{n1} & y_{n2} & \cdots & y_{nn} \end{bmatrix} \begin{bmatrix} \alpha_1 \\ \alpha_2 \\ \vdots \\ \vdots \\ \alpha_n \end{bmatrix} = \begin{bmatrix} \tau_1 \\ \tau_2 \\ \vdots \\ \vdots \\ \tau_n \end{bmatrix} \quad (4\text{-}33)$$

where the diagonal terms y_{kk} denote the sum of all admittances connected to node k, the node that is associated with source α_k. The terms y_{ik} denote the sum of all admittances common to nodes i and k. The quantities τ_k denote the algebraic sum of the net equivalent through sources connected to node k, with a positive sign if the source is in the positive reference direction in the branch.

4-8 EQUILIBRIUM EQUATIONS ON THE LOOP BASIS

The procedure in this section will parallel that given in Section 4-7. We again begin with Eq. (4-21) and premultiply the second of this set by the fundamental circuit matrix M. This gives,

$$M(\mathscr{A}+A_s) = MZ_b(\mathscr{T}+T_s), \quad (4\text{-}34)$$

which is now rewritten in the form,

$$M\mathscr{A} = M(Z_b\mathscr{T}+Z_bT_s-A_s). \quad (4\text{-}35)$$

The quantity $M\mathscr{A}$ specifies the net branch across variables in traversing a closed loop. By the KVL this quantity must vanish. Using this result, we rearrange Eq. (4-35) to the form,

$$MZ_b\mathscr{T} = M(A_s-Z_bT_s). \quad (4\text{-}36)$$

This expression is combined with Eq. (4-10) which relates the loop variables of the fundamental circuits with the tree branch variables. The result is,

$$MZ_b M^T \mathcal{T}_L = M(A_s - Z_b T_s). \tag{4-37}$$

This set of equations is the general form of the KVL equilibrium law expressed in terms of loop variables, and is a generalization of Eq. (2-25). In parallel with the feature noted in connection with Eq. (4-25), M depends on the selection of the tree of the network, and because of this, Eq. (4-37) represents a complete set of variables, but not, in general, the only set. That is, many acceptable loop variables are possible.

The righthand side of this matrix equation represents the net equivalent across sources that feed the loops. The sources A_s represent the across sources associated with the passive branches. $Z_b T_s$ represent the transformation of the through sources associated with the passive branches into equivalent across sources. The minus sign arises from the specified reference conditions in Fig. 4-10. The triple matrix product $MZ_b M^T$ yields a square matrix whose elements give the net impedance of each loop, including mutual impedances, if any, between loops, where mutual coupling exists.

We proceed to transform Eq. (4-37) into more recognizable form. This is done by partitioning the matrix M into three parts,

$$M = [M_l \mid M_r \mid M_c], \tag{4-38}$$

where M_l refers to the l submatrix of Z_b, M_r refers to the submatrix r, and M_c refers to the submatrix $1/c$. With this choice of M, the triple matrix product becomes,

$$MZ_b M^T = [M_l \mid M_r \mid M_c] \begin{bmatrix} lp & & \\ & r & \\ & & \frac{1}{cp} \end{bmatrix} \begin{bmatrix} M_l^T \\ M_r^T \\ M_c^T \end{bmatrix} \tag{4-39}$$

which may be expanded into the form,

$$MZ_b M^T = M_l l M_l^T p + M_r r M_r^T + M_c \frac{1}{c} M_c^T p^{-1}. \tag{4-40}$$

The loop impedance submatrixes are defined,

$$L = M_l l M_l^T, \qquad R = M_r r M_r^T, \qquad \frac{1}{C} = M_c \frac{1}{c} M_c^T, \tag{4-41}$$

148 System Geometry and Constraint Equations

which are respectively, the inductance, the resistance, and the inverse capacitance (or elastance) parameter matrixes, on the loop basis. Eq. (4-37) thus assumes the form,

$$\left(Lp + R + \frac{1}{Cp}\right)\mathcal{T}_L = M(A_s - Z_b T_s), \qquad (4\text{-}42)$$

which is, in more compact form,

$$\left(Lp + R + \frac{1}{Cp}\right)\mathcal{T}_L = A_L \qquad (4\text{-}43)$$

where A_L represents the net equivalent across sources feeding the loops. In expanded form, this equation defines the set,

$$\begin{bmatrix} \left(L_{11}p + R_{11} + \frac{1}{C_{11}p}\right) & \cdots & \left(L_{1l}p + R_{1l} + \frac{1}{C_{1l}p}\right) \\ \cdot & \cdots & \cdot \\ \cdot & \cdots & \cdot \\ \cdot & \cdots & \cdot \\ \left(L_{l1}p + R_{l1} + \frac{1}{C_{l1}p}\right) & \cdots & \left(L_{ll}p + R_{ll} + \frac{1}{C_{ll}p}\right) \end{bmatrix} \begin{bmatrix} \tau_1 \\ \cdot \\ \cdot \\ \cdot \\ \tau_l \end{bmatrix} = \begin{bmatrix} \alpha_1 \\ \cdot \\ \cdot \\ \cdot \\ \alpha_l \end{bmatrix} \qquad (4\text{-}44)$$

This expression is written in the usual form, as already discussed in Eq. (2-24)

$$\begin{bmatrix} z_{11}(p) & \cdots & z_{1l}(p) \\ z_{21}(p) & \cdots & z_{2l}(p) \\ \cdot & \cdots & \cdot \\ \cdot & \cdots & \cdot \\ \cdot & \cdots & \cdot \\ z_{l1}(p) & \cdots & z_{ll}(p) \end{bmatrix} \begin{bmatrix} \tau_1 \\ \tau_2 \\ \cdot \\ \cdot \\ \cdot \\ \tau_l \end{bmatrix} = \begin{bmatrix} \alpha_1 \\ \alpha_2 \\ \cdot \\ \cdot \\ \cdot \\ \alpha_l \end{bmatrix} \qquad (4\text{-}45)$$

Here z_{kk} denotes the sum of all elements on the contour of loop k, the loop that is associated with the loop variable τ_k. The elements z_{ik} is the sum of all impedances that are common to loops i and k. The quantities α_k denote the net equivalent loop sources and specify the algebraic sum of the across and the transformed net through sources on the contour of each loop.

4-9 THE CANONIC *LC* NETWORK

There is some value in considering the formulation of state equations from a somewhat more general point of view. We first consider a network that is composed only of L and C elements. Further, it is assumed, (1) that there are no circuits of capacitors, with or without voltage sources, and (2) that there are no cutsets of inductors, with or without current sources. The circuit may contain cutsets of capacitors with voltage sources only, and circuits of inductors with current sources only. A network of this type is called a canonic LC network.

To establish the topological structure, the graph of the network is drawn. A tree is so selected that the capacitors are included in the branches and the inductors are included in the chords. We shall write V_{bc} to represent capacitor voltages and V_{cl} to represent inductor voltages. V_c and V_1 will denote the voltages of voltage sources and current sources, respectively. The circuit equations defined by the complement of the tree are then written by Eq. (4-5), which we here write in the form

$$\begin{bmatrix} F_{11} & F_{12} & | & I & 0 \\ F_{21} & F_{22} & | & 0 & I \end{bmatrix} \cdot \begin{bmatrix} V_0 \\ V_{bc} \\ V_{cl} \\ V_1 \end{bmatrix} = 0 \qquad (4\text{-}46)$$

Correspondingly, the cutset equations given by Eq. (4-2) are,

$$\begin{bmatrix} I & 0 & | & S_{11} & S_{12} \\ 0 & I & | & S_{21} & S_{22} \end{bmatrix} \cdot \begin{bmatrix} I_0 \\ I_{bc} \\ --- \\ I_{cl} \\ I_1 \end{bmatrix} = 0 \qquad (4\text{-}47)$$

Furthermore, by Eq. (4-9) this may be written,

$$\begin{bmatrix} I & 0 & | & -F_{11}^T & -F_{21}^T \\ 0 & I & | & -F_{12}^T & -F_{22}^T \end{bmatrix} \cdot \begin{bmatrix} I_0 \\ I_{bc} \\ --- \\ I_{cl} \\ I_1 \end{bmatrix} = 0 \qquad (4\text{-}48)$$

where the current vectors are the appropriate edge currents.

To establish the algebraic structure, refer to the component terminal

150 System Geometry and Constraint Equations

equations, which are,

$$\begin{bmatrix} C_b & 0 \\ 0 & L_c \end{bmatrix} \frac{d}{dt} \begin{bmatrix} V_{bc} \\ I_{cl} \end{bmatrix} = \begin{bmatrix} I_{bc} \\ V_{cl} \end{bmatrix} \quad (4\text{-}49)$$

This component terminal equation is to be combined with the constraint equations obtained from the network graph. We first write the constraint equations in expanded form, from Eqs. (4-46) and (4-48). These are,

$$\begin{aligned} I_0 - F_{11}^T I_{cl} - F_{21}^T I_1 &= 0, \\ I_{bc} - F_{12}^T I_{cl} - F_{22}^T I_1 &= 0, \end{aligned} \quad (4\text{-}50)$$

and

$$\begin{aligned} F_{11} V_0 + F_{12} V_{cl} + V_1 &= 0, \\ F_{21} V_0 + F_{22} V_{bc} + V_1 &= 0. \end{aligned}$$

From the second and third of these equations,

$$\begin{bmatrix} I_{bc} \\ V_{cl} \end{bmatrix} = \begin{bmatrix} 0 & F_{12}^T \\ -F_{12} & 0 \end{bmatrix} \cdot \begin{bmatrix} V_{bc} \\ I_{cl} \end{bmatrix} + \begin{bmatrix} 0 & F_{22}^T \\ -F_{11} & 0 \end{bmatrix} \cdot \begin{bmatrix} V_0 \\ I_1 \end{bmatrix} \quad (4\text{-}51)$$

This is combined with Eq. (4-49) and yields,

$$\begin{bmatrix} C_b & 0 \\ 0 & L_c \end{bmatrix} \frac{d}{dt} \begin{bmatrix} V_{bc} \\ I_{cl} \end{bmatrix} = \begin{bmatrix} 0 & F_{12}^T \\ -F_{12} & 0 \end{bmatrix} \cdot \begin{bmatrix} V_{bc} \\ I_{cl} \end{bmatrix} + \begin{bmatrix} 0 & F_{22}^T \\ -F_{11} & 0 \end{bmatrix} \cdot \begin{bmatrix} V_0 \\ I_1 \end{bmatrix} \quad (4\text{-}52)$$

We also write the first and fourth of Eq. (4-50), which leads to,

$$\begin{bmatrix} I_0 \\ V_1 \end{bmatrix} = \begin{bmatrix} 0 & F_{11} \\ -F_{22} & 0 \end{bmatrix} \cdot \begin{bmatrix} V_{bc} \\ I_{cl} \end{bmatrix} + \begin{bmatrix} 0 & F_{21} \\ -F_{21} & 0 \end{bmatrix} \cdot \begin{bmatrix} V_0 \\ I_1 \end{bmatrix} \quad (4\text{-}53)$$

Equation (4-53) is the complementary input-output vector, and is given in terms of the state vector. Equations (4-52) and (4-53) together constitute a state model of the network. Note moreover, that a knowledge of this state model immediately gives the circuit matrix, i.e., the topology of the LC network and the element values.

Suppose that Eq. (4-52) is multiplied on the left by,

$$\begin{bmatrix} C_b & 0 \\ 0 & L_c \end{bmatrix}^{-1} = \begin{bmatrix} C_b^{-1} & 0 \\ 0 & L_c^{-1} \end{bmatrix} \quad (4\text{-}54)$$

The result is,

$$\frac{d}{dt} \begin{bmatrix} V_{bc} \\ I_{cl} \end{bmatrix} = \begin{bmatrix} 0 & C_b^{-1} F_{12}^T \\ -L_c^{-1} F_{12} & 0 \end{bmatrix} \cdot \begin{bmatrix} V_{bc} \\ I_{cl} \end{bmatrix} + \begin{bmatrix} 0 & C_b^{-1} F_{22}^T \\ L_c^{-1} F_{11} & 0 \end{bmatrix} \cdot \begin{bmatrix} V_0 \\ I_1 \end{bmatrix} \quad (4\text{-}55)$$

This equation, together with Eq. (4-53) constitute a state model of the system. It is customary to write these equations in the form,

$$\frac{dx}{dt} = Ax + Bu,$$
$$\bar{u} = Tx + Wu, \qquad (4\text{-}56)$$

which is one of the usual forms for the state model of the system.

4-10 THE GENERAL *LC* NETWORK

The canonic requirement is now relaxed. We choose a tree to include the maximum possible number of capacitor branches and the minimum number of inductor chords†. We again choose the capacitor voltages corresponding to the tree branches and the inductor currents corresponding to the tree complement to constitute the components of the state vector. In this case the cutset equations defined by the tree are

$$\begin{bmatrix} I & 0 & 0 & | & -F_{11}^T & -F_{21}^T & -F_{31}^T \\ 0 & I & 0 & | & -F_{12}^T & -F_{22}^T & -F_{32}^T \\ 0 & 0 & I & | & 0 & -F_{23}^T & -F_{33}^T \end{bmatrix} \cdot \begin{bmatrix} I_0 \\ I_{bc} \\ I_{bl} \\ I_{cc} \\ I_{cl} \\ I_1 \end{bmatrix} = 0 \qquad (4\text{-}57)$$

The selection of the tree (each capacitive link of the tree defines a circuit containing only voltage sources and capacitors, i.e., no inductors) is such that the matrix element $B_{13} = 0$. Also, the circuit equations defined by the tree complement are,

$$\begin{bmatrix} F_{11} & F_{12} & 0 & | & I & 0 & 0 \\ F_{21} & F_{22} & F_{23} & | & 0 & I & 0 \\ F_{31} & F_{32} & F_{33} & | & 0 & 0 & I \end{bmatrix} \cdot \begin{bmatrix} V_0 \\ V_{bc} \\ V_{bl} \\ V_{cc} \\ V_{cl} \\ V_1 \end{bmatrix} = 0 \qquad (4\text{-}58)$$

†A tree that contains all voltage sources as tree branches, all current sources as links, and as many capacitive tree branches and inductive links as possible, is called a *proper* or *normal* tree. Stated another way, a normal tree is one that contains all of the voltage sources, none of the current sources, the maximum number of capacitors and the minimum number of inductors.

The component terminal equations, Eq. (4-49), which is also valid in the present case, are combined with the revised constraint equations of the network graph. We obtain the following expressions,

$$\begin{bmatrix} C_b + F_{12}^T C_c F_{12} & 0 \\ 0 & L_c + F_{23} L_b F_{23}^T \end{bmatrix} \frac{d}{dt} \begin{bmatrix} V_{bc} \\ I_{cl} \end{bmatrix} = \begin{bmatrix} 0 & F_{22}^T \\ -F_{22} & 0 \end{bmatrix} \cdot \begin{bmatrix} V_{bc} \\ I_{cl} \end{bmatrix}$$

$$- \begin{bmatrix} F_{12}^T C_c F_{11} & 0 \\ 0 & F_{23} L_b F_{33} \end{bmatrix} \frac{d}{dt} \begin{bmatrix} V_0 \\ I_1 \end{bmatrix} + \begin{bmatrix} 0 & F_{32}^T \\ -F_{21} & 0 \end{bmatrix} \begin{bmatrix} V_0 \\ I_1 \end{bmatrix} \quad (4\text{-}59)$$

and

$$I_0 = -F_{11}^T C_c F_{11} \frac{d}{dt} V_0 - F_{11}^T C_c F_{12} \frac{d}{dt} V_{bc} + F_{21}^T I_{cl} + F_{31}^T I_1, \quad (4\text{-}60)$$

$$V_1 = -\left(F_{31} V_0 + F_{32} V_{bc} + \frac{d}{dt} F_{33} L_b F_{23}^T I_{cl} + \frac{d}{dt} F_{33} L_b F_{33}^T I_1 \right) \quad (4\text{-}61)$$

These equations show that when the networks contain cutsets consisting of capacitors and current sources, or loops consisting of inductors and voltage sources, there is a linear dependence among currents through capacitors or voltages across inductors, or both, and a linear dependence exists among the state variables.

Now premultiply all terms in Eq. (4-59) by the inverse of the coefficient matrix of the left hand term. This gives

$$\frac{d}{dt} \begin{bmatrix} V_{bc} \\ I_{cl} \end{bmatrix} = \begin{bmatrix} 0 & (C_b + F_{12}^T C_c F_{22})^{-1} F_{22}^T \\ -(L_c + F_{23} L_b F_{23}^T) & 0 \end{bmatrix} \begin{bmatrix} V_{bc} \\ I_{cl} \end{bmatrix}$$

$$- \begin{bmatrix} (C_b + F_{12}^T C_c F_{12})^{-1} F_{12}^T C_c F_{11} & 0 \\ 0 & -(L_c + F_{23} L_b F_{23}^T)^{-1} F_{23} L_b F_{33}^T \end{bmatrix} \frac{d}{dt} \begin{bmatrix} V_0 \\ -I_1 \end{bmatrix}$$

$$+ \begin{bmatrix} 0 & -(C_b + F_{12}^T C_c F_{12})^{-1} F_{32}^T \\ -(L_c + F_{23} L_b F_{23}^T)^{-1} F_{21} & 0 \end{bmatrix} \begin{bmatrix} V_0 \\ -I_1 \end{bmatrix} \quad (4\text{-}62)$$

This equation is used for derivatives of the state variables, and is combined with Eqs. (4-60) and (4-61). The results are,

$$V_1 = -\{ F_{33} L_b F_{33}^T - F_{33} L_b F_{23}^T (L_c + F_{23} L_b F_{23}^T)^{-1} F_{23} L_b F_{33}^T \} \frac{d}{dt} I_1$$

$$- \{ F_{32} - F_{33} L_b F_{23}^T (L_c + F_{23} L_b F_{23}^T) \} V_{bc} \quad (4\text{-}63)$$

$$+ \{ F_{33} L_b F_{23}^T (L_c + F_{23} L_b F_{23}^T)^{-1} F_{21} - F_{31} \} V_0$$

and

$$-I_0 = \{F_{11}^T C_c F_{11} - F_{11}^T C_c F_{12}(C_b + F_{12}^T C_c F_{12})^{-1} F_{12}^T C_c F_{11}\} \frac{d}{dt} V_0$$
$$+ \{F_{11}^T C_c F_{12}(C_b + F_{12}^T C_c F_{12})^{-1} F_{22}^T - F_{21}^T\} I_{cl} \qquad (4\text{-}64)$$
$$+ \{F_{11}^T C_c F_{12}(C_b + F_{12}^T C_c F_{12})^{-1} F_{32}^T - F_{31}^T\} I_1$$

Equations (4-62), (4-63), (4-64) constitute a state vector of the network, and is seen to be of the general form,

$$\frac{dx}{dt} = Ax + Bu + G\frac{du}{dt}$$
$$\bar{u} = Tx + Wu + X\frac{du}{dt} \qquad (4\text{-}65)$$

A comparison of these equations with Eq. (4-56) for the canonic LC network shows that the presence of chord capacitors and branch inductors reflects itself in the appearance of the terms involving the time derivatives of the source vector.

4-11 The CANONIC LC NETWORK CONTAINING R

We now wish to extend the considerations of Section 4-9 to include resistors. To avoid unnecessary complexity, it will be assumed that all excitation sources are zero. Thus the present case is essentially that illustrated in Fig. 4-11, which shows the R, L, C as being essentially withdrawn. The fundamental circuit equations are,

$$\begin{bmatrix} F_{11} & 0 & I & 0 \\ F_{21} & F_{22} & 0 & I \end{bmatrix} \cdot \begin{bmatrix} V_{bc} \\ V_{br} \\ V_{cr} \\ V_{cl} \end{bmatrix} = 0 \qquad (4\text{-}66)$$

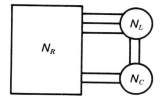
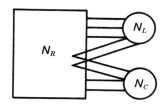

Fig. 4-11. General network forms under consideration.

and the cutset or complementary circuit equations are,

$$\begin{bmatrix} I & 0 & -F_{11}^T & -F_{21}^T \\ 0 & I & 0 & -F_{22}^T \end{bmatrix} \cdot \begin{bmatrix} I_{bc} \\ I_{br} \\ I_{cr} \\ I_{cl} \end{bmatrix} = 0 \qquad (4\text{-}67)$$

where V_{bc} = voltage vector of capacitors, V_{br} = voltage vector of branch resistors, V_{cr} = voltage vector of chord resistors, and V_{cl} = voltage vector of inductors. Similar meanings are attached to the corresponding currents. Note that B_{12} in the circuit matrix has been taken as zero in order to simplify the problem.

The terminal equations of the components are written, for C_b and L_c

$$\begin{bmatrix} C_b & 0 \\ 0 & L_c \end{bmatrix} \frac{d}{dt} \begin{bmatrix} V_{bc} \\ I_{cl} \end{bmatrix} = \begin{bmatrix} I_{bc} \\ V_{cl} \end{bmatrix} \qquad (4\text{-}68)$$

and for the branch and chord resistors

$$\begin{bmatrix} V_{br} \\ I_{cr} \end{bmatrix} = \begin{bmatrix} R_b & 0 \\ 0 & G_c \end{bmatrix} \cdot \begin{bmatrix} I_{br} \\ V_{cr} \end{bmatrix}$$

The system of equations, Eqs. (4-66), (4-67) and (4-68) can be reduced to state form by eliminating the capacitor current variables, inductor voltage variables, and resistor component variables, using the circuit and cutset equations. This leads to

$$\begin{bmatrix} C_b & 0 \\ 0 & L_c \end{bmatrix} \frac{d}{dt} \begin{bmatrix} V_{bc} \\ I_{cl} \end{bmatrix} = \begin{bmatrix} -F_{11}^T G_c B_{11} & F_{21}^T \\ -F_{21} & -F_{22} R_b F_{22}^T \end{bmatrix} \cdot \begin{bmatrix} V_{bc} \\ I_{cl} \end{bmatrix} \qquad (4\text{-}69)$$

A comparison of this expression with the corresponding terms of Eq. (4-52) for the case when $F_{12} = 0$ shows that the coefficient matrix on the right in Eq. (4-69) is the terminal matrix of the resistance network. To express this equation in the more usual state variable form, premultiply both sides of the equation by the inverse of the C, L coefficient matrix. The result is,

$$\frac{d}{dt} \begin{bmatrix} V_{bc} \\ I_{cl} \end{bmatrix} = \begin{bmatrix} -C_b^{-1} F_{11}^T G_c F_{11} & C_b^{-1} F_{21}^T \\ -L_c^{-1} F_{21} & -L_c^{-1} F_{22} R_b F_{22}^T \end{bmatrix} \cdot \begin{bmatrix} V_{bc} \\ I_{cl} \end{bmatrix} \qquad (4\text{-}70)$$

This equation shows that in the more general case of a network containing RLC elements that the operator matrix A will contain nonzero entries in all positions.

We can generalize these results without carrying out the details. Thus

The Canonic LC Network Containing R

for the case of the network with drivers we would expect Eqs. (4-56) still to be valid, but now there would be entries in all positions of the matrixes A, B, T, W. The general form would thus be of the following form:

$$\frac{d}{dt}\begin{bmatrix} V_{bc} \\ I_{cl} \end{bmatrix} = \begin{bmatrix} S_{11} & S_{12} \\ S_{21} & S_{22} \end{bmatrix} \cdot \begin{bmatrix} V_{bc} \\ I_{cl} \end{bmatrix} + \begin{bmatrix} F_{11} & F_{12} & F_{13} & F_{14} \\ F_{21} & F_{22} & F_{23} & F_{24} \end{bmatrix} \cdot \begin{bmatrix} V_1 \\ V_2 \\ I_3 \\ I_4 \end{bmatrix}$$

$$\begin{bmatrix} I_1 \\ I_2 \\ V_3 \\ V_4 \end{bmatrix} = \begin{bmatrix} T_{11} & T_{12} \\ T_{21} & T_{22} \\ T_{31} & T_{32} \\ T_{41} & T_{42} \end{bmatrix} \cdot \begin{bmatrix} V_{bc} \\ I_{cl} \end{bmatrix} + \begin{bmatrix} W_{11} & W_{12} & W_{13} & W_{14} \\ W_{21} & W_{22} & W_{23} & W_{24} \\ W_{31} & W_{32} & W_{33} & W_{34} \\ W_{41} & W_{42} & W_{43} & W_{44} \end{bmatrix} \cdot \begin{bmatrix} V_1 \\ V_2 \\ I_3 \\ I_4 \end{bmatrix} \quad (4\text{-}71)$$

where V_1, V_2, I_3, I_4 correspond to the drivers.

Example 4-11.1

Write the state equations for the circuit shown.

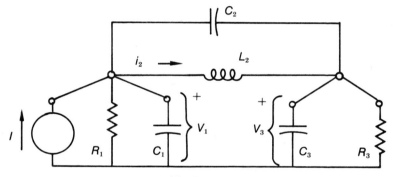

Fig. 4-11.1-1

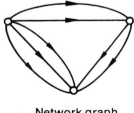

Network graph

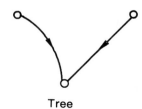

Tree

Fig. 4-11.1-2

156 System Geometry and Constraint Equations

Solution: To select the state variables, draw the network graph and select a tree that includes the maximum number of capacitors and the minimum number of inductors. These are shown in the figures. From the tree, the state variables are chosen to be v_1, v_3, i_2. The appropriate equations in terms of these are the following,

$$I - \frac{v_1}{R_1} - C_1 \frac{dv_1}{dt} - i_2 - C_2 \frac{d}{dt}(v_1 - v_3) = 0,$$

$$C_2 \frac{d}{dt}(v_1 - v_3) + i_2 - C_3 \frac{dv_3}{dt} - \frac{v_3}{R_3} = 0,$$

$$v_1 - L_2 \frac{di_2}{dt} - v_3 = 0.$$

These equations are rearranged in the order of the state variables,

$$(C_1 + C_2)\frac{dv_1}{dt} = -\frac{v_1}{R_1} - i_2 + C_2 \frac{dv_3}{dt} + I,$$

$$(C_2 + C_3)\frac{dv_3}{dt} = C_2 \frac{dv_1}{dt} + i_2 - \frac{v_3}{R_3},$$

$$L_2 \frac{di_2}{dt} = v_1 - v_3.$$

Now combine the first and second equations. This gives,

$$\left(C_1 + C_2 - \frac{C_2^2}{C_2 + C_3}\right)\frac{dv_1}{dt} = -v_1 - \left(1 - \frac{C_2}{C_2 + C_3}\right)i_2 - \frac{C_2}{C_2 + C_3}\frac{v_3}{R_3} + I$$

$$\left(C_2 + C_3 - \frac{C_2^2}{C_1 + C_2}\right)\frac{dv_3}{dt} = -\frac{C_2}{C_1 + C_2}\frac{v_1}{R_1} + \left(1 - \frac{C_2}{C_1 + C_2}\right)i_2 - \frac{v_3}{R_3} + \frac{C_2}{C_1 + C_2}I$$

$$L_2 \frac{di_2}{dt} = v_1 - v_3$$

From these we write the state equation,

$$\frac{d}{dt}\begin{bmatrix} v_1 \\ v_3 \\ \end{bmatrix} = \begin{bmatrix} \frac{C_2 + C_3}{C_\pi R_1} & -\frac{C_2}{C_\pi} & -\frac{C_3}{C_\pi} \\ -\frac{C_2}{C_\pi R_1} & -\frac{C_1 + C_2}{C_\pi R_3} & \frac{C_1}{C_\pi} \\ \frac{1}{L_2} & -\frac{1}{L_2} & 0 \end{bmatrix} \cdot \begin{bmatrix} v_1 \\ v_3 \\ i_2 \end{bmatrix} + \begin{bmatrix} I \\ \frac{C_2}{C_\pi}I \\ 0 \end{bmatrix}$$

where

$$C_\pi = C_1 C_2 + C_2 C_3 + C_3 C_1$$

4-12 THE GENERAL *RLC* NETWORK

For the general *RLC* network the state equations are in the form given in Eq. (4-65) but with entries in almost all positions of the coefficient matrixes. The results can be shown to have the form,

$$\frac{d}{dt}\begin{bmatrix}V_{bc}\\I_{cl}\end{bmatrix} = \begin{bmatrix}A_{11} & A_{12}\\A_{21} & A_{22}\end{bmatrix}\begin{bmatrix}V_{bc}\\I_{cl}\end{bmatrix} + \begin{bmatrix}B_{11} & B_{12} & B_{13} & B_{14}\\B_{21} & B_{22} & B_{23} & B_{24}\end{bmatrix}\begin{bmatrix}V_1\\V_2\\V_3\\V_4\end{bmatrix}$$

$$+ \begin{bmatrix}G_{11} & G_{12} & 0 & 0\\0 & 0 & G_{23} & G_{24}\end{bmatrix}\frac{d}{dt}\begin{bmatrix}V_1\\V_2\\I_3\\I_4\end{bmatrix}$$

$$\begin{bmatrix}I_1\\I_2\\V_3\\V_4\end{bmatrix} = \begin{bmatrix}T_{11} & T_{12}\\T_{21} & T_{22}\\T_{31} & T_{32}\\T_{41} & T_{42}\end{bmatrix} \cdot \begin{bmatrix}V_{bc}\\I_{cl}\end{bmatrix} \begin{bmatrix}W_{11} & W_{12} & W_{13} & W_{14}\\W_{21} & W_{22} & W_{23} & W_{24}\\W_{31} & W_{32} & W_{33} & W_{34}\\W_{41} & W_{42} & W_{43} & W_{44}\end{bmatrix} \cdot \begin{bmatrix}V_1\\V_2\\I_3\\I_4\end{bmatrix} \quad (4\text{-}72)$$

$$+ \begin{bmatrix}X_{11} & X_{12} & & 0\\X_{21} & X_{22} & &\\ & \text{---} & \text{---} & \text{---}\\0 & & X_{33} & X_{34}\\ & & X_{43} & X_{44}\end{bmatrix}\frac{d}{dt}\begin{bmatrix}V_1\\V_2\\I_3\\I_4\end{bmatrix}$$

Attention is called to the fact that a similarity transformation is possible which will transform these equations into a form like Eq. (4-65) without the time derivative factors of the sources, as already shown in Section 2-9.

Example 4-12.1

Formulate a state model of the circuit shown.

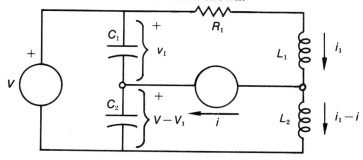

Fig. 4-12.1-1

158   System Geometry and Constraint Equations

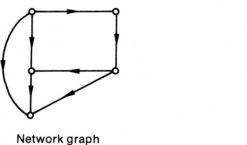
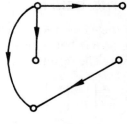

Network graph   Tree

Fig. 4-12.1-2

Solution: Clearly, the capacitors form a closed loop across the source. Also, the inductors form a cutset. Hence only one capacitor voltage and one inductor current will be used in establishing the state vector. Refer now to the graph of the network and select a tree with the minimum number of inductors. The graph and the selected tree are illustrated. We select as state variables only v_1 and i_1, even though four variables appear to be possible state variables. We now write the equations

$$v = R_1 i_1 + L_1 \frac{di_1}{dt} + L_2 \frac{d}{dt}(i_1 - i)$$

$$C_1 \frac{dv_1}{dt} + i = C_2 \frac{d}{dt}(v - v_1)$$

These equations are rearranged to the form,

$$(L_1 + L_2)\frac{di_1}{dt} = -R_1 i_1 + v + L_2 \frac{di}{dt},$$

$$(C_1 + C_2)\frac{dv_1}{dt} = C_2 \frac{dv}{dt} - i.$$

From these we write the state equation,

$$\frac{d}{dt}\begin{bmatrix} v_1 \\ i_1 \end{bmatrix} = \begin{bmatrix} 0 & 0 \\ 0 & \frac{-R_1}{L_1+L_2} \end{bmatrix}\begin{bmatrix} v_1 \\ i_1 \end{bmatrix} + \begin{bmatrix} 0 & \frac{-1}{C_1+C_2} \\ \frac{1}{L_1+L_2} & 0 \end{bmatrix}\begin{bmatrix} v \\ i \end{bmatrix}$$

$$+ \begin{bmatrix} \frac{C_2}{C_1+C_2} & 0 \\ 0 & \frac{L_2}{L_1+L_2} \end{bmatrix}\frac{d}{dt}\begin{bmatrix} v \\ i \end{bmatrix}$$

4-13 DUALITY

We call specific attention to Fig. 2-3 and its describing equations, Eqs. (2-13) and Fig. 2-4 and its describing equations, Eqs. (2-21). Observe that the describing equations are identical in form, but with the role of the dependent and independent variables interchanged. Thus Figs. 2-3 and 2-4 are dually related. We wish to add to the significance of duality from topological considerations.

In addition to the quantitative aspects of duality which were contained in Fig. 1-19, there are qualitative aspects of duality. These are contained in Table 4-2.

TABLE 4-2. Qualitative aspects of duality.

Loop	Node Pair
Tree Branch	Chord
Across variable	Through variable
Series	Parallel
Open circuit	Short circuit
Tie set	Cutset

The dual character of the loop and the node-pair suggests a rule for drawing the dual of a given *mappable graph* (one which can be laid on a plane or a sphere without essential cross-overs; these are also known as *flat* networks. This rule is specified by a series of steps for obtaining the dual of a mappable graph.

1. Place a dot in each loop of the mapped graph, and include also a dot for the outside of the network.
2. Connect two dots by one line through each edge. A degenerate case requires special care. Thus, an open branch in series with a closed network has, as its dual, a branch in parallel with the dual of the original network.
3. Draw the dual graph. Observe that all edges that are cut, as in 2, are common to the nodes appropriate to the dots.

Clearly from the manner of the construction, the loops of a dual graph will correspond to the node pairs of the original graph, and vice versa. Also, the number of edges remains invariant in this transformation.

Another feature of the dual of an oriented graph is that the dual is also an oriented graph. Hence, it is important to establish a method which ensures that the proper orientation of the two graphs is established. This is possible by establishing a duality of sense, with converging sense being the dual of clockwise sense. Refer to Fig. 4-12 which illustrates the situa-

160 System Geometry and Constraint Equations

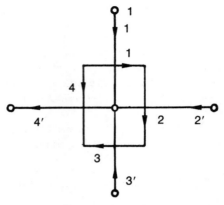

Fig. 4-12. Duality of sense.

tion. Here edges 1,2,3 are in a clockwise sense with edge 4 in a counterclockwise sense. The corresponding dual edges 1',2',3' are converging towards the node, with edge 4' away from the node.

Example 4-13.1

Draw the dual of the oriented graph shown.

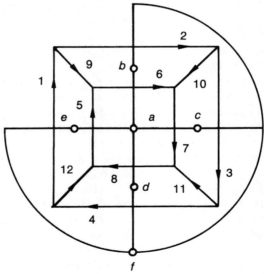

Fig. 4-13.1-1

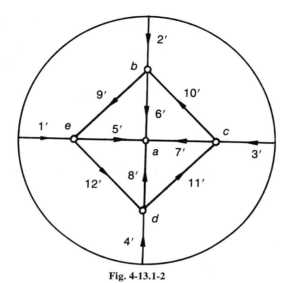

Fig. 4-13.1-2

Solution: The dual graph is drawn by following the rules given above. The result is shown.

Since duality is a functional property of network graphs, it carries no obvious information concerning numerical coefficients. It is possible, of course, to draw two networks which are dually related and which have the same numerical values for dually related elements. Such dually related networks do play an important role in the design of filters; they are called *inverse* or *reciprocal* networks. For example, Fig. 4-13 shows a network, its dual, and its reciprocal. The topology of the networks of Fig. 4-13b and 4-13c are the same, but the numerical coefficients differ. Clearly, because the networks in Fig. 4-13a and 4-13b or 4-13c are duals, the form of the equilibrium equations will be the same (compare Eqs. (2-13) and (2-21)). Note that the equations that describe Fig. 4-13a and 4-13c will have the same form and also equal numerical coefficients.

An important consequence of duality is, since the mathematical form of the network equations of the circuit and its dual are the same, that the general mathematical solution of one set of equations automatically provides a solution to the other. However, the numerical factors will be different in general. For reciprocal networks, the numerical factors will be the same. A further consequence of duality is that network theorems developed for systems which are analyzed on a loop basis will have counterpart network theorems for systems which are analyzed on a node-pair basis. These considerations will receive later attention.

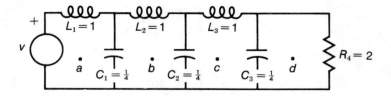

a. Given network

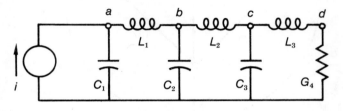

b. Dual network

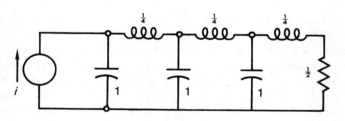

c. Inverse network

Fig. 4-13. A network, its dual, and its reciprocal.

PROBLEMS

4-1. Write the connection matrixes for the given networks.

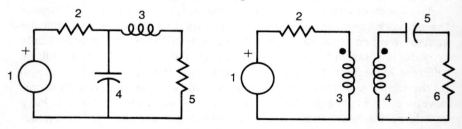

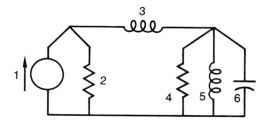

4-2. Deduce the network graphs which are specified by the following connection matrixes:

$$\begin{bmatrix} 1 & 1 & 0 & 1 & 0 & 0 & 0 \\ -1 & 0 & 0 & 0 & 0 & 0 & 1 \\ 0 & 0 & 0 & -1 & 1 & 1 & -1 \\ 0 & 0 & -1 & 0 & 0 & -1 & 0 \\ 0 & -1 & 1 & 0 & -1 & 0 & 0 \end{bmatrix} \quad \begin{bmatrix} -1 & -1 & 1 & 1 & 0 & 0 & 0 & 0 \\ 0 & 0 & -1 & -1 & 0 & 0 & -1 & -1 \\ 1 & 1 & 0 & 0 & 1 & 1 & 0 & 0 \\ 0 & 0 & 0 & 0 & -1 & -1 & 1 & 1 \end{bmatrix}$$

4-3. The terminal graph of a four node system is given in the accompanying figure. Draw at least 10 trees for this system.

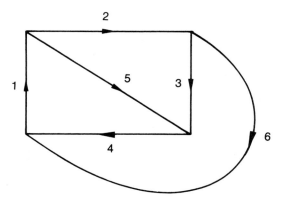

4-4. Refer to the system illustrated.
   a. Draw a system graph.
   b. Choose a tree of the graph and write the fundamental cutset equations.
   c. Write the fundamental circuit equations from the cutset equations. Check your results.

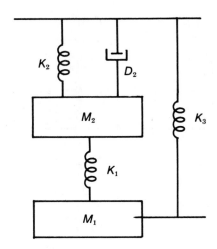

4-5. Consider the network illustrated.

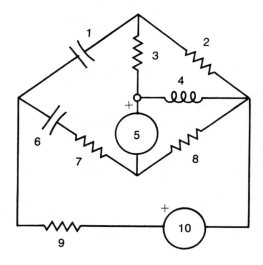

a. Draw the graph of the network.

b. Select a tree and determine a set of independent loop currents and independent node-pair voltages.

c. Consider the tree 1,2,3,5,7,9. Determine the cutset matrix; the fundamental circuit matrix.

4-6. In the network graph shown, choose the tree 5 to 13 inclusive. Find the cutset and the fundamental circuit matrixes appropriate to this tree.

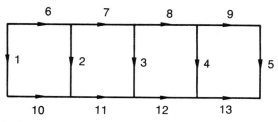

4-7. Refer to the following systems.
  a. Draw three trees for each.
  b. Determine the A and B matrices for each.

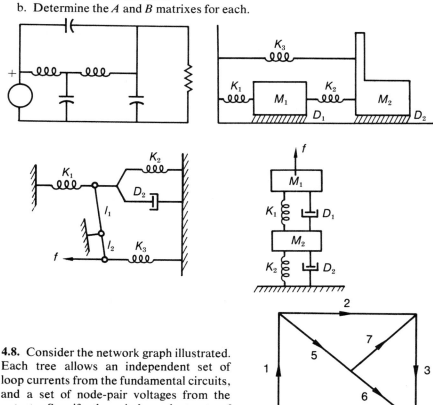

4.8. Consider the network graph illustrated. Each tree allows an independent set of loop currents from the fundamental circuits, and a set of node-pair voltages from the cutsets. Specify three independent sets of each for the given network.

4-9. Refer to the figure shown. Here the R network is assumed to be a 4-terminal device; hence the terminal graph must be retained intact. The terminal equations for this device are contained in the expression,

$$\begin{bmatrix} I_1 \\ I_2 \\ I_3 \end{bmatrix} = \begin{bmatrix} 1 & 1 & 0 \\ 1 & 2 & -1 \\ 0 & -1 & 3 \end{bmatrix} \cdot \begin{bmatrix} V_1 \\ V_2 \\ V_3 \end{bmatrix}$$

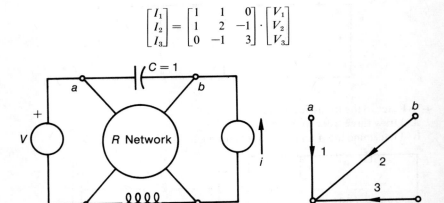

Use a branch formulation to express the voltage variables for the inductor and the capacitor.

4-10. Construct the duals of the graphs shown.

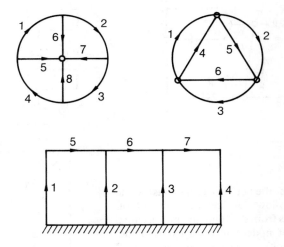

4-11. Draw duals of the networks of Problem 3-2.

Part III System Response

a. Time Domain

5

System Response

The study in Part II was concerned with methods for writing the differential equations, in the case of continuous systems; or the difference equations, in the case of discrete time systems, which describe the dynamics of the interconnected system. The Kirchhoffian methods lead to sets of integrodifferential equations, and the state variable method leads to sets of first order differential equations. It was shown both analytically and through the signal flow graph that the Kirchhoff formulation and the state variable formulations can be obtained from each other.

Once a description of the equilibrium conditions of the system are established, then it is possible to determine the response of the system to a wide variety of excitation functions. A number of methods exist for carrying out such calculations, including analytic, numerical and machine methods. Part a of this chapter will concern itself with the important features of the analytic solution of linear differential equations with constant coefficients. This is a restricted class of differential equations, but many important systems are described by such equations. Part b will consider numerical methods for the solution of differential equations, not necessarily linear, and Part c will discuss machine methods of solution.

5-a CLASSICAL DIFFERENTIAL EQUATIONS

5-1 FEATURES OF LINEAR DIFFERENTIAL EQUATIONS

We note initially that the Kirchhoff formulation of dynamic equations leads, in the simple case of one loop or one node-pair systems, to an integrodifferential equation of the form,

$$L\frac{di}{dt} + Ri + \frac{1}{C}\int i\,dt = v(t). \tag{5-1}$$

When sets of such equations are involved, these may be combined, and in this case the system is described by a series of nonhomogeneous differential equations of the form,

$$a_n\frac{d^n y}{dt^n} + a_{n-1}\frac{d^{n-1} y}{dt^{n-1}} + \cdots + a_0 y = b_m\frac{d^m u}{dt^m} + b_{m-1}\frac{d^{m-1} u}{dt^{m-1}} + \cdots + b_0 u. \tag{5-2}$$

The state formulation leads to sets of first order differential equations, and since special methods exist for handling these, we shall defer detailed attention to these until Chapter 6 when we shall show that the entire set can be handled together. Of course, if the state equations are combined, they lead to higher order equations of the form in Eq. (5-2), and the discussion here to be undertaken will be applicable.

Equation (5-2) is a linear, nonhomogeneous differential equation of order n with constant coefficients. Such classes of equations possess important properties. These are here noted since these properties will be of subsequent interest to us. Suppose that we write Eq. (5-2) in the functional form,

$$T(p)y = f(u), \tag{5-3}$$

where $T(p)y$ denotes the total differential operational form of the expression on the left, and $f(u)$ is the functional form on the right, and where y denotes the output or response variable, and u denotes the input or excitation variable. If the differential equation is linear, then we may write that,

$$T(p)(c_1 y_1) = c_1 T(p)y_1 = c_1 f(u_1). \tag{5-4}$$

Also

$$T(p)(c_2 y_2) = c_2 T(p)y_2 = c_2 f(u_2), \tag{5-5}$$

where c_1 and c_2 are constants, and where y_1 and y_2 are solutions of Eq. (5-3). In the light of these results, we can write,

$$T(p)(c_1 y_1 + c_2 y_2) = c_1 T(p)y_1 + c_2 T(p)y_2 = c_1 f(u_1) + c_2 f(u_2). \tag{5-6}$$

This expression shows that for the linear differential equation, the response $T(p)y_1 + T(p)y_2$ to the input $f(u_1) + f(u_2)$ is simply the sum of the response due to each excitation function separately applied. This property is known as the *superposition* property, and can, in fact, be used as a basic definition for a linear system.

An important feature of linear differential equations, especially in connection with the solution of Eq. (5-2), is that the pth order response of a system is the solution of the system to the pth order of the initial excitation function. This feature of linear differential equations was employed in Section 2-8 to reduce an nth order differential equation to the normal form. In our present studies we shall use these results to find a solution to Eq. (5-2). In accomplishing this, we begin with the reduced equation,

$$a_n \frac{d^n \eta}{dt^n} + a_{n-1} \frac{d^{n-1} \eta}{dt^{n-1}} + \cdots + a_0 \eta = u(t), \qquad (5\text{-}7)$$

which is assumed to have as its solution the function η, and then by Eq. (2-36), the solution to Eq. (5-2) is given as,

$$y(t) = b_m \frac{d^m \eta}{dt^m} + b_{m-1} \frac{d^{m-1} \eta}{dt^{m-1}} + \cdots + b_0 \eta. \qquad (5\text{-}8)$$

Observe that Eq. (5-8) has made use of the superposition property of Eq. (5-6).

5-2 GENERAL FEATURES OF SOLUTIONS OF DIFFERENTIAL EQUATIONS

The general problem that we wish to consider is concerned with the details of solving Eqs. (5-1) and (5-2) when $v(t)$ and $u(t)$ are known time functions. In principle, these may be any analytic function of the time. From a practical point of view, however, there are a relatively limited number of excitation functions that form the basis for most subsequent analyses. Typical of these are the waveforms illustrated in Fig. 5-1.

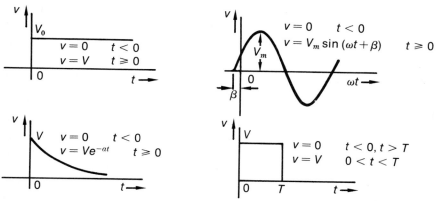

Fig. 5-1. Typical waveforms for through or across time functions.

Observe that all of these have been drawn to indicate a discontinuity, which is to denote that the wave is switched into the system at a given time t_0, usually taken as $t = 0$. From an analytic point of view, even though these waveforms may be continuous after $t = 0$, nevertheless they must be considered to be discontinuous in time. This is important since it places in focus the need for initial statements of the state of the system, which, together with specified switching conditions, will be needed to completely specify the subsequent behavior of the system.

We shall begin our considerations with Eq. (5-1) since the pattern that is employed in studying this equation is typical of that for linear differential equations in general. We proceed by noting that the solution to an equation of this type consists of two parts. One part of the solution is entirely due to the nature of the network, in the absence of the excitation. The second part of the solution is determined by the character of the applied excitation function. That part of the solution that is due to the nature of the network alone is obtained from the homogeneous or *complementary equation*, which is the original integrodifferential equation with the right hand terms set to zero,

$$L\frac{di}{dt} + Ri + \frac{1}{C}\int i\, dt = 0. \tag{5-9}$$

The solution to this equation is known as the *complementary function*, and will be written as i_t, where the subscript denotes that this is the *transient* portion of the total solution, as we shall soon see.

The *particular solution* will be designated i_s and is often called the *steady state* solution, although in general, this term will be time varying. The complete solution is,

$$i(t) = i_t(t) + i_s(t), \tag{5-10}$$

and explicitly shows it to be made up of the complementary function and the particular integral. The particular integral depends explicitly on the form of the excitation function $v(t)$. We shall consider below the important techniques for finding i_s — all of Part III-b is concerned with this essential problem.

The essential features of systems analysis can be displayed graphically as shown in Fig. 5-2. As discussed above, for linear systems and employing classical methods for the solution of ordinary differential equations, the sequence is precisely that illustrated, with each step being separately accomplished. When using Laplace transform methods, as shall be done in Chapter 7, the several steps are accomplished together. Whatever the

General Features of Solutions of Differential Equations 173

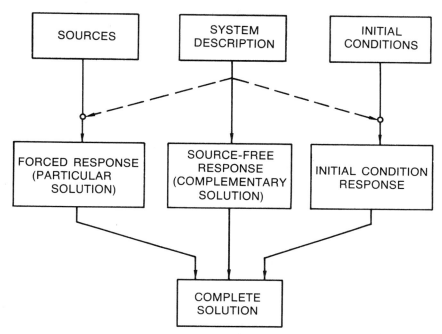

Fig. 5-2. The factors involved in determining the response of a system.

method, the complete solution is the superposition of the continuously acting forced response, the response of the originally relaxed system (zero initial conditions—this is referred to as the normal mode response), and the effects of excitations caused by energies stored within the system itself.

We shall limit our discussion to those passive physical systems for which the output response is zero for $t < t_0$, if the input excitation is zero for $t < t_0$. This means that if,

$$u(t) = 0 \quad (t < t_0)$$

then, (5-11)

$$y(t) = \{\text{Linear operation on}\}\, u(t) = 0 \quad (t < t_0).$$

Systems which satisfy these conditions are called *causal* systems. In general, a function $f(t)$ is called causal if it has zero value for $t < 0$, i.e., $f(t) = 0$ for $t < 0$. It can be shown that all physically realizable systems are causal.

174 System Response

Suppose that we now combined Eq. (5-10) with Eq. (5-1). The result is,

$$L\frac{d}{dt}(i_s+i_t)+R(i_s+i_t)+\frac{1}{C}\int(i_s+i_t)\,dt$$
$$=\left(L\frac{di_s}{dt}+Ri_s+\frac{1}{C}\int i_s dt\right)+\left(L\frac{di_t}{dt}+Ri_t+\frac{1}{C}\int i_t dt\right)=v(t).$$

Now, because of Eq. (5-9), this expression may be separated into two parts,

$$L\frac{di_t}{dt}+Ri_t+\frac{1}{C}\int i_t dt = 0,$$
$$L\frac{di_s}{dt}+Ri_s+\frac{1}{C}\int i_s dt = v(t). \quad (5\text{-}12)$$

The first is the complementary equation; the second is the particular equation. It is convenient to write these equations in normalized form by defining the quantities,

$$\text{Characteristic time} \quad \tau_n = \sqrt{LC}$$
$$\text{Damping ratio} \quad \zeta = \sqrt{\frac{R^2 C}{4L}} = \frac{R}{2}\sqrt{\frac{C}{L}} \quad (5\text{-}13)$$

Using these, Eqs. (5-12) become, using the p notation,

$$\tau_n^2 p i_t + 2\zeta\tau_n i_t + \frac{i_t}{p} = 0,$$
$$\tau_n^2 p i_s + 2\zeta\tau_n i_s + \frac{i_s}{p} = Cv. \quad (5\text{-}14)$$

We wish to study techniques for the solution of both of these equations.

5-3 THE COMPLEMENTARY FUNCTION

It is customary to adopt as a tentative solution to the complementary equation an exponential function of the form e^{st}, since the form of this function does not change with repeated differentiations or integrations. Thus we shall test as a solution the function,

$$i_t = Be^{st}, \quad (5\text{-}15)$$

with the thought that it may be possible to choose the constant B and the value of s to satisfy the equation. If Eq. (5-15) is combined with Eq. (5-14), we find that,

$$\tau_n^2 s B e^{st} + 2\zeta\tau_n B e^{st} + \frac{1}{s} B e^{st} = 0. \tag{5-16}$$

It follows from this that,

$$\tau_n^2 s + 2\zeta\tau_n + \frac{1}{s} = 0. \tag{5-17}$$

This shows that the exponential function is a solution of the complementary function provided that s satisfies Eq. (5-17), independent of the value of B. Let us write Eq. (5-17) in the form,

$$\tau_n^2 s^2 + 2\zeta\tau_n s + 1 = 0. \tag{5-18}$$

This shows that s must satisfy a quadratic function. There are two roots to this equation, namely,

$$\begin{aligned} s_1 &= -\frac{\zeta}{\tau_n} + \frac{1}{\tau_n}\sqrt{\zeta^2 - 1}, \\ s_2 &= -\frac{\zeta}{\tau_n} - \frac{1}{\tau_n}\sqrt{\zeta^2 - 1}. \end{aligned} \tag{5-19}$$

The roots s_1 and s_2 are known as the *normal modes* or the *eigenvalues* (characteristic values) of the system. Since $e^{s_1 t}$ and $e^{s_2 t}$ are solutions of the given equation, the sum is also a solution, as per the superposition principle already discussed. Thus the complementary function is,

$$i_t = B_1 e^{s_1 t} + B_2 e^{s_2 t}. \tag{5-20}$$

This solution satisfies a basic requirement of the theory of differential equations that there be as many arbitrary constants in the solution as the highest order of the derivative appearing in the equation. To ascertain this order requires that the integrodifferential equation be differentiated as many times as is necessary to convert it into the lowest order differential equation form. In this particular case, Eq. (5-15) must be differentiated once, with a resulting second order differential equation.

The detailed form of the complementary solution may assume different appearances, depending on the particular values of s, which depend on the values of ζ. Three forms are possible, and we shall examine these.

1. $\zeta^2 > 1$. In this case, the roots s_1 and s_2 in Eq. (5-19) are real and different. Hence the form in Eq. (5-20) is appropriate in the given form. Often, however, it is differently written by making use of the identity that

176 System Response

relates the exponential with the hyperbolic forms. This is,

$$e^{\pm x} = \cosh x \pm \sinh x. \tag{5-21}$$

Using this form, Eq. (5-20) may be written,

$$i_t = B_1 \exp\left(-\zeta + \sqrt{\zeta^2 - 1}\right)\frac{t}{\tau_n} + B_2 \exp\left(-\zeta - \sqrt{\zeta^2 - 1}\right)\frac{t}{\tau_n},$$

or

$$i_t = e^{-\zeta t/\tau}\left[B_1 \exp\left(\sqrt{\zeta^2 - 1}\,\frac{t}{\tau_n}\right) + B_2 \exp\left(-\sqrt{\zeta^2 - 1}\,\frac{t}{\tau_n}\right)\right]. \tag{5-22}$$

This result may be expanded to the form,

$$i_t = e^{-\zeta t/\tau_n}\left[B_1\left(\cosh\sqrt{\zeta^2 - 1}\,\frac{t}{\tau_n} + \sinh\sqrt{\zeta^2 - 1}\,\frac{t}{\tau_n}\right)\right.$$
$$\left. + B_2\left(\cosh\sqrt{\zeta^2 - 1}\,\frac{t}{\tau_n} - \sinh\sqrt{\zeta^2 - 1}\,\frac{t}{\tau_n}\right)\right],$$

or

$$i_t = e^{-\zeta t/\tau_n}\left[(B_1 + B_2)\cosh\sqrt{\zeta^2 - 1}\,\frac{t}{\tau_n} + (B_1 - B_2)\sinh\sqrt{\zeta^2 - 1}\,\frac{t}{\tau_n}\right]. \tag{5-23}$$

Since B_1 and B_2 are arbitrary constants, their sum and difference are also arbitrary. Hence we write this in the form,

$$i_t = e^{-\zeta t/\tau_n}\left[B_3 \cosh\sqrt{\zeta^2 - 1}\,\frac{t}{\tau_n} + B_4 \sinh\sqrt{\zeta^2 - 1}\,\frac{t}{\tau_n}\right] \tag{5-24}$$

which is seen to include two arbitrary constants.

2. $\zeta^2 = 1$. This is a special case since it requires a special selection of system parameters to meet this condition. However, it is a form that is often achieved. In this case, Eqs. (5-19) show that the two roots are equal. This common value is written as s_0, with,

$$s_0 = -\frac{1}{\tau_n}. \tag{5-25}$$

Now, however, Eq. (5-20) reduces to a single term. As a result, we must seek a second solution, so as to satisfy the required conditions for two arbitrary constants in the solution.

The Complementary Function

The approach in this case is to examine the function,

$$i_t = Bte^{s_0 t}. \tag{5-26}$$

It will be shown that this expression is also a solution to our differential equation. To combine this with Eq. (5-15) will require the terms,

$$p(Bte^{s_0 t}) = Be^{s_0 t} + Bs_0 te^{s_0 t}$$

$$\frac{1}{p}(Bte^{s_0 t}) = \frac{Bte^{s_0 t}}{s_0} - \frac{Be^{s_0 t}}{s_0^2} \tag{5-27}$$

Now combine these with Eq. (5-15) to get, after canceling the common factor $Be^{s_0 t}$ from each term in the equation,

$$\left(\tau_n^2 - \frac{1}{s_0^2}\right) + \left(\tau_n^2 s_0 + 2\zeta\tau_n + \frac{1}{s_0}\right)t = 0. \tag{5-28}$$

The quantity on the right is identically zero, as required by Eq. (5-17). The quantity on the left is zero by Eq. (5-25). Hence Eq. (5-15) is satisfied by the trial function, Eq. (5-26), and the complementary function is then,

$$i_t = (B_1 + B_2 t)e^{-t/\tau_n}, \tag{5-29}$$

where B_1 and B_2 are arbitrary constants.

We note without proof that a differential equation with three equal roots will have the complementary function,

$$i_t = (B_1 + B_2 t + B_3 t^2)e^{s_0 t}. \tag{5-30}$$

This form can be appropriately extended for n equal roots.

3. $\zeta^2 < 1$. Under these conditions the quantity $\sqrt{\zeta^2 - 1}$ is imaginary, and may be written $j\sqrt{1 - \zeta^2}$. The roots in Eq. (5-19) assume the form,

$$s_1 = -\frac{\zeta}{\tau_n} + j\sqrt{\frac{1-\zeta^2}{\tau_n}}, \tag{5-31}$$

$$s_2 = -\frac{\zeta}{\tau_n} - j\sqrt{\frac{1-\zeta^2}{\tau_n}}.$$

Use is now made of the Euler identity,

$$e^{\pm jx} = \cos x \pm j \sin x, \tag{5-32}$$

and Eq. (5-20) assumes the form,

$$i_t = e^{-\zeta t/\tau}\left[B_1 \exp\left(j\sqrt{1-\zeta^2}\,\frac{t}{\tau_n}\right) + B_2 \exp\left(-j\sqrt{1-\zeta^2}\,\frac{t}{\tau_n}\right)\right]. \tag{5-33}$$

By following the same procedure as in going from Eq. (5-22) to Eq. (5-24) the result is,

$$i_t = e^{-\zeta t/\tau}\left(B_3 \cos\sqrt{1-\zeta^2}\,\frac{t}{\tau_n} + B_4 \sin\sqrt{1-\zeta^2}\,\frac{t}{\tau_n}\right). \qquad (5\text{-}34)$$

In fact, this expression can be deduced from Eq. (5-24) by making use of the fact that,

$$\begin{aligned}\cosh jx &= \cos x, \\ \sinh jx &= j \sin x.\end{aligned} \qquad (5\text{-}35)$$

5-4 THE PARTICULAR SOLUTION

As already noted, the particular solution will depend on the excitation function $v(t)$. No acceptable general method exists for finding this function, although the method of variation of parameters to be discussed below, is worthy of attention.

We shall adopt a rather heuristic viewpoint; we shall try to find a solution by whatever means may be possible, recognizing that the validity of the solution is readily checked by direct recourse to the differential equation. We then rely on the uniqueness theorem to support the fact that such a solution is a unique solution. Fortunately, most problems that arise in systems with linear coefficients can be handled by one of the following methods.

1. $v(t)$, *a polynomial in t*. Refer to the second part of Eq. (5-14), which is written here for convenience,

$$Z(p)i_s(t) = v(t), \qquad (5\text{-}36)$$

where $Z(p)$ is the impedance operator

$$Z(p) = Lp + R + \frac{1}{Cp}.$$

Suppose, in particular, that $v(t)$ is a polynomial in t of the form

$$v(t) = a_0 + a_1 t + a_2 t^2 + \cdots + a_r t^r. \qquad (5\text{-}37)$$

A usual approach in determining an expression for $i_s(t)$ is to adopt a similar power series as a trial solution. Hence we write tentatively,

$$i_s(t) = c_0 + c_1 t + c_2 t^2 + \cdots + c_n t^n. \qquad (5\text{-}38)$$

Equation (5-38) is combined with Eq. (5-36). This involves differentiation and integration of the power series, which can be done on a term by term basis, and then adding the results, in the manner called for in Eq. (5-36). The result of such an operation is a power series in t. This leads to a power series, appropriate to the left hand member of Eq. (5-36), which is equal to the power series expression for $v(t)$ given in Eq. (5-37). For equality, the coefficients of like powers of t must be equal. This leads to a set of equations involving the coefficients which can be solved for the various c-s in terms of known factors.

The solution may also be found using a mathematical procedure that involves operations that require subsequent verification. This procedure is based on writing Eq. (5-36) in the operational form,

$$i_s = \frac{1}{Z(p)} v(t). \qquad (5\text{-}39)$$

We recall that we have established meanings for p and $1/p$ when associated with a variable. Here, however, we are treating $Z(p)$, a rational function in p, a functional that has not been defined, as an algebraic quantity. By simple algebraic manipulation we are able to cast this function into a form that we can interpret. This is done by expanding $1/Z(p)$ by long division to deduce an expression of the form,

$$\frac{1}{Lp + R + (1/Cp)} = b_0 + b_1 p + b_2 p^2 + \cdots + b_r p^r + \frac{r(p)}{Lp + R + (1/Cp)}, \qquad (5\text{-}40)$$

where $r(p)$ denotes the remainder after the rth power. Now we combine Eqs. (5-39) and (5-40) with Eq. (5-37) to get,

$$i_s = \left[b_0 + b_1 p + b_2 p^2 + \cdots + b_r p^r + \frac{r(p)}{Z(p)} \right] (a_0 + a_1 t + a_2 t^2 + \cdots + a_r t^r). \qquad (5\text{-}41)$$

Note that each term in the bracket except the remainder term denotes a differentiation, and hence has meaning. Further, we see that all terms in the expansion of $1/Z(p)$ beyond the rth contribute nothing to the final result. In consequence we drop the remainder term and assert that the particular solution is given by completing the operational expression,

$$i_s = (b_c + b_1 p + b_2 p^2 + \cdots + b_r p^r)(a_0 + a_1 t + a_2 t^2 + \cdots + a_r t^r). \qquad (5\text{-}42)$$

The result can be verified by combining it with Eq. (5-36).

Example 5-4.1

A series RLC circuit for which $L = R = C = 1$ is excited by $v(t) = 1 + t^2$. Find the particular integral.

Solution. By simple division,

$$\frac{1}{(1/p) + 1 + p} = p - p^2 + p^4 + \cdots.$$

It follows from this that,

$$i_s = (p - p^2)(1 + t^2) = 2t - 2.$$

To verify this solution, combine i_s with the original differential equation,

$$Z(p)i_s = \left(\frac{1}{p} + 1 + p\right)(2t - 2) = (t^2 - 2t + k) + (2t - 2) + 2 = t^2 + k.$$

If the constant of integration k is chosen as 1, then the result is valid.

2. *$v(t)$, an exponential function of t.* In this case we assume that

$$v(t) = A e^{st}. \tag{5-43}$$

By a trial process which assumes that $i_s = B e^{st}$, it is found that the solution to Eq. (5-39) retains the exponential form, but with p replaced by s. The solution is,

$$i_s(t) = \frac{A e^{st}}{Z(s)}. \tag{5-44}$$

3. *$v(t)$, a constant.* This is an important case, since it corresponds to a system that is subjected to a constant excitation. In electric circuit theory, it is the d-c response. The solution is just a special case under (1) and (2). Case (1) applies by setting $a_0 = 1$ and all other a-s $= 0$. Case (2) applies by setting $s = 0$.

4. *$v(t)$, a sinusoidal function.* This too is a very important case, and gives the so-called sinusoidal steady state solution. It will receive considerable attention in the remainder of this book. Suppose that the excitation function is,

$$v(t) = A \cos(\omega t + \varphi), \tag{5-45}$$

where A and φ are constants. Using the Euler relationship, this expression can be written as,

$$v(t) = A \frac{e^{j(\omega t + \varphi)} + e^{-j(\omega t + \varphi)}}{2}, \tag{5-46}$$

which shows it as the sum of two exponential functions. We can use Eq. (5-44) with $s = j\omega$ to write,

$$i_s = \frac{A}{2}\left[e^{j\varphi}\frac{e^{j\omega t}}{Z(j\omega)} + e^{-j\varphi}\frac{e^{-j\omega t}}{Z(-j\omega)}\right]. \tag{5-47}$$

Often the result is conveniently left in this form. The result can be recast in other useful forms.

Since $Z(j\omega)$ and $Z(-j\omega)$ are complex numbers, in general, they may be written in the following polar forms,

$$Z(j\omega) = j\omega L + R + \frac{1}{j\omega C} = |Z(\omega)|e^{j\eta},$$

$$Z(-j\omega) = -j\omega L + R + \frac{1}{-j\omega C} = |Z(\omega)|e^{-j\eta}, \tag{5-48}$$

where the magnitude quantity is $|Z(\omega)|$ and the angle function is η, with

$$|Z(\omega)| = \sqrt{R^2 + \left(\omega L - \frac{1}{\omega C}\right)^2}, \quad \eta = \tan^{-1}\frac{\omega L - (1/\omega C)}{R}. \tag{5-49}$$

A complex function $Z(-j\omega)$ that is obtained by setting $-j$ for $+j$ in $Z(j\omega)$ is called the complex conjugate of $Z(j\omega)$, and often is written $Z*(j\omega)$. Equation (5-47) can now be written as,

$$i_s = \frac{A}{2}\frac{e^{j(\omega t+\varphi-\eta)} + e^{-j(\omega t+\varphi-\eta)}}{|Z(\omega)|},$$

which is the expression,

$$i_s = \frac{A}{|Z(\omega)|}\cos(\omega t + \varphi - \eta). \tag{5-50}$$

The same result is possible by recognizing that,

$$v(t) = A\cos(\omega t + \varphi) = Re[Ae^{j(\omega t+\varphi)}], \tag{5-51}$$

where Re denotes that the real part of the expansion is being designated. It follows that,

$$i_s = Re\left[\frac{Ae^{j(\omega t+\varphi)}}{Z(j\omega)}\right] = Re\left[A\frac{e^{j(\omega t+\varphi-\eta)}}{|Z(\omega)|}\right],$$

from which,

$$i_s = \frac{A}{Z(\omega)}\cos(\omega t + \varphi - \eta), \tag{5-52}$$

which is the same as above.

5. $v(t)$, a time function multiplied by an exponential function. For the case when,

$$v(t) = \psi(t)e^{st}, \tag{5-53}$$

the solution, given without proof, is expressed symbolically,

$$i_s = \frac{1}{Z(p)}[\psi(t)e^{st}] = e^{st}\left[\frac{\psi(t)}{Z(p+s)}\right]. \tag{5-54}$$

This essentially specifies a shifting procedure, and moves the exponential function out of the influence of the operational function, after modifying $Z(p)$. The function $Z(p+s)$ remains a rational function of p, the resulting form for i_s depending on the particular form of $\psi(t)$. The next step in the solution might come under rules (1) through (4).

Example 5-4.2

A simple series RLC network has the constants: $L = 1$, $R = 2$, $C = \frac{1}{2}$. The applied excitation is $v(t) = 3e^{-t}\sin 5t$. Find the form for i_s.

Solution. An application of Eq. (5-54) leads to the expression,

$$i_s = \frac{1}{Z(p)}[3e^{-t}\sin 5t] = 3e^{-t}\,Im\left[\frac{1}{Z(p-1)}\right]e^{j5t}.$$

We now proceed as in Eq. (5-51), which gives,

$$i_s = 3e^{-t}\,Im\left[\frac{1}{(p-1)+2+(2/p-1)}\right]e^{j5t}$$

$$= 3e^{-t}\,Im\left[\frac{e^{j5t}}{(j5-1)+2+(2/j5-1)}\right]$$

$$= 3e^{-t}\,Im\left[\frac{26\,e^{j5t}}{120j+24}\right] = 3\times\frac{26}{24}\,Im\left[\frac{e^{j(5t-\eta)}}{\sqrt{26}}\right]$$

$$= \frac{\sqrt{26}}{8}\sin(5t-\tan^{-1}5).$$

5-5 VARIATION OF PARAMETERS

In Sections 5-3 and 5-4, the complementary function and the particular solution are deduced by independent methods. This would seem to imply that these operations are essentially separate and unrelated. This is not really the case, as the method of variation of parameters will show. This is an elegant though somewhat artificial method that uses a knowledge of the complementary function to find the particular solution. The method permits finding the complete solution of a linear equation whose complementary function is known. Actually, this is not a practical approach, but the conclusion that a knowledge of the solution of the system equations to one set of conditions—here to zero excitation—allows us to find the complete solution to any excitation function is a most important concept. This same general concept will appear again in our study, but under different requirements.

To illustrate the method, let us consider the general linear differential equation of the second order,

$$p^2 y + Ppy + Qy = u, \tag{5-55}$$

where, as previously noted, $p \triangleq d/dt \triangleq \cdot$ are equivalent time derivative operations. We begin by supposing that the complementary function, the solution to,

$$p^2 y + Ppy + Qy = 0, \tag{5-56}$$

is known, and is of the form,

$$y_t = B_1 y_1 + B_2 y_2, \tag{5-57}$$

where B_1 and B_2 are arbitrary constants.

We proceed by assuming that the differential equation Eq. (5-55) is satisfied by a solution of the form given by Eq. (5-57) but where B_1 and B_2 are no longer constants, but depend on t. That is, we assume as a solution,

$$y = B_1(t) y_1 + B_2(t) y_2. \tag{5-58}$$

To find the conditions for the solution, differentiate this expression with respect to time. This gives, now using the dot notation,

$$\dot{y} = B_1 \dot{y}_1 + B_2 \dot{y}_2 + \dot{B}_1 y_1 + \dot{B}_2 y_2.$$

Now let B_1 and B_2 be so chosen that

$$\dot{B}_1 y_1 + \dot{B}_2 y_2 = 0, \tag{5-59}$$

from which,

$$\dot{y} = B_1 \dot{y}_1 + B_2 \dot{y}_2.$$

Differentiate again to get,

$$\ddot{y} = B_1 \ddot{y}_1 + B_2 \ddot{y}_2 + \dot{B}_1 \dot{y}_1 + \dot{B}_2 \dot{y}_2.$$

This result is combined with Eq. (5-56) to give,

$$B_1\{\ddot{y}_1 + P\dot{y}_1 + Qy_1\} + B_2\{\ddot{y}_2 + P\dot{y}_2 + Qy_2\} + \dot{B}_1\dot{y}_1 + \dot{B}_2\dot{y}_2 = u.$$

But since we have chosen y_1 and y_2 as solutions of Eq. (5-56), then we have,

$$\dot{B}_1\dot{y}_1 + \dot{B}_2\dot{y}_2 = u. \tag{5-60}$$

We now employ Eqs. (5-59) and (5-60) to determine $\dot{B}_1$ and $\dot{B}_2$. The solutions are,

$$\dot{B}_1 = \frac{y_2 u}{y_1 \dot{y}_2 - \dot{y}_1 y_2} \quad \text{and} \quad \dot{B}_2 = \frac{-y_1 u}{y_1 \dot{y}_2 - \dot{y}_1 y_2}.$$

Integrating these expressions gives,

$$B_1 = \int \frac{y_2 u}{y_1 \dot{y}_2 - \dot{y}_1 y_2} \, dt + E,$$

and

$$B_2 = -\int \frac{y_1 u}{y_1 \dot{y}_2 - \dot{y}_1 y_2} \, dt + F,$$

(5-61)

where E and F are arbitrary constants.

Equation (5-61) is combined with Eq. (5-58) to yield the complete solution,

$$y = Ey_1 + Fy_2 + \left[y_1 \int \frac{y_2 u}{y_1 \dot{y}_2 - \dot{y}_1 y_2} \, dt - y_2 \int \frac{y_1 u}{y_1 \dot{y}_2 - \dot{y}_1 y_2} \, dt \right]. \tag{5-62}$$

Clearly, the term in the bracket is the particular integral of Eq. (5-55). This expression may be written in more convenient form by making use of the fact that,

$$\ddot{y}_1 + P\dot{y}_1 + Qy_1 = 0,$$
$$\ddot{y}_2 + P\dot{y}_2 + Qy_2 = 0.$$

Now multiply the first equation by y_2 and the second by y_1 and subtract the resulting expressions to get,

$$\ddot{y}_1 y_2 - y_1 \ddot{y}_2 = -P[\dot{y}_1 y_2 - y_1 \dot{y}_2].$$

But the form of the left hand member can be rewritten as shown,

$$p(\dot{y}_1 y_2 - y_1 \dot{y}_2) = -P(\dot{y}_1 y_2 - y_1 \dot{y}_2),$$

where p again denotes d/dt. Now integrate this expression to get,

$$\dot{y}_1 y_2 - y_1 \dot{y}_2 = C \exp\left\{-\int P \, dt\right\}, \tag{5-63}$$

where the constant C may be evaluated from the particular choice of y_1 and y_2.

Using Eq. (5-63) in conjunction with Eq. (5-62) allows us to write an expression for the particular solution. The result is,

$$y_s = \frac{1}{C}\left[y_1 \int y_2 u(t) \exp\left\{\int P \, dt\right\} dt - y_2 \int y_1 u(t) \exp\left\{\int P \, dt\right\} dt\right]. \tag{5-64}$$

Thus we have as a general solution of Eq. (5-55), with a slight change in notation of Eq. (5-64),

$$y = E y_1 + F y_2 + \frac{1}{C}\int^t u(\xi) \exp\left\{\int P(\xi) \, d\xi\right\}\left\{y_1(t) y_2(\xi) - y_1(\xi) y_2(t)\right\} d\xi. \tag{5-65}$$

5-6 EVALUATION OF INTEGRATION CONSTANTS—INITIAL CONDITIONS

Once the complementary function and the particular solution have been deduced for any particular example, the next step is to determine the constants of integration that arise in the complementary function. These must be so chosen as to yield a response function that satisfies all conditions of the particular problem. These conditions, as previously discussed, are the known charges on all capacitors (usually given as initial voltages across the capacitors), and known current through all inductors, at some specified initial time, usually taken as $t = 0$. If these initial voltages and currents are zero, the system is said to be *initially relaxed*. In the more general case, the initial conditions would be specified by initial through and across variables pertinent to the system elements. In all cases, the specified initial conditions must be combined with the complete solution for the determination of the constants of integration. In this way the character of the system disturbance is taken into account. Of course, if the system disturbance involves more than the switching of excitation functions, as, for example, the switching of system elements, such changes must be taken into account also.

A variety of switching operations are possible, and these are of such importance that we wish to discuss them. Some of the important switching operations are:

1. The most common switching operation is that of introducing an excitation source into a circuit. The more common input waveforms are given in Fig. 5-1. If the circuit is initially relaxed, then this excitation function establishes the form of the particular solution, and the fact that initial voltages across capacitors and initial currents through inductors are zero, will provide sufficient relations for the evaluation of the constants of integration.

2. If initial voltages across capacitors are not zero, and if during the switching operation the total system capacitance remains unchanged, then the voltage across the capacitor will be the same before and after the instant of switching. This condition assumes the absence of switching impulses to the capacitor or capacitors. This result follows from the fact that for the capacitor $i = C\,dv/dt$, when C is a constant, and in the switching operation that occurs during the interval from $t = 0-$ to $t = 0+$,

$$\int_{v(0-)}^{v(0+)} C\,dv = \int_{0-}^{0+} i\,dt, \qquad (5\text{-}66)$$

the value of the integral on the right is zero unless i is an impulse function. This is a statement of conservation of charge (conservation of momentum in mechanics) and states that if no impulse is applied during the switching interval, then,

$$Cv\,(0+) - Cv\,(0-) = 0. \qquad (5\text{-}67)$$

3. If initial currents exist in inductors and if during the switching operation the total system inductance remains unchanged, then the current through the inductor will remain unchanged over the switching instant. This condition is the dual of that for the voltage across the capacitor, and assumes the absence of switching voltage impulses. This result follows directly from the fact that the terminal relation for the inductor is $v = L\,di/dt$, and with L constant over the switching interval,

$$\int_{i(0-)}^{i(0+)} L\,di = \int_{0-}^{0+} v\,dt,$$

the right hand side will be zero in the absence of voltage impulses. This is a statement of conservation of flux linkages, and states that in the absence of voltage impulses during the switching interval,

$$Li(0+) - Li(0-) = 0. \tag{5-68}$$

4. In the light of the discussion in 2 and 3, for circuits with constant excitation:
  a. capacitors behave as open circuits in the steady state in networks with constant excitation.
  b. inductors behave as short circuits in the steady state in networks with constant excitation.

5. In the event that L or C or both are changed instantaneously during a switching operation, and in the absence of switching impulses, Eqs. (5-67) and (5-68) must be modified in form to the following:

for capacitors, conservation of charge $q(0+) = q(0-)$,
for inductors, conservation of flux linkages $\psi(0+) = \psi(0-)$.

As a practical matter, circuits with switched L or C are easily accomplished in electrical circuits by placing switches across all or part of the L or C of a circuit.

More will be said about initial conditions in Chapter 7 in connection with the Laplace transform.

5-7 THE SERIES *RL* CIRCUIT AND ITS DUAL

We wish to carry out the solutions in detail of a number of important problems. We first consider the simple series RL circuit and its dual, as illustrated in Fig. 5-3. The excitation function is called a step function, and is conveniently written $Vu_{-1}(t)$, where the symbol $u_{-1}(t)$ denotes the

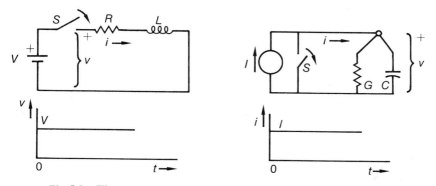

Fig. 5-3. The series RL circuit and its dual with step function excitation.

188   **System Response**

unit step function which is defined as,

$$u_{-1}(t) = \begin{cases} 0 & t < 0 \\ 1 & t \geq 0 \end{cases}. \tag{5-69}$$

We note that except for a change of symbols, the solution in the present case applies for both the circuit and its dual.

The dynamic equation for Fig. 5-3 is,

$$L\frac{di}{dt} + Ri = v(t). \tag{5-70}$$

To find the complementary solution, we choose an exponential function of the form,

$$i_t = Be^{st}, \tag{5-71}$$

as a trial solution. This leads to the complementary solution, if,

$$Ls + R = 0,$$

from which we have that,

$$s = -\frac{R}{L}. \tag{5-72}$$

The particular solution is,

$$i_s = \frac{V}{R}. \tag{5-73}$$

This is simply the d-c current in the circuit, since in the steady state, the inductor has no effect, i.e., $Ldi/dt = 0$ when $i =$ constant. The complete solution is then,

$$i = i_s + i_t = \frac{V}{R} + Be^{st}. \tag{5-74}$$

To evaluate the constant B in this equation, we require the availability of one initial condition. Here we shall assume that the initial current through the inductor is zero, i.e., $i(0-) = 0$; that is, we assume that the switch in the series circuit had been open for a long time prior to $t = 0$. In the dual circuit, the corresponding condition is that the switch across the parallel circuit had been closed for a long time prior to $t = 0$, and that the voltage across the capacitor is zero, or $v(0-) = 0$. Now we invoke the conservation laws discussed in Section 5-6, which require, respectively, that $i(0+) = 0$ for the series circuit and $v(0+) = 0$ for the dual circuit. We thus write from Eq. (5-74) for $t = 0$,

$$0 = \frac{V}{R} + B,$$

from which,

$$B = -\frac{V}{R}. \tag{5-75}$$

The final solution for the current is thus,

$$i = \frac{V}{R}(1 - e^{-Rt/L}). \tag{5-76}$$

The nature of this function is illustrated in Fig. 5-4. This figure shows that the current rises from zero, the value imposed by the condition $i(0-) = i(0+) = 0$, with a decreasing slope as time progresses, and ultimately attains its asymptotic value.

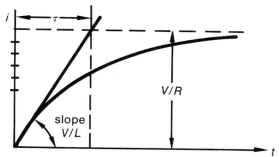

Fig. 5-4. The response of an initially relaxed RL circuit to a step function excitation.

The initial slope is an important feature of the exponential function. The slope of the current curve at any point is,

$$\frac{di}{dt} = \frac{V}{L} e^{-Rt/L}, \tag{5-77}$$

which has the value V/L at $t = 0$. Suppose that we use this initial slope to calculate the intercept τ, the point that the tangent to the response curve at the origin makes with the final value of the current. This is,

$$\tau = \frac{V/R}{\text{slope}} = \frac{L}{R}. \tag{5-78}$$

The quantity $\tau = L/R$, which is known as the *time constant* of the circuit (τ is in seconds when L is in Henrys and R is in Ohms), can be interpreted as the time that it would take the current to reach the steady-state value, if it were to continue at its original rate. The time constant may also be interpreted from Eq. (5-76) to show that for $t = \tau$, the current will reach 0.632 of its final value. We further note that the current will reach 99 percent of its final value in 5 time constants. We shall find that these interpretations apply even when the initial current is not zero.

5-8 THE SERIES *RL* CIRCUIT WITH AN INITIAL CURRENT

We now wish to examine the features of a circuit behavior with an initial current subject to a variety of switching conditions. First we consider the circuit that is given in Fig. 5-5, which shows the switching of the

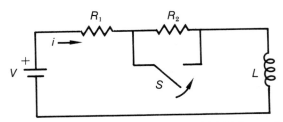

Fig. 5-5. Switching of R in a circuit with initial current.

circuit resistance. In this circuit the total resistance is $R_1 + R_2$ prior to switching, and becomes R_1 after switching at $t = 0$.

Clearly, there is an initial current prior to the switching time which is assumed to be constant, and of value,

$$i_0 = i(0-) = \frac{V}{R_1 + R_2}. \tag{5-79}$$

After switching, the controlling differential equation is given by the form of Eq. (5-70), with the general solution,

$$i = \frac{V}{R_1} + Be^{-R_1 t/L}. \tag{5-80}$$

To evaluate the constant B requires that we impose the initial condition, which is that $i(0-) = i(0+)$ since L remains constant in the circuit, and by

The Series RL Circuit with an Initial Current

Eq. (5-68) no instantaneous change in current may occur. We now write,

$$\frac{V}{R_1+R_2} = \frac{V}{R_1} + B,$$

from which we find that,

$$B = V\left(\frac{1}{R_1+R_2} - \frac{1}{R_1}\right) = \frac{-VR_2}{R_1(R_1+R_2)}. \tag{5-81}$$

The final solution is given by the expression,

$$i = \frac{V}{R_1}\left(1 - \frac{R_2}{R_1+R_2} e^{-R_1 t/L}\right). \tag{5-82}$$

The nature of this function is illustrated in Fig. 5-6. A little thought will show that the meaning of the time constant here remains precisely as in Section 5-7, but current changes are now measured from the initial value of current rather than from zero, which was previously the initial value.

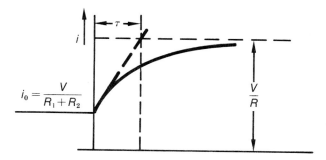

Fig. 5-6. Response of *RL* circuit with initial current.

As a second example, we refer to Fig. 5-7 which shows the switching of L in a circuit with an initial current. Here it is assumed that the switch is perfect so that when it is closed all of the current passes through the closed-switch. In this example the circuit inductance prior to switching is L_1 and becomes L_1+L_2 after $t=0$. As a result, the current prior to switching, for assumed steady conditions, is,

$$i(0-) = \frac{V}{R_1}. \tag{5-83}$$

To find the current after switching requires the careful use of the law of conservation of flux linkages, discussed in Section 5-6. We can write, over

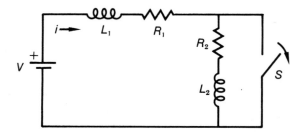

Fig. 5-7. Switching of L in a circuit with initial current.

the switching period,

$$L_1 i(0-) = (L_1 + L_2) i(0+), \tag{5-84}$$

from which,

$$i(0+) = \frac{L_1}{L_1+L_2} i(0-) = \frac{L_1}{L_1+L_2} \frac{V}{R_1}. \tag{5-85}$$

Now, because there has been a change in the circuit parameters, a transient current will result. Again the circuit dynamics are described by Eq. (5-70), with the general solution,

$$i = \frac{V}{R_1+R_2} + Be^{-(R_1+R_2)t/(L_1+L_2)}. \tag{5-86}$$

But it is required that at the switching instant, $t = 0$,

$$\frac{V}{R_1+R_2} + B = \frac{L_1}{L_1+L_2} \frac{V}{R_1},$$

from which,

$$B = V\left[\frac{L_1}{R_1(L_1+L_2)} - \frac{1}{R_1+R_2}\right] = \frac{L_1 R_2 - L_2 R_1}{R_1(R_1+R_2)(L_1+L_2)}. \tag{5-87}$$

Thus the final solution is given by the expression,

$$i = \frac{V}{R_1+R_2}\left[1 + \frac{L_1 R_2 - L_2 R_1}{R_1(L_1+L_2)} e^{-(R_1+R_2)t/(L_1+L_2)}\right]. \tag{5-88}$$

The nature of this function is illustrated in Fig. 5-8. This figure shows that at the switching instant the current may rise or fall, with or without a time delay until the circuit current assumes its final changed value. Observe

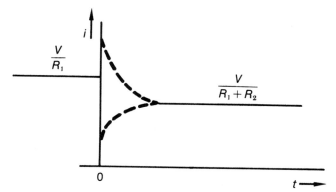

Fig. 5-8. Response of RL circuit when L is switched.

that if $R_2 = 0$, there will be a drop in current, and following a transient, the current will recover to its original value.

5-9 THE SERIES RC CIRCUIT AND ITS DUAL

This section is the counterpart of Section 5-7. The circuits under survey are shown in Fig. 5-9. We shall assume that there is an initial charge on

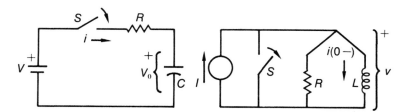

Fig. 5-9. The series RC circuit and its dual.

the capacitor so that an initial voltage V_0 exists thereon. Conversely, in the dual case, it is assumed that an initial current $i(0-)$ exists in the inductor.

The controlling equilibrium equation for the circuit, after switching, is

$$Ri + \frac{1}{C}\int i\,dt = v(t). \tag{5-89}$$

As before, a simple exponential form is assumed for the complementary solution. This leads to the requirement on s that,

$$R + \frac{1}{Cs} = 0,$$

which establishes the value for s,

$$s = -\frac{1}{RC}. \tag{5-90}$$

We note that since the capacitor cannot support a d-c through it, the steady state, or particular solution, is zero. Hence the complete solution is the expression,

$$i = B\, e^{-t/RC}. \tag{5-91}$$

To find the value of B requires an appropriate initial condition, $i(0+)$. This can be found from the fact that the voltage across the capacitor cannot change instantaneously, hence $v_c(0+) = v_c(0-) = V_0$, and the initial current is then,

$$i_0 = i(0+) = \frac{V - V_0}{R}. \tag{5-92}$$

This value is combined with Eq. (5-91) to yield, as the final expression for the current,

$$i = \frac{V - V_0}{R} e^{-t/RC}. \tag{5-93}$$

The nature of this function is shown in Fig. 5-10. Here, as in Fig. 5-4, the

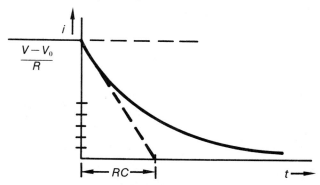

Fig. 5-10. The current in an RC circuit.

time constant $= RC$ (τ is in seconds when R is in Ohms and C is in Farads) can be related both to the initial slope of the curve and to the 0.368 value of the change in current, or the time to reach $1/e$th of its final value.

Often the voltage across the capacitor is of interest rather than the current in the circuit. This may be found in several different ways. If the current is known, as in the present case, then the voltage across the capacitor is given by the equation,

$$v_c = \frac{1}{C} \int i \, dt.$$

The constant of integration K that arises is evaluated from a knowledge of v_c when $t = 0$. That is, by performing the indicated integration, we have,

$$v_c = -(V - V_0) e^{-t/RC} + K. \tag{5-94}$$

But at time $t = 0$,

$$V_0 = -(V - V_0) + K,$$

and Eq. (5-94) becomes,

$$v_c = V - (V - V_0) e^{-t/RC}. \tag{5-95}$$

Another approach would be to write the dynamics of the system in terms of v_c, recalling that $i = C \, dv_c/dt$. Thus Eq. (5-89) would be written in the form,

$$RC \frac{dv_c}{dt} + v_c = v(t), \tag{5-96}$$

which defines the equilibrium conditions in terms of v_c as the dependent variable. The solution proceeds precisely as before, with the complementary function being an exponential provided that s satisfies the equation,

$$RCs + 1 = 0, \tag{5-97}$$

which again leads to Eq. (5-90) for the system time constant. The particular solution of Eq. (5-96) is the steady state value of v_c, which is V, as seen from an inspection of the circuit (no d-c current can be supported by the capacitor) or from Eq. (5-96) since in the steady state $dv_c/dt = 0$. The general solution to Eq. (5-96) is,

$$v_c = V + Be^{-t/RC}. \tag{5-98}$$

B is evaluated using the initial condition $v_c(0-) = v_c(0+) = V_0$;

thus,
$$V_0 = V + B,$$
and leading again to Eq. (5-95). A sketch of the solution for v_c is given in Fig. 5-11. Also shown in this figure is the value v_R, the voltage across the

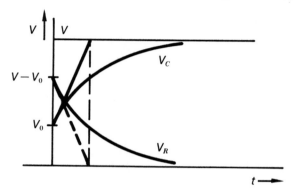

Fig. 5-11. The voltage in an RC circuit.

resistor. This is immediately written down, since from Eq. (5-89) we have that,
$$V = v_c + v_R.$$
Then v_R has the analytic description,
$$v_R = (V - V_0)e^{-t/RC}, \tag{5-99}$$
which may be written, of course, from Eq. (5-93).

Example 5-9.1

The accompanying figure is supposed to represent a heating furnace, of a type that might be used in continuous heat treating. A gas-air mixture is burned in the furnace, with a rate of heat supplied of Q_1. It is supposed that metal boxes are carried continuously through the furnace on a conveyor belt. Assuming that the process is a continuous one, establish the controlling equation, and discuss the results.

Solution. One arrives at the desired equation by an application of conservation of energy. The resulting expression is,

$$(C_p M L)\dot{\theta} - Q_1 + (C_p M v)\theta + hA(\theta - \theta_a) + Q_s - Q_i = 0,$$

The Series RC Circuit and its Dual

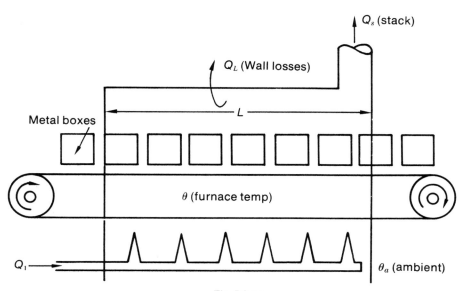

Fig. 5-9.1-1

where C_p = specific heat of metal boxes and contents, Joules/Kg-deg,
M = mass of material on conveyor, Kg/m,
L = length of furnace, m,
v = velocity of conveyor, m/sec,
h = heat loss coefficient, Joules/m²-deg,
A = heat loss area, m²,
θ_a = ambient temperature, degK,
Q_i = heat carried in by metal boxes,
Q_s = stack heat loss, Joules/sec.

The stack heat losses are assumed to be a function of the gas-air flow rate and the furnace temperature, i.e.,

$$Q_s = Q_s(Q_1,\theta),$$

although these losses are not assumed to vary over wide limits. We will suppose that the variations are relatively small, and hence we expand this implicit function in a Taylor expansion, and retain only the lowest order terms. This gives,

$$dQ_s = \frac{\partial Q_s}{\partial Q_1} dQ_1 + \frac{\partial Q_s}{\partial \theta} d\theta.$$

198   System Response

But for small fluctuations in heat input, the partial derivatives are essentially constant, which suggests that we write,

$$Q_s = K_1 Q_1 + K_2 \theta + Q_{ss}.$$

Combining this expression with the first equation gives,

$$(C_p ML)\dot{\theta} + (C_p M v + hA + K_2)\theta = (1 - K_1)Q_1 - (hA\theta_a + Q_i - Q_{ss}).$$

This process is seen to be of a single time constant type, with time constant,

$$T = \frac{C_p ML}{C_p M v + hA + K_2}.$$

It is observed that in this process that M, the amount of material on the conveyor, alters the time constant of the process.

5-10 THE SERIES RLC CIRCUIT AND ITS DUAL

A detailed study of the series RLC circuit, or its dual, is of interest because it introduces some additional ideas. Thus we now wish to consider Fig. 5-12 for study. The dynamic description of this circuit is the equation,

$$L\frac{di}{dt} + Ri + \frac{1}{C}\int i\,dt = v, \qquad (5\text{-}100)$$

which is exactly the equation discussed in Sections 5-3 and 5-4. The complete solution is thus written,

$$i = i_s + B_1 e^{s_1 t} + B_2 e^{s_2 t}, \qquad (5\text{-}101)$$

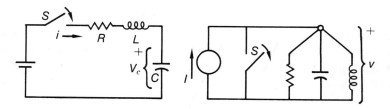

Fig. 5-12. The series RLC circuit and its dual.

where s_1 and s_2 have the forms,

$$s_1 = -\frac{\zeta}{\tau_n} + \frac{1}{\tau_n}\sqrt{\zeta^2 - 1},$$
$$s_2 = -\frac{\zeta}{\tau_n} - \frac{1}{\tau_n}\sqrt{\zeta^2 - 1},$$
(5-102)

where,

$$\tau_n = \sqrt{LC} \qquad \zeta = \sqrt{\frac{R^2 C}{4L}} = \frac{R}{2}\sqrt{\frac{C}{L}}.$$

As discussed in some detail in Section 5-3, the form of the solution depends on the relation among the parameters of the system. We wish to consider each case in some detail, subject to the following initial conditions,

$$i(0-) = 0,$$
$$v_c(0-) = V_0.$$
(5-103)

That is, there is zero initial current through the inductor, but the capacitor is assumed to have an initial charge.

From the conservation laws, the initial conditions after closing the switch are

$$i(0+) = 0,$$
$$v_c(0+) = V_0.$$
(5-104)

Since our solution is in $i(t)$, we must translate the second of these into an equivalent initial condition involving a function of i. This is done by making use of the integrodifferential equation, Eq. (5-100) which is valid of course from $t = 0$ for all time. We use this equation at the instant $t = 0+$. This yields directly,

$$L\frac{di}{dt}(0+) + 0 + v_c(0+) = V,$$

from which we write,

$$\frac{di}{dt}(0+) = \frac{V - V_0}{L}.$$
(5-105)

To carry out the solution, we use the known fact that i_s must be zero, since d-c is applied to a circuit containing a series capacitor C. Thus we must study the equation,

$$i = B_1 e^{s_1 t} + B_2 e^{s_2 t},$$
(5-106)

200   System Response

subject to the initial conditions specified by Eqs. (5-104) and (5-105). We proceed by evaluating the constants B_1 and B_2. At $t = 0+$,

$$i(0+) = 0 = B_1 + B_2,$$

and

$$\frac{di}{dt}(0+) = s_1 B_1 + s_2 B_2 = \frac{V - V_0}{L}. \tag{5-107}$$

The simultaneous solution of these two equations yields,

$$B_1 = -B_2 = \frac{V - V_0}{L(s_1 - s_2)} = \frac{V - V_0}{(2L/\tau_n)\sqrt{\zeta^2 - 1}}. \tag{5-108}$$

The final solution for the current is,

$$i(t) = \frac{V - V_0}{(2L/\tau_n)\sqrt{\zeta^2 - 1}} e^{-\zeta t/\tau_n} \left[\exp\left(\sqrt{\zeta^2 - 1}\,\frac{t}{\tau_n}\right) - \exp\left(-\sqrt{\zeta^2 - 1}\,\frac{t}{\tau_n}\right) \right]. \tag{5-109}$$

We now examine the details of Eq. (5-109) for the three ranges of values for ζ^2.

1. $\zeta^2 > 1$. In this case, by Eq. (5-24) the final solution has the form

$$i(t) = \frac{V - V_0}{(L/\tau_n)\sqrt{\zeta^2 - 1}} e^{-\zeta t/\tau_n} \sinh \sqrt{\zeta^2 - 1}\,\frac{t}{\tau_n} \tag{5-110}$$

The form of this equation is illustrated in Fig. 5-13, which also includes representations of Eq. (5-109), for which $|s_1| < |s_2|$, so that $e^{s_1 t}$ has a longer time constant than $e^{s_2 t}$. We observe that as C becomes very large in value, the circuit approaches the series RL case, and Fig. 5-13 approaches Fig. 5-4 in form. Likewise, if L becomes very small, the circuit approaches the series RC case, and Fig. 5-13 approaches Fig. 5-10 in form.

2. $\zeta^2 = 1$. In accordance with the discussion in Section 5-3.2 the equation for the current is,

$$i = \frac{V - V_0}{L/\tau_n} t\, e^{-\zeta t/\tau_n}. \tag{5-111}$$

The general form of the variation shown in Fig. 5-13 is valid in this case, but because the circuit is now *critically damped*, the time for decay is a minimum, without overshoot.

3. $\zeta^2 < 1$. This is the *underdamped* case, and the solution is, by Section 5-3.3,

$$i = \frac{V - V_0}{(L/\tau_n)\sqrt{1 - \zeta^2}} e^{-\zeta t/\tau_n} \sin \sqrt{1 - \zeta^2}\,\frac{t}{\tau_n}. \tag{5-112}$$

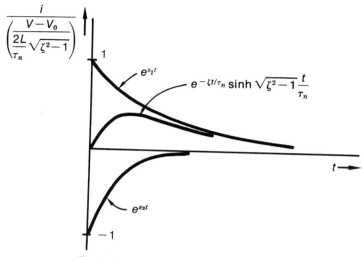

Fig. 5-13. Graphs of Eqs. (5-109) and (5-110).

This expression also follows directly from Eq. (5-110) by using the second of Eq. (5-35). This equation can exhibit rather different variations depending on the quantities ζ and τ_n. When ζ/τ_n is large and $\sqrt{1-\zeta^2}/\tau_n$ is small, there will be an overshoot of the current curve, and the sinusoidal variation specified by the equation will be masked. For intermediate values there will be a generally sinusoidal character, but with a noticeable decrement from one cycle to the next. For very light damping, the function begins to resemble a true sinusoid over a short time interval. These variations are illustrated in Fig. 5-14.

From Figs. 5-13 and 5-14, if R is considered as an adjustable parameter, Case (1) corresponds to large R. As R decreases, the quantity $\zeta = (R/2)\sqrt{(C/L)}$ decreases, with the curve approaching the axis in shorter and shorter times. A point is reached, Case (2) when the time is a minimum without overshoot. Continued reduction of R results in overshoot, with the results as shown in Fig. 5-14. Often a circuit of the type studied here is used to generate finite groups of lightly damped sine waves by applying a square pulse instead of the step function, and arranging, when the excitation is removed to add heavy damping to the circuit to cause the waves to damp rapidly to zero. Such a circuit is often referred to as a shock-excited oscillator.

Finally, if there were an initial current in the inductor, this would reflect itself in the constants B_1 and B_2 in Eq. (5-107), but otherwise would not affect the subsequent discussion.

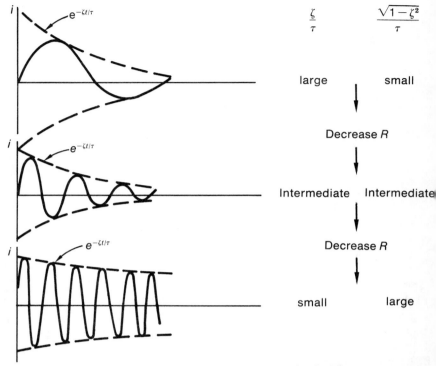

Fig. 5-14. The damped sinusoid, with decreasing damping.

Example 5-10.1

A fluid manometer is illustrated. Develop the equilibrium equation for this device, and from this specify the characteristic time and the damping ratio.

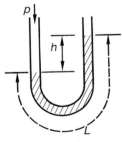

Fig. 5-10.1-1

Solution. An application of the Newton force law yields the following equation,

$$(AL\rho)\ddot{h} = -(RA^2\rho g)\dot{h} - (A\rho g)h + pA,$$

or

$$L\ddot{h} + RAg\dot{h} + gh = \frac{p}{\rho},$$

where L = length of the liquid column through the liquid, m,
A = area of the manometer tube, m²,
h = displacement difference,
ρ = mass density of manometer fluid, g/m³,
R = fluid resistance of manometer tube, sec/m²,
p = pressure, Newton/m² = Kg/m-sec²,
g = acceleration due to gravity, m/sec².

Using the results in Eq. (5-102) properly adapted to the above equation, we have that,

$$\tau = \sqrt{\frac{L}{g}},$$

$$\zeta = \frac{RAg}{2}\sqrt{\frac{1}{gL}} = \frac{RA}{2}\sqrt{\frac{g}{L}}.$$

Example 5-10.2

Refer to the accompanying figure which shows a process that is being controlled. In this system a level h_0 is established for the level of tank 2. The level control LC constrains the inflow Q_1 at a rate that is proportional to the difference between the set level h_0 and the level h_2 (this difference is often called the error, and is designated ϵ). Specify the characteristic time, the damping ratio and the steady state error (often called the offset) for a step-function input. Draw a SFG of the interconnected system.

Solution. The continuity relation for Tank 1 is,

$$C_1\dot{h}_1 = Q_1 - Q_2,$$

where, for laminar flow,

$$h_1 = R_1 Q_2.$$

These equations combine to,

$$R_1 C_1 \dot{h}_1 + h_1 = Q_1 R_1. \tag{1}$$

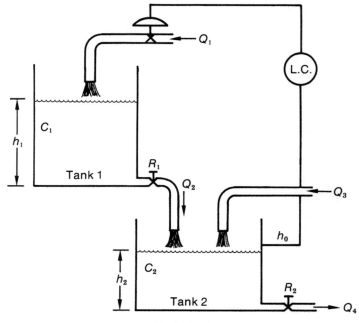

Fig. 5-10.2-1

This expression shows that R_1C_1 is the time constant for Tank 1.

The continuity equation for Tank 2 is,

$$C_2 \dot{h}_2 = Q_2 + Q_3 - Q_4,$$

where, for laminar flow,

$$h_2 = R_2 Q_4.$$

These expressions are combined to yield,

$$R_2 C_2 \ddot{h}_2 + h_2 = (Q_2 + Q_3) R_2. \tag{2}$$

For a proportional controller, the level control is specified by the relation,

$$Q_1 = K_c (h_0 - h_2). \tag{3}$$

The equations above are written in operational form,

$$(T_1 p + 1) h_1 = Q_1 R_1 \qquad T_1 = R_1 C_1, \tag{4}$$

$$(T_2p+1)h_2 = R(Q_2+Q_3) = h_1R_2/R_1 + R_2Q_3. \tag{5}$$

A SFG of Eqs. (3) through (5) is given in Fig. 5-10.2-2.

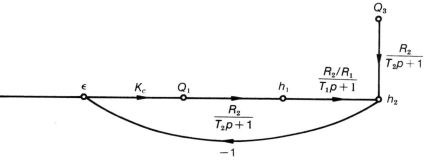

Fig. 5-10.2-2

We now combine the above equations in the following manner. By Eqs. (4) and (5),

$$(T_2p+1)h_2 = R_2Q_3 + \frac{R_2Q_1}{(T_1p+1)},$$

so that,

$$(T_1p+1)(T_2p+1)h_2 - R_2Q_3(T_1p+1) = R_2Q_1.$$

When this expression is combined with Eq. (3), we have,

$$R_2K_c(h_0 - h_2) = (T_1p+1)(T_2p+1)h_2 - R_2Q_3(T_1p+1). \tag{6}$$

This expression is solved for h_2, yielding,

$$h_2 = \frac{R_2K_ch_0}{(T_1p+1)(T_2p+1)+R_2K_c} - \frac{R_2Q_3(T_1p+1)}{(T_1p+1)(T_2p+1)+R_2K_c}. \tag{7}$$

This equation is combined with Eq. (6) to yield an expression for the error $\epsilon = (h_0 - h_2)$. The equation is,

$$\epsilon = (h_0 - h_2) = \frac{(T_1p+1)(T_2p+1)}{(T_1p+1)(T_2p+1)+R_2K_c}h_0 \\ - \frac{R_2(T_1p+1)}{(T_1p+1)(T_2p+1)+R_2K_c}Q_3.$$

This equation is written in the following form,

$$\epsilon = \frac{1}{R_2K_c+1}\frac{(T_1p+1)(T_2p+1)}{T^2p^2+2\zeta Tp+1}h_0 - \frac{R_2}{R_2K_c+1}\frac{(T_1p+1)}{T^2p^2+2\zeta Tp+1}Q_3,$$

where

$$T = \sqrt{\frac{T_1T_2}{R_2K_c+1}} \qquad \zeta = \sqrt{\frac{(T_1+T_2)^2}{4T_1T_2(R_2K_c+1)}}.$$

The offset to an applied step function is,

$$\epsilon_{ss} = \frac{-R_2}{(R_2K_c+1)}.$$

5-11 SWITCHING OF SINUSOIDAL SOURCES

While the general procedure is not dependent on the form of the source that is being switched into a circuit, we wish to consider the switching of a sinusoidal source in the RL circuit. The reason for this is that the transient term, and hence the transient performance under such an excitation source, is critically dependent on the circuit parameters, and it is possible to effect the switching without a switching transient.

We study the series RL circuit or its dual shown in Fig. 5-15. It is assumed that the excitation source is specified as,

$$v = V_m \cos(\omega t + \theta). \tag{5-113}$$

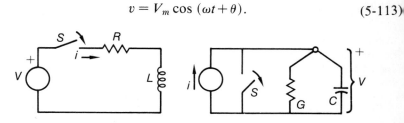

Fig. 5-15. The series RL circuit and its dual.

We proceed in the usual way by writing the equilibrium equation, Eq. (5-70) and choose as the complete solution,

$$i = i_s + i_t,$$

where, by the results of Section 5-4.4 we write for the particular solution,

$$i_s = \frac{V_m}{\sqrt{R^2 + \omega^2 L^2}} \cos(\omega t + \theta - \eta) \qquad \eta = \tan^{-1}\frac{\omega L}{R}, \qquad (5\text{-}114)$$

and as in Section 5-7 we write for the complementary function,

$$i_t = Be^{-Rt/L}. \qquad (5\text{-}115)$$

The complete solution is,

$$i = \frac{V_m}{\sqrt{R^2 + \omega^2 L^2}} \cos(\omega t + \theta - \eta) + Be^{-Rt/L}. \qquad (5\text{-}116)$$

For convenience, we assume that the circuit is initially relaxed, so that $i(0+) = 0$. Therefore we have that,

$$0 = \frac{V_m}{\sqrt{R^2 + \omega^2 L^2}} \cos(\theta - \eta) + B,$$

from which it follows that,

$$B = -\frac{V_m}{\sqrt{R^2 + \omega^2 L^2}} \cos(\theta - \eta). \qquad (5\text{-}117)$$

The final expression for the current is,

$$i = \frac{V_m}{\sqrt{R^2 + \omega^2 L^2}} [\cos(\omega t + \theta - \eta) - \cos(\theta - \eta)e^{-Rt/L}]. \qquad (5\text{-}118)$$

Observe from this equation that if the parameters of the circuit are so chosen that $\theta - \eta = \pi/2$, then the amplitude of the transient term is zero, and switching has occurred without any transient term, and the current is immediately at its steady state value. In the general case, the situation will be somewhat like that illustrated in Fig. 5-16.

5-b NUMERICAL METHODS

Much of the foregoing discussion relates to analytic methods for deducing the response of simple systems to normally well-behaved excitation functions. Further, the considerations have been limited to linear systems. When the systems are more extensive than of second order, or if the system is nonlinear, then finding the solution becomes increasingly more difficult. In fact, if the system is nonlinear, the likelihood of finding a closed-form solution is rather slim, and usually numerical methods of

208 System Response

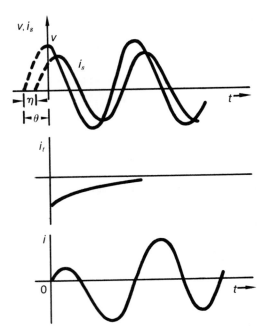

Fig. 5-16. Response of an *RL* circuit to a sinusoidal driving source.

integration must be used. It is the purpose of this section to discuss some of the methods of numerical analysis which are particularly important in the solution of system problems.

When employing numerical methods, one does not seek a continuous solution, but rather, the equations are solved at a sequence of closely spaced, and usually uniform intervals. The selection of the integration interval is a critical one and for linear time invariant systems depends on the largest and smallest eigenvalues of the system. The numerical integration of differential equations is quite slow when the network has a wide spread of eigenvalues. If the network is nonlinear, the numerical integration of the resulting differential equation may be quite burdensome. We wish to consider numerical methods for solving differential equations. These are of two general types, (1) those which develop an iterative procedure for the solution, and (2) those which replace the differential equation by a difference equation approximation, and then complete the solution recursively.

5-12 THE NEWTON-RAPHSON METHOD

Before considering the problem of the numerical solution of differential equations, we consider the matter of deducing the normal modes or eigenvalues of the system, a question that arose in Section 5-4. For the simple quadratic expression in Eq. (5-18) the procedure is straightforward and well known. For more extensive problems, the root-finding problem is not a simple one.

Consider a function $f(x)$ which is a continuous polynomial or transcendental function of x with real numerical coefficients, and suppose that it is required to find the real roots of the equation,

$$f(x) = 0. \tag{5-119}$$

The Newton-Raphson method is one of the more convenient methods for approximating a real root of such an equation. In this method, which is shown graphically in Fig. 5-17, the zero of the function is approached by

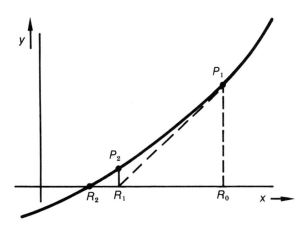

Fig. 5-17. The Newton-Raphson Method.

generating a sequence of successive approximations which converge to the real root in the limit. That is, an initial approximation to the desired real root must first be found; say this is R_0. This establishes point P_1 on the curve. Now, as shown in the figure, a tangent line is drawn at point P_1. This intersects the X-axis at R_1, which is a better estimate of the root than is R_0. The root R_1 establishes point P_2, and a tangent line to the curve at

P_2 provides a new estimate, R_2. This iterative procedure is repeated until the root R is approached, to some specified degree of accuracy.

The Newton-Raphson method proceeds in three steps:

1. Obtain an initial guess x_i to the desired root by graphing or other means.
2. Use the Newton-Raphson recurrence relation to obtain a new estimate of the root. This relationship results by setting the first two terms in a Taylor series to zero. The result is,

$$x_{i+1} = x_i - \frac{f(x_i)}{f'(x_i)}. \qquad (5\text{-}120)$$

3. Repeat the previous step as many times as is required until the desired accuracy has been obtained.

    It is noted that carrying out such a solution is readily programmed for a digital computer.

5-13 NUMERICAL SOLUTION OF DIFFERENTIAL EQUATIONS

The Euler method is a straightforward one for solving differential equations numerically. This is a step by step method with the dependent variable being determined for successive incremental steps of the independent variable. The shortcoming of the method is that it is subject to cumulative errors, and the solution is inaccurate when the method is used over wide ranges. The details of the method are best understood by reference to a particular example.

Example 5-13.1

Solve the nonlinear differential equation,

$$\frac{d^2y}{dx^2} + A\frac{dy}{dx} + Bxy = 0,$$

subject to: $x_0 = 0$, $y_0 = 1.0$, $(dy/dx)_0 = 1$, $A = -2$, $B = -3$. Use the Euler method.

Solution. Suppose that the functional relationship between y and x has the form illustrated in the figure. We employ a step-wise or staircase approximation, as illustrated, between successive points; that is, we select,

$$x_{n+1} = x_n + \Delta x,$$

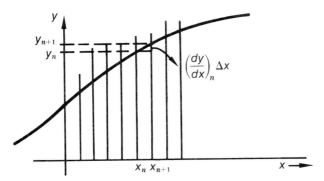

Fig. 5-13.1-1

and

$$y_{n+1} = y_n + \left(\frac{dy}{dx}\right)_n \Delta x,$$

which are the first two terms in the Taylor series approximation. Now, by differentiation, we have that,

$$\left(\frac{dy}{dx}\right)_{n+1} = \left(\frac{dy}{dx}\right)_n + \left(\frac{d^2y}{dx^2}\right)_n \Delta x.$$

These three expressions will be combined with the original differential equation at the sample points,

$$\left(\frac{d^2y}{dx^2}\right)_{n+1} = -A\left(\frac{dy}{dx}\right)_{n+1} - Bx_{n+1}y_{n+1}.$$

With these, we can set up an iterative procedure. We begin with,

$$\left(\frac{d^2y}{dx^2}\right)_0 = -A\left(\frac{dy}{dx}\right)_0 - Bx_0y_0,$$

to specify the initial $(d^2y/dx^2)_0$. With this, we can develop the following sequence of calculations,

$n = 0$, calculate x_1 y_1 $(dy/dx)_1$ $(d^2y/dx^2)_1$
$n = 1$, calculate x_2 y_2 $(dy/dx)_2$ $(d^2y/dx^2)_2$ etc.

The solution is given by $y = f(x)$.

5-14 DIFFERENCE EQUATION APPROXIMATION

The numerical methods of integration currently in use are, for the most part, based on the use of finite difference approximations to the actual differential equation to be solved. In this approximation, as discussed in Section 2-10, the derivatives are replaced by differences, using the substitutions,

$$\frac{dy}{dt} \simeq \frac{y(t+T)-y(t)}{T} = \frac{y_{n+1}-y_n}{T},$$
$$\frac{d^2y}{dt^2} \simeq \frac{y_{n+2}-2y_{n+1}+y_n}{T^2}, \quad (5\text{-}121)$$
$$\frac{d^3y}{dt^3} \simeq \frac{y_{n+3}-3y_{n+2}+3y_{n+1}-y_n}{T^3}.$$

Here the continuous time t is sliced into equal increments T seconds long, and we concern ourselves only with the values of y at $t = 0, T, 2T, \ldots, nT$. For the values of y for t between any pair of values of T, interpolation can be employed. Of course, smaller time intervals T may be used also, but this will necessitate an increased number of calculations.

Let us apply these ideas to a simple first order differential equation of the form,

$$\frac{dy}{dt} = f(y,t). \quad (5\text{-}122)$$

In finite difference form this equation becomes,

$$\frac{y_{n+1}-y_n}{T} = f(y_n,n),$$

which is written,

$$y_{n+1} = y_n + Tf(y_n,n). \quad (5\text{-}123)$$

This is a *recursive* formula which allows a solution on a step-by-step basis. From this equation we write the sequence of steps,

$$\begin{aligned} y_1 &= y_0 + Tf(y_0,0) \\ y_2 &= y_1 + Tf(y_1,1) \\ y_3 &= y_2 + Tf(y_2,2) \\ &\vdots \quad \vdots \quad \vdots \end{aligned} \quad (5\text{-}124)$$

Clearly, if we know $y(0)$ and the functional form $f[y(n),n]$ then we can find y_1, then y_2, then y_3, etc. The general finite difference expression is of the implicit integration form (3)

$$y_n = \sum_{i=1}^{M} a_i y_{n-1} + T \sum_{j=0}^{M} b_j f_{n-j} \qquad (5\text{-}125)$$

where the coefficients a_i and b_i are determined by the number of terms of the Taylor expansion of $y(nT)$ which are to be matched by the integration formula so defined.

Equation (5-125) when combined with Eq. (5-122) is written,

$$y_n = \sum_{i=1}^{M} a_i y_{n-i} + T \sum_{j=0}^{M} b_j \dot{y}_{n-j}. \qquad (5\text{-}126)$$

This form is the numerical analysis technique that is used in ECAP-II (An Electronic Circuit Analysis Program) in the IBM 1130 computing system and for machines supporting the IBM operating system (S/360) (4).

Note that if $b_0 = 0$ in Eq. (5-126), then this expression represents an *explicit* method of integration since all the data on the right hand side are known from previous integration steps. The relation of step size T and the system time constant is important in these calculations, as will be discussed below. If $b_0 \neq 0$, then determining a new value of y_n requires a knowledge of $\dot{y}_n$, a quantity that has not yet been found. In this case Eq. (5-126) represents an *implicit* method of integration. To effect a solution, it is necessary to set up a system of equations relating $\dot{y}_n$ to y_n at every integration step. If these equations are nonlinear, an iterative process of solution must be used. This requirement for solving simultaneous equations would ordinarily make such a method quite impractical, except that new "sparse matrix" techniques allow these calculations to be carried out so quickly that implicit integration becomes quite feasible. An advantage of the implicit integration method is that the calculation eliminates the time constant problem, the integration step size being dictated by the allowed truncation error. Another desirable feature of implicit integration is that the algebraic-differential equations can be solved simultaneously, which simplifies the coding.

Example 5-14.1

Consider for numerical solution, the differential equation,

$$\frac{dy}{dt} + y = 0 \qquad y(0) = y_0 = 1,$$

at the integration time intervals $T = 1, 0.5, 0.1$. Discuss the results.

Solution. It is observed that this differential equation can be solved exactly, with the solution $y(t) = y_0 e^{-t}$. This solution can be used as a reference for comparison of the approximate solutions. The finite difference approximation to the given equation is,

$$\frac{y_{n+1} - y_n}{T} + y_n = 0,$$

from which,

$$y_{n+1} = (1 - T) y_n.$$

a. $T = 1$. From the recursive formula we find that,

$$y_{n+1} = (1 - 1) y_n = 0.$$

Since the function is zero after one increment, we conclude that $T = 1$ is too large.

b. $T = 0.5$. The present calculation is based on the recursive formula,

$$y_{n+1} = 0.5 y_n.$$

We now find the following sequence,

$y_1 = 0.5 y_0 = 0.5$
$y_2 = 0.5 y_1 = 0.25$ (one time constant in time).
$y_3 = 0.5 y_2 = 0.125$
$y_4 = 0.5 y_3 = 0.0625$

. .
. .
. .

c. $T = 0.1$. The recursive formula becomes,

$$y_{n+1} = 0.9 y_n.$$

We calculate the sequence,

$y_1 = 0.9 y_0 = 0.9$ $y_6 = 0.9 y_5 = 0.531441$
$y_2 = 0.9 y_1 = 0.81$ $y_7 = 0.4782969$
$y_3 = 0.729$ $y_8 = 0.43046901$
$y_4 = 0.6561$ $y_9 = 0.387422109$
$y_5 = 0.59049$ $y_{10} = 0.3486798981$ (one time constant).

A sketch of these results is informative. We note that since the exact value at $t = 1$ (one time constant) is $y = 0.368$, we see that the error is about 5 percent when $T = 0.1$ sec. For better accuracy T must be chosen less than 0.1, with a corresponding increase in the number of calculations, e.g., for $T = 0.01$ there would be 100 calculations per time

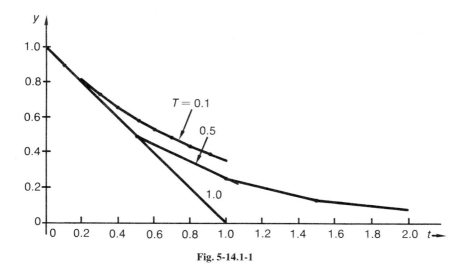

Fig. 5-14.1-1

constant; for $T = 0.001$ there would be 1000 calculations per time constant.

As already noted, a complication that arises in the numerical solution of differential equations stems from the fact that the system eigenvalues may be widely spaced. The largest eigenvalue (or equivalently, its reciprocal, the smallest time constant) controls the permissible size of the integration step T. But small eigenvalues (large time constants) control the network response, and so determine the total length of time over which the integration must be carried out to characterize the response. For a network with a 1000 to 1 ratio of largest to smallest eigenvalue, it might be necessary to take 1000 times as many integration steps using a method that was sensitive to this factor as in one which is not. As a consequence, response waveforms that would be adequately described by a few hundred points may necessitate finding several thousand integration points.

It is the "minimum time constant" barrier that is the most serious obstacle in the transient analysis of networks. This is so because the minimum time constant barrier severely limits the speed of numerical integration in the transient analysis, even for small networks. The small time constant barrier can be breached for linear networks by expanding the solution in terms of eigenfunctions. Unfortunately no comparable scheme exists for nonlinear systems, and the problem remains a serious one. Attention is being directed to this problem (5).

5-15 NONLINEAR SYSTEMS

There is nothing inherent in the discussion in the foregoing section which restricts the techniques of numerical integration to linear systems; the application to nonlinear systems follow in precisely the same way as for the linear system although now one cannot speak of eigenvalues. Again the derivatives are replaced by differences, and the solution is carried out from the resulting recursive formula. For example, if we were to consider the nonlinear differential equation,

$$\frac{d^2y}{dt^2} + (2y+1)\frac{dy}{dt} + y = 0, \qquad (5\text{-}127)$$

we would write the equation in difference equation form,

$$\frac{y_{n+2} - 2y_{n+1} + y_n}{T^2} + (2y_n + 1)\frac{y_{n+1} - y_n}{T} + y_n = 0. \qquad (5\text{-}128)$$

The recursive formula is,

$$y_{n+2} = (2 - 2Ty_n - T)y_{n+1} + [T(2y_n + 1) - T^2 - 1]y_n = 0. \qquad (5\text{-}129)$$

For $T = 0.1$, this expression becomes,

$$y_{n+2} = (1.9 - 0.2y_n)y_{n+1} + (0.2y_n - 0.91)y_n = 0. \qquad (5\text{-}130)$$

From this, we can proceed in a systematic way to find the value of y for successive values of n, given y_0 and y_1.

The same ideas may be employed if the systems elements and the driving function are given in graphical form rather than in analytic form. This is shown in the following example.

Example 5-15.1

The characteristics of a nonlinear inductor and the applied driving force are given in the accompanying figures. Determine the current in the series RL circuit shown, for $i(0+) = 2$ amps.

Solution. The controlling equation for this circuit is

$$\frac{d\psi}{dt} + 0.2i = v,$$

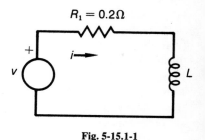

Fig. 5-15.1-1

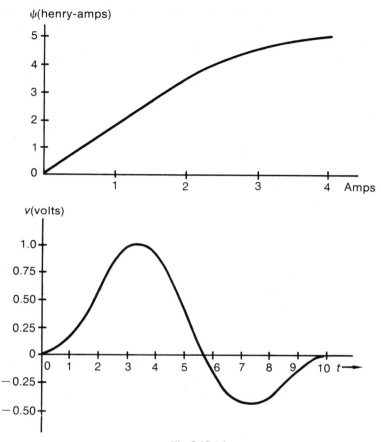

Fig. 5-15.1-2

where $\psi = \psi(i)$ is the nonlinear characteristic of the inductor. This equation is written in difference equation form, thus,

$$\frac{\psi_{n+1} - \psi_n}{T} + 0.2\, i_n = v_n,$$

$$\psi_n = \psi(i_n).$$

The recursive formula is given as,

$$\psi_{n+1} = \psi_n - 0.2\, T i_n + T v_n.$$

218   System Response

To establish a reasonable sampling time, we note that over the substantially linear portion of the flux-linkage curve, $L = 1.8$ Henrys. Thus the approximate time constant of the circuit is $L/R = 9$ secs. A value of $T = 1$ thus appears reasonable. Consequently, the equation under study is,

$$\psi_{n+1} = \psi_n - 0.2\, i_n + v_n.$$

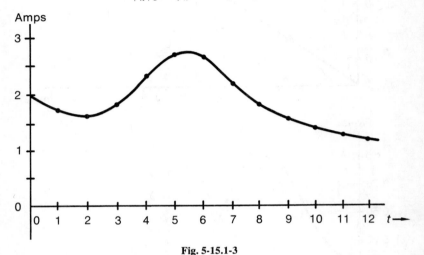

Fig. 5-15.1-3

The solution is contained in the following table, noting that the initial condition $i(0+) = 2$ amps corresponds to $\psi(0+) = 3.6$. A curve showing the current-time response is given in the accompanying figure.

n	v_n	i_n	ψ_n	$0.2 i_n$	ψ_{n+1}	i_{n+1}
0	0	2	3.6	0.4	3.2	1.7
1	0.19	1.7	3.2	0.34	3.1⁺	1.7⁻
2	0.63	1.7⁻	3.2⁻	0.34⁻	3.5	1.82
3	0.98	1.82	3.5	0.36	4.1	2.35
4	0.90	2.35	4.1	0.47	4.5	2.7
5	0.38	2.7	4.5	0.54	4.4	2.7
6	−0.015	2.7	4.4	0.54	3.8	2.2
7	−0.041	2.2	3.8	0.44	3.4	1.8
8	−0.039	1.8	3.4	0.36	3.0	1.6
9	−0.010	1.6	3.0	0.32	2.7	1.45
10	0	1.45	2.7	0.29	2.4	1.3
11	0	1.3	2.4	0.26	2.1	1.15
12	0	1.15	2.1	0.23	1.8	0.95
13	0	0.95	1.8	0.19	1.6	0.85

5-16 VARIOUS METHODS FOR NUMERICAL INTEGRATION (6)

The method discussed in the foregoing section for obtaining recursion formulas for the numerical solution of differential equations is only one of a number of ways for solving differential equations. Most methods begin by writing the differential equation in normal form. This gives a set of first order equations that are connected by derivative relations, but otherwise are independent. Thus these solutions are concerned with first order equations. We shall now consider the essential features of a number of methods for solving a differential equation of the general form given in Eq. (5-122).

The essential question to which each of the various methods addresses itself is, given the equation,

$$\frac{dy}{dt} = f(y,t), \tag{5-131}$$

which specifies the slope of the curve at each position in space and time, what approximation should be made for the slope over any given time interval? The approximation by each method is discussed.

a. Euler Method. This method begins by replacing the differential equation by a difference equation, as in Eq. (5-123), namely,

$$y_{n+1} = y_n + Tf(y_n,n). \tag{5-132}$$

This method assumes that the slope of the curve $f(y_n,n)$ at nT remains constant over the interval. Thus a simple linear stepwise calculation yields y_{n+1} from y_n. The method is inefficient, and for reasonable accuracy requires that the integration interval must be small. The inaccuracy stems from the estimate of the slope of the curve at nT, and that it is used throughout the succeeding interval. Another disadvantage is that Eq. (5-132) is an implicit formula because it involves $f(y_n,n)$, which depends on the estimated y_n.

b. Improved Euler Method. The difference equation in this method is written,

$$y_{n+1} = y_n + \frac{T}{2}\{f(y_n,n) + f[y_n + Tf(y_n,n), n+1]\}. \tag{5-133}$$

It is noted that the terms within the braces, $y_n + Tf(y_n,n)$, as discussed in connection with Eq. (5-132), is an estimate of the value of y_{n+1}. Thus this method uses an estimated y_{n+1} to establish a slope dy/dt at the end of the interval. The known slope at the beginning of the interval together with the estimated slope at the end of the interval is used to give an average slope over the interval. This average slope is assumed constant over the interval, and is used to compute the value of y_{n+1}.

This calculation, as for Eq. (5-132), is of low order accuracy and the integration interval must often be shortened to insure sufficient accuracy. Further, it too is an implicit formula since it involves f_{n+1}, which cannot be computed until y_{n+1} is known.

c. Modified Euler Method. Here the recursive equation is written,

$$y_{n+1} = y_n + Tf\left[y_n + \frac{T}{2}(y_n, n), n + \frac{1}{2}\right]. \qquad (5\text{-}134)$$

In this method an estimate is made of the slope of the function at the midpoint of the interval, which is then used over the interval for finding y_{n+1}. The inherent disadvantages in using this formula are essentially those discussed under b.

d. Runge-Kutta Method. This method employs a relatively sophisticated method for finding the average slope during the integration interval. The recursive equation is written,

$$y_{n+1} = y_n + \frac{T}{6}(K + 2L + 2M + N), \qquad (5\text{-}135)$$

where,

$$K = f(y_n, n),$$
$$L = f\left(y_n + \frac{T}{2}K, n + \frac{1}{2}\right),$$
$$M = f\left(y_n + \frac{T}{2}L, n + \frac{1}{2}\right),$$
$$N = f(y_n + TM, n + 1).$$

It is observed that,

$K =$ the slope at the beginning of the interval.
$L =$ an estimate of the slope at the midpoint of the interval, assuming that K is the slope over the entire half interval.
$M =$ an estimate of the slope at the midpoint of the interval, assuming that L is the slope over the entire half interval.
$N =$ an estimate of the slope at the end of the interval.

Clearly in Eq. (5-135) the average slope that is used is the weighted combination of the estimated slopes K, L, M, N. The complication of the method is that it requires four slope evaluations for each step forward. A disadvantage of this method is the lack of a measure of accuracy so that it is not possible to tell when the integration interval should be changed.

e. Predictor-Corrector Methods. This method involves the following two recursive formulas:

$$y_{n+1} = y_{n-1} + 2Ty'_n \qquad \text{Predictor equation},$$
$$y_{n+1} = y_n + \frac{T}{2}(y'_{n+1} + y'_n) \quad \text{Corrector equation}. \qquad (5\text{-}136)$$

The first equation is called a predictor, since the new value y_{n+1} is predicted in terms of the previous value y_{n-1} and the previous derivative y'_n. This predicted value of y_{n+1} is used in the corrector equation to compute an approximate value of the "new" derivative, which is then used in the corrector equation for a separate estimate of y_{n+1}. The difference between the predicted and the corrected values provides a measure of the error.

Let us reexamine the cycle. Denote the y_{n+1} from the predictor equation as p_{n+1}. This value is combined with Eq. (5-136), the difference equation representation of the differential equation, to evaluate $p'_{n+1} = f(p_{n+1}, n+1)$, which is the predicted slope at the end of the interval. This value of p'_{n+1} is used in the corrector equation in place of y'_{n+1}, and the corrected value of y_{n+1}, now called c_{n+1}, is found. If c_{n+1} does not differ appreciably from p_{n+1}, then continue with the next step, otherwise use c_{n+1} in Eq. (5-136) to find a new slope $c'_{n+1} = f(c_{n+1}, n+1)$. This new slope c'_{n+1} is used in place of y'_{n+1} to compute a new y_{n+1}. This cycle can be continued until satisfactory results are achieved. Clearly, if the accuracy is not satisfactory after the first cycle, then it is probably sensible to reduce the value of the time interval T and again use the predictor-corrector equations. As a comparison with the Runge-Kutta method, it is noted that here only one, or perhaps two, evaluations of derivatives need to be computed at each step, against the four slopes required in the Runge-Kutta method.

A feature of the predictor-corrector method is that it is not self-starting, that is, it is necessary to know y_0 and y'_1 in order to begin operations. y_0 is given, but y'_1 must be calculated. One may use the Runge-Kutta or one of the Euler methods (or other methods) to find y'_1.

5-c MACHINE SOLUTIONS

Two important machine methods exist for the solution of the equations that arise in system studies. These involve the analog computer and the

222 System Response

digital computer. The analog computer is a simulation device which provides continuous information. In it, the differential equations that describe a system are duplicated on the machine, and the excitation function is duplicated exactly or approximately by function generators. The solution of the system equations appears on a strip chart as a time varying result.

In the digital computer the differential equation is solved numerically using methods like those discussed in Section 5-b. It is possible also to program the digital machine so that it will simulate the analog machine. The simulation programming is accomplished using precisely the same type of program graph used for the analog computer so that outwardly, at least, the two machines perform the same calculations in the same way. Internally, of course, the machine operations are entirely different. Another difference lies in the ability to time-scale the analog machine, thereby allowing "real" time calculations. In those cases where real time data are required the analog machine is indicated, whereas for results in numerical or graphical form, both machines would serve. This matter receives additional attention in Section 5-20.

Analog computers must perform the basic mathematical operations specified by a differential equation, and involves addition, subtraction, differentiation, integration, multiplication, division, etc. Consequently, electronic circuits which can perform these basic mathematical operations must be available. In some cases these circuits depend for their operation on the special features of transistor or tube characteristics. In other cases feedback is applied in special ways to achieve the desired results. Often complicated circuit arrangements are used to accomplish given operations. We shall not be concerned here with the details of such circuits, although we do wish to consider certain aspects of the operational amplifier, which is the essential element of the analog computer, as we shall discuss.

5-17 THE OPERATIONAL AMPLIFIER

The operational amplifier is a high gain amplifier with high stability. It is provided with terminals for applying feedback by appropriately connecting resistors and capacitors into the circuit. The amplifier may be a single high gain state, or it may be a cascade of stages to achieve high gain. For an input $v(t)$, the output is $-Av(t)$, where A, the nominal gain without feedback, is quite large, say 1000 or more. Commercially available integrated circuit operational amplifiers usually have gains in excess of 100,000.

a. The Summing Amplifier—Scale and Sign Changing. An amplifier without feedback is usually symbolized by either a triangular symbol or a SFG, as in Fig. 5-18. The representation of the amplifier with feedback is given in Fig. 5-19. In this case, a scale change or multiplication by a con-

Fig. 5-18. Amplifier and its SFG representation.

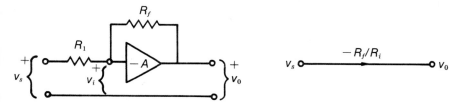

Fig. 5-19. A feedback amplifier with scale change.

stant results. An implicit assumption in the analysis of the operation of this circuit is that the amplifier input impedance is very large, and does not load any circuit connected to its input terminals. This is a valid approximation since the design provides this property. Observe in the figure that a distinction is made between v_s, the input to the circuit, and v_i, the input to the amplifier terminals. An important feature of the operational amplifier follows from the simple gain expression,

$$v_0 = -Av_i. \qquad (5\text{-}137)$$

For a reasonable output signal (say a fraction of a volt), v_i will be very small if A is large. In the limit as A becomes infinitely large, v_i approaches zero for the finite v_0. Because of this, point G is often referred to as a *virtual ground*. Under these conditions, the analysis of Fig. 5-19 is exceedingly simple.

Under the conditions that the current to the input terminals of the amplifier is zero (this assumes an infinite input impedance; it can be 10^6 ohms with vacuum tubes, and with proper circuits can be 10^6 with transistors, and 10^8 or higher with FET-s) we find, by an application of the KCL at node G,

$$i_1 + i_f = i_A = 0,$$

which is,

$$\frac{v_s}{R_1} + \frac{v_0}{R_f} = 0,$$

so that,

$$\frac{v_0}{v_s} = \frac{-R_f}{R_1}. \tag{5-138}$$

The resistance values can be chosen to make the resistance ratio assume any desired factor, within the limits of the amplifier. Using the convention of sign-changing through the operational amplifier, it may be necessary to use a second stage with unit gain as a sign changer, if a positive output is required for positive input.

An important extension of Fig. 5-19 is shown in Fig. 5-20. Different

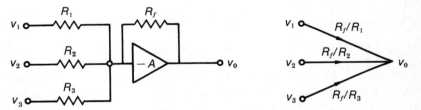

Fig. 5-20. The summing amplifier.

voltages are applied to each input. The result, using the KCL, is,

$$\frac{v_1}{R_1} + \frac{v_2}{R_2} + \frac{v_3}{R_3} + \frac{v_0}{R_f} = 0,$$

from which

$$v_0 = -\left(\frac{R_f v_1}{R_1} + \frac{R_f v_2}{R_2} + \frac{R_f v_3}{R_3}\right). \tag{5-139}$$

This shows that the output is the negative of the sum of the input voltages, each of which is separately scaled by its own resistance ratio. When the resistances are chosen to be,

$$R_1 = R_2 = R_3 = R_f,$$

then,

$$v_0 = -(v_1 + v_2 + v_3), \qquad (5\text{-}140)$$

and v_0 is simply the negative of the sum of the inputs. As a practical matter the resistance values of R_1, R_2, R_3, should be high in order to avoid mutual interaction among the sources. This requires that the input resistance to the amplifier must be high, if loading of the sources is to be avoided.

b. The integrator. Refer to Fig. 5-21, which is a generalization of Fig. 5-19, with Z_1 and Z_f replacing the resistors R_1 and R_f, where the Z-s may

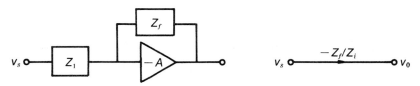

Fig. 5-21. The operational amplifier.

denote single elements or combinations thereof. One proceeds in the manner that leads to Eq. (5-138) to find that, in the present case,

$$v_0 = -\frac{Z_f}{Z_1} v_s. \qquad (5\text{-}141)$$

This expression is now to be viewed as an operational equation, with the output being specified as $-(Z_f/Z_1)v_s$. Clearly, if each Z is a resistor, then Eq. (5-138) results. We wish to examine particularly the case when $Z_f = 1/Cp$ (the operational form for the capacitor) and $Z_1 = R$. Equation (5-141) now assumes the form,

$$v_0 = -\frac{1}{CRp} v_s, \qquad (5\text{-}142)$$

which indicates an integration process. Moreover, if the circuit constants are so chosen that $RC = 1$, then,

$$v_0 = -\frac{1}{p} v_s. \qquad (5\text{-}143)$$

In this case the transmittance of the SFG representation is $-1/p$, a simple integrator, as shown in Fig. 5-22. From practical considerations, the

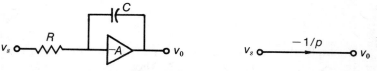

Fig. 5-22. The simple integrator.

series R is made large (of the order of 10^6 ohms or larger) to avoid loading the source v_s.

Figure 5-22 may be extended in much the same way that Fig. 5-19 was extended to Fig. 5-20. We thus consider Fig. 5-23, which is the summing

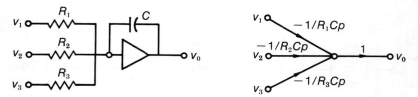

Fig. 5-23. The summing integrator.

integrator. The output from this amplifier circuit is readily shown to be,

$$v_0 = -\left(\frac{1}{R_1 C p}v_1 + \frac{1}{R_2 C p}v_2 + \frac{1}{R_3 C p}v_3\right). \tag{5-144}$$

For equal time constants, such that $R_1 C = R_2 C = R_3 C = 1$, the output is,

$$v_0 = -\left(\frac{1}{p}v_1 + \frac{1}{p}v_2 + \frac{1}{p}v_3\right) = -\frac{1}{p}(v_1 + v_2 + v_3). \tag{5-145}$$

c. The differentiator. Differentiation may be achieved in much the same way that integration was achieved. This involves an interchange of the R and C in Fig. 5-22, with the circuit representation in Fig. 5-24. As a practical matter, because of the inherent property of a differentiator to aggravate any noise content of an input signal, the use of the differentiator is avoided in analog computer practice. The differentiator may be used

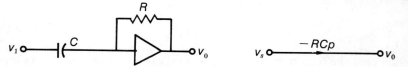

Fig. 5-24. The differentiator.

when a digital computer is used to simulate the analog computer, since the noise problem does not exist, unless it is specifically programmed into the computer program. Ordinarily, however, the differentiator is not used, and its use is avoided by properly programming the computer to accomplish the desired results without requiring differentiators. We shall see how this is accomplished in our further studies.

d. Function generation. By using somewhat more elaborate network combinations than simple R and C in the operational amplifier, somewhat more elaborate output functions may be obtained. Specifically, refer to Fig. 5-25 which shows a not too involved feedback impedance. In this

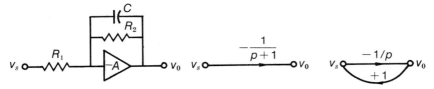

Fig. 5-25. A feedback function generator.

amplifier we have that,

$$Z_f = \frac{R_2}{R_2 Cp + 1} \qquad Z_1 = R_1.$$

The output of the operational amplifier is then,

$$v_0 = -\frac{R_2}{R_1} \frac{1}{R_2 Cp + 1} v_s. \tag{5-146}$$

Note that if the element values are chosen so that $R_2 = R_1$ and $R_2 C = 1$, this output function is,

$$v_0 = -\frac{1}{p+1} v_s = \frac{-1/p}{1-(-1/p)} v_s. \tag{5-147}$$

Observe that this result is consistent with the second SFG which shows the integrator with a simple feedback line. It is generally desirable, in most cases, to use a simple integrator with an appropriate feedback line to accomplish a given function than it is to seek special RC combinations for Z_1 and Z_f for achieving the same result.

5-18 COMPUTER SIMULATION OF DIFFERENTIAL EQUATIONS

The availability of operational amplifiers makes possible a ready simulation of linear, ordinary differential equations with constant coefficients. The extension to nonlinear differential equations is straightforward, but it is necessary that appropriate nonlinear functions be generated. This is frequently accomplished with diode function generators which are often included with the analog computer. The diode function generator may be programmed to produce a piecewise linear approximation to a specified nonlinear function.

Reference has been made in Chapter 4 that the SFG serves extremely well as a program graph for the analog computer. We now wish to carry out the details of such programming. It is well to note that experience and care will often yield an optimum program, that is, one that requires a minimum number of amplifiers for simulating a given differential equation. There are a few basic rules to guide one in preparing an analog computer program. These are:

1. Differentiators must not be used.
2. A sign reversal is to be associated with each operational amplifier. (As a practical matter, sign reversing amplifiers are often included with the operational amplifier to allow an output of either sign.)
3. The use of summing amplifiers and phase reversing amplifiers (sign changers) in program graph preparation can often result in a saving in the number of operational amplifiers that are required.

We illustrate the procedure with several examples.

Example 5-18.1

Draw a SFG program graph and the analog computer program for simulating the differential equation,

$$\frac{d^2y}{dt^2} + K_1 \frac{dy}{dt} + K_2 y = f(t),$$

subject to $y(0) = \dot{y}(0) = 0$.

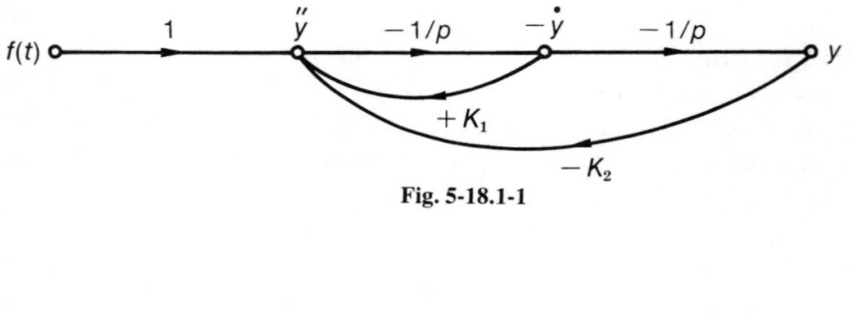

Fig. 5-18.1-1

Solution. This differential equation is solved for the highest derivative, to give,

$$\frac{d^2y}{dt^2} = f(t) - K_2 y - K_1 \frac{dy}{dt}.$$

We now draw a SFG for this equation according to the methods discussed in Chapter 4. This leads to the SFG shown. The appropriate analog computer program is a direct translation of this SFG to analog computer graphical symbolism. The result is given in the accompanying figure. Observe the inclusion of one sign-changer in order to account for one of the required signs.

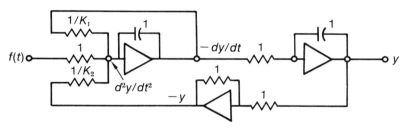

Fig. 5-18.1-2

Example 5-18.2

Draw the program graph and the analog computer program for simulating the simultaneous equations,

$$\frac{d^2y}{dt^2} + 2y - x = f(t),$$

$$\frac{dx}{dt} - 2x = \frac{dy}{dt} - 3y.$$

Solution. Each equation is solved for the highest order derivative in a different variable. Thus we write,

$$\ddot{y} = f(t) - 2y + x,$$
$$\dot{x} = \dot{y} - 3y + 2x.$$

SFG-s are drawn for each equation and the interconnected graph is then drawn; the analog computer program follows from this. These steps are given in the associated figures.

230 System Response

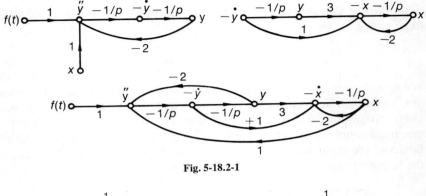

Fig. 5-18.2-1

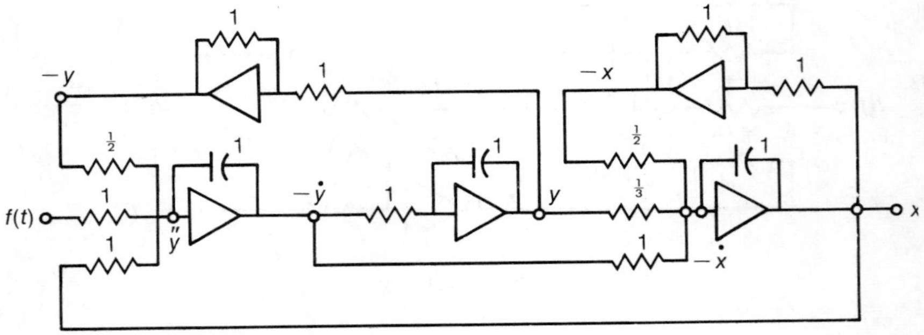

Fig. 5-18.2-2

5-19 INTRODUCING INITIAL CONDITIONS

We have already discussed that the solution of a particular problem that is described by an nth order differential equation requires the specification of n quantities which describe the state of the system at some particular time, usually chosen as $t = 0$. Correspondingly, of course, if the system is described by n equivalent first order state equations, the initial state vector also requires a knowledge of n initial quantities. As a result, to meet the initial conditions of the problem, each integrator output variable must be set at its proper initial value. This means, therefore, that initial voltages across the integrator capacitors must be preset to certain values.

The procedure consists in having a controlled source of voltage which can be switched across each capacitor, thus charging each capacitor to a prescribed voltage appropriate to the initial conditions. A typical method for injecting initial conditions is shown in Fig. 5-26. Note specifically that in this figure a positive voltage relative to ground is applied across the capacitor, owing to the location of the virtual ground of the circuit. At the start of the run the switches are opened, thereby removing the

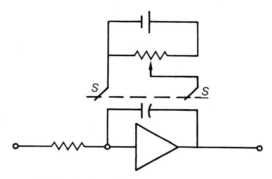

Fig. 5-26. Insertion of an initial condition.

source from the circuit. Clearly, if the input to the integrator is $\dot{y}$, then the output from the integrator is y, which is,

$$y(t) = -\left[\int_0^t \dot{y}\,dt + y(0)\right]. \qquad (5\text{-}148)$$

It is noted that in commercial analog computers provision is made both to introduce initial conditions, and also, by means of systems of relays, to disconnect the capacitors from their amplifiers, thereby "holding" the solution on the capacitors at the time of disconnect. The capacitors can be switched back into their operating positions and the solution continued, if desired, or the initial conditions can be reestablished for another run.

5-20 TIME AND MAGNITUDE SCALING OF ANALOG COMPUTERS

The real time response of a given simulation problem on an analog computer may prove to be an impractical one for study. This can result from a variety of reasons:
1. The problem may be such that the time required is so long that speed-up is desirable so as to save time. 2. The solution time may be so short that time slow-down is desirable to make study possible. 3. There may be a best range of time events for operating a computer, and it is

desired to establish this. In such cases, the time scale of the actual problem can be scaled, as desired. The procedure in time scaling is quite straightforward. It is best to discuss this matter by reference to a particular differential equation.

Consider the differential equation for a simple series RLC circuit, which is here written as,

$$L\frac{d^2i}{dt^2} + R\frac{di}{dt} + \frac{i}{C} = f(t). \tag{5-149}$$

It is desired to change the time variable from t to τ such that,

$$\tau = \alpha t, \tag{5-150}$$

where τ = machine time,
α = time coefficient (scaling factor),
t = real time.

Equation (5-150) is combined with Eq. (5-149) to yield,

$$\alpha^2 L\frac{d^2i}{d\tau^2} + \alpha R\frac{di}{d\tau} + \frac{i}{C} = f\left(\frac{\tau}{\alpha}\right). \tag{5-151}$$

This equation shows that time scaling requires the alteration of the coefficients of the differential equation, and hence in the analog computer program.

It may be necessary to do magnitude scaling in the case of a particular problem, perhaps to avoid factors that are too small, or to avoid saturation or other nonlinear operating factors. Magnitude scaling is accomplished by writing,

$$I = \beta i, \tag{5-152}$$

where I = machine volts,
β = magnitude coefficient,
i = real variable.

Now substitute this scaling into Eq. (5-151), the previously developed time-scaled equation. The result is,

$$\alpha^2 L\frac{d^2I}{d\tau^2} + \alpha R\frac{dI}{d\tau} + \frac{I}{C} = \beta f\left(\frac{\tau}{\alpha}\right). \tag{5-153}$$

It is observed that the only effect of magnitude scaling is an alteration of the magnitude of the input signal.

As a practical matter, analog computers may be used either in a nonrepetitive mode or in a repetitive mode. In the nonrepetitive mode, the problem must be time-scaled for operation over a convenient time, say from 1 to 10 minutes. The input is actuated once and the solution is plotted with a mechanical pen recorder. In the repetitive mode the problem is scaled for operation from approximately 10 msec to 1 sec, with the input being a repetitive signal with a frequency from 1 to 100 Hz. The solution in this case is usually displayed on an oscilloscope, and often is recorded photographically. It is important in the repetitive mode that the integrators be returned to the proper initial state at the start of each cycle.

5-21 SIMULATION LANGUAGES FOR THE DIGITAL COMPUTER

A number of programs have been written which permit the ready handling of simulation problems. Important simulation programs are the General Purpose System Simulator (GPSS) which is designed to investigate phenomena relating to information flow in business organizations. These are discrete processes and require the use of discrete system simulation programs. By contract, our interest is in continuous system simulation, and the analog computer is a natural tool for such studies. The analog computer type operation has been simulated on large-scale digital machines. The IBM Continuous System Modeling Program (CSMP) is an example of such a program. Other well-known continuous system simulation programs are MIDAS, MIMIC, and PACTOLUS (a forerunner of CSMP) which solve systems of ordinary differential equations, and which check out analog computer programs.

For many problems, CSMP obviates the need to use an analog computer facility. It allows simulation problems to be prepared directly and simply from either a SFG representation or a set of ordinary differential equations. It provides a basic set of functional blocks with which the components of a continuous system may be represented, and accepts application-oriented statements for defining the connections between these functional blocks. CSMP also accepts FORTRAN statements, thereby allowing the user to handle readily nonlinear and time varying problems of considerable complexity. Included in the basic function set are conventional analog computer components such as integrators, relays, plus many special purpose functions like delay time and limiter functions. This basic library is augmented by the FORTRAN library. Of course, special functions can be defined by the user, thereby allowing CSMP to take on the characteristics of a language oriented to any particular special purpose field in continuous system simulation. Input and output are facilitated by means of user-oriented control statements. A fixed format is provided for printing (tabular format), print plotting (graphic format), and preparation of a data set at selected increments of the independent variable. Convenient means are available for terminating a simulation run with a sequence of computations and logical tests. These can be designed to test run responses and define run control conditions for accomplishing iterative simulation of the type required in parameter optimization studies. Further, with few exceptions, parameter data, run control data, and connection statements can be prepared in any order for automatic sequencing by the program. Clearly, CSMP permits the user to concentrate upon the phenomenon being simulated.

As already noted, the one essential difference between the analog computer and the digital computer provided with CSMP or some other simulation language is the time scaling features of the analog machine. Where real time data are required, the analog machine would be used.

Owing to the speed with which the digital machine will perform computations, it is often convenient to couple an analog computer with a digital computer to allow the digital machine to perform many calculations and to provide these data to the analog machine for real time control. Such hybrid computers, which comprise an analog machine, analog/digital and digital/analog converters and the digital machine, often perform in a way that would not be possible by either machine operating alone.

5-22 PROBLEM ORIENTED LANGUAGES

CSMP is one of a broad class of problem oriented languages. It is a symbolic source program in which individual components or subsystem computations are implemented as subroutines. The advantage of such problem oriented programs is that the need for flow-charting solution algorithms is largely eliminated, the essential algorithms required to solve a particular problem being written by the processor in accordance with the description of the problem. An important feature of such an approach is that if the program is successfully processed and executed, the algorithm generated will be without errors.

A number of problem oriented programs have been developed for the solution of circuit problems from considerations of the circuit topology and the circuit parameters. Among the more general of these are ECAP (Electronic Circuit Analysis Program) and NET-1. Unlike CSMP, the considerations here begin with the circuit diagram itself and not with the dynamic description of the response in differential equation form. In the early IBM 1620 version of ECAP, accommodation is possible for a circuit with as many as 20 nodes, 60 branches, 10 dependent sources, and 20 switches (these may be used in developing nonlinear elements). The modern version is substantially more extensive than this. This program provides for a-c, d-c, and transient analysis, as follows:

a. The d-c analysis program provides a steady-state solution (voltages, current, power) of linear networks, the partial derivatives and sensitivity coefficients of the network voltages with respect to parameter changes, a worst-case analysis for specified tolerances, and a standard deviation analysis for specified parameter tolerances.

b. The a-c analysis program provides steady-state solutions (voltage, current, power) of linear networks for sinusoidal excitation at fixed frequency. Provision exists for the automatic modification of parameters. An automatic modification of the frequency (in logarithmic scale) allows a frequency response analysis to be effected.

c. The transient analysis program is used to obtain the time response (voltage, current) subject to one or more specified driving forces. This program can handle circuits containing nonlinearities by switching branch parameter values using switches as part of the input data to the program.

The steps in the use of ECAP are essentially the following:

1. *Design the circuit.* This is the engineer's task, if he is developing a device for a prescribed performance.

2. *Sketch the equivalent circuit.* This is particularly required if the circuit involves electronic devices (transistors, vacuum tubes, diodes, etc.) which must be replaced by equivalent circuit models.

3. *Number the nodes and branches.*

4. *Define the circuit topology using the data statements.* This specifies the circuit constants and the details of the interconnection of the circuit elements in the network.

5. *Add appropriate control statements.* This defines the type of analysis that is requested, e.g., an a-c analysis for specified frequency increments over a prescribed frequency range.

6. *Run, Evaluate, and Modify.* In this step the program will use the data statements to set up the circuit equations, and from the control cards, it will determine which solutions to these equations to perform. It will then carry out the calculations.

PROBLEMS

5-1. Find the general solution to each of the following differential equations:
$$\frac{d^2y}{dt^2} + 3\frac{dy}{dt} + 2y = 0,$$

$$\frac{d^2y}{dt^2}+3\frac{dy}{dt}+7y=0,$$

$$\frac{dy}{dt}+2y=te^{-t},$$

$$\frac{d^2y}{dt^2}+3\frac{dy}{dt}+2y=e^{-t}+3e^{-3t}.$$

5-2. Solve the set of simultaneous differential equations given, for the case of zero initial conditions,

$$\left(L_1 p+R_1+\frac{1}{C_1 p}\right)i_1-Mpi_2=v,$$

$$-Mpi_1+(L_2 p+R_2)i_2=0.$$

5-3. Write the equilibrium equations in both Kirchhoff and state form for each of the following networks. Specify the initial conditions, assuming that the situation shown existed for a long time before the switching operation was effected. The solutions are *not* required.

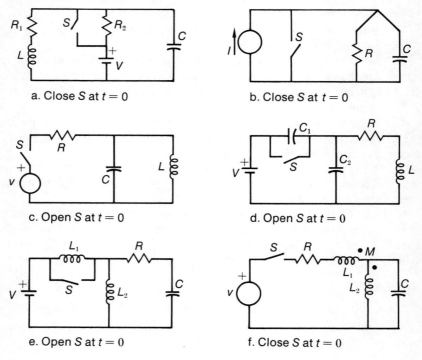

a. Close S at t = 0

b. Close S at t = 0

c. Open S at t = 0

d. Open S at t = 0

e. Open S at t = 0

f. Close S at t = 0

5-4. Illustrated is a ballistic pendulum that is initially at rest. A bullet of mass m traveling with a speed v is fired into the pendulum mass M at time $t = 0$ and remains lodged therein. Determine the initial value $\omega(0+)$. Draw the analogous electrical circuit.

5-5. Step function excitations are applied to the several systems shown. If all through and across variables are known at time $t = 0-$, determine these quantities at time $t = 0+$.

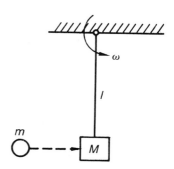

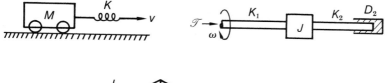

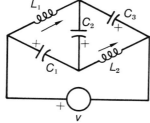

5-6. The circuit shown is initially in its quiescent steady-state when the switch is opened, say at $t = 0$. Deduce an expression for the inductor currents for $t > 0$.

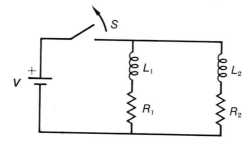

5-7. In the sketch shown, the switch is closed at time $t = 1$ sec. Initially $v_c(0-) = 0$. Determine $v_c(t)$ and $i_c(t)$, and sketch the time variations.

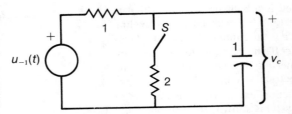

5-8. The circuit shown is initially relaxed. Find an expression for $i_L(t)$ and sketch the variation with time. The switch is closed at $t = 2$ sec.

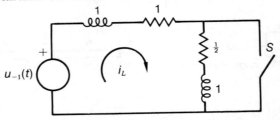

5.9. The circuit shown is initially in steady-state equilibrium. Find an expression for the voltage $v_c(t)$, and sketch the variation with time. The switch is closed at $t = 0$.

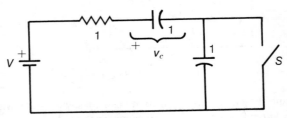

5.10. Initially switch S is closed and the circuit is in its steady-state. The switch is opened at time $t = 0$.
 a. Find an expression for the voltage across the switch after it is opened.
 b. If the parameters are such that $i(0+) = 5$ amp, $di(0+)/dt = -25$ amp/sec, find the voltage $v(0+)$ and the voltage gradient $dv(0+)dt$ across the switch.

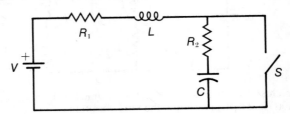

5-11. The voltage applied at $t = 0$ is $v = \cos \omega t$; the initial charge on the capacitor is zero.
   a. Find an expression for the current $i_c(t)$.
   b. Is there some value $v_c(0-)$ for which the transient term is zero?

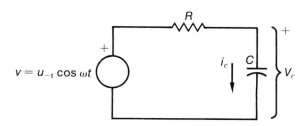

5-12. a. Show that the output voltage of the circuit shown to the recurring pulse train is given by the expressions,

$$v(n) = v(n-1)e^{-[n-(n-1)]T/RC},$$
$$v(n+1) = [v(n) - V]e^{-[(n+1)-n]T/RC}.$$

b. Calculate $v(n)$ for $n = 1, 2, 3, 4$ for an initially relaxed circuit, with $T/RC = 0.5$.

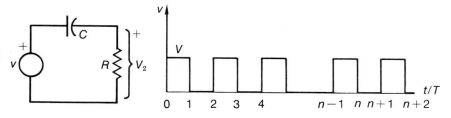

5-13. The switch in the circuit shown is periodically closed and opened. The circuit is initially relaxed. S is closed at $t = 0$, opened at $t = 1$, closed at $t = 2$, etc. Write a difference equation for v_2 at the end of the nth close-open cycle.

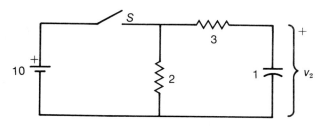

5-14. A three pulse wavetrain is applied to the initially relaxed RC circuit shown. From a knowledge of the step function response of the RC circuit, deduce the output voltage v_2.

HINT: Use a step-by-step procedure beginning with $t = 0$ and then determine the results at successive switching times.

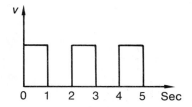
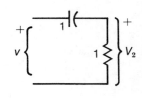

5-15. The state equations of a given nonlinear system are,

$$\dot{x}_1 = 2x_1 + x_2^2 + t,$$
$$\dot{x}_2 = x_1^2 t + x_2,$$
$$x_1(0) = x_2(0) = 1.$$

Evaluate the first term in the recursive series using the following methods: Euler, Runge-Kutta, predictor-corrector.

5-16. A system is described by the difference equation,

$$y(n+3) - 5y(n+2) + 3y(n+1) - 9y(n) = u(n+2),$$

where $u(n) = 0$, for all $n \geq 0$, $y(1) = 0$, $y(2) = 1$, first difference $\Delta y(1) = y(1) - y(0) = 7$. Determine the output for all $n \geq 1$.

5-17. Calculate the transfer functions of the operational amplifiers shown.

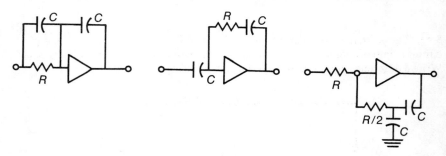

5-18. Draw SFG program graphs and the analog computer program for simulating the following differential equations, for zero initial conditions. The desired output is $+y$.

$$\frac{d^3y}{dt^3} + \frac{5d^2y}{dt^2} + \frac{3dy}{dt} + 7y = 8,$$

$$\frac{d^2y}{dt^2} + \frac{3dy}{dt} + 5y = \frac{4dx}{dt} + 7x,$$

$$\frac{d^3y}{dt^3} + 4y = \frac{7dx}{dt} + 3x^2.$$

5-19. A system SFG is shown. Note that this is not in a form that involves only simple integrators.
  a. Find an expression for $y = f(u_1, u_2)$.
  b. Draw a program graph for this system.
  c. Write the corresponding state description of the system.
  d. Draw the computer program.

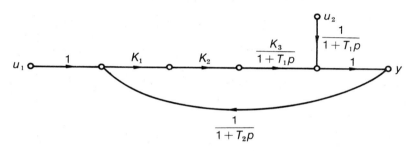

5-20. Refer to the state equations given,

$$\frac{d}{dt}\begin{bmatrix} x_1 \\ x_2 \\ x_3 \end{bmatrix} = \begin{bmatrix} -1 & -1 & 0 \\ 0.3 & -3 & 1 \\ -3 & -1 & -2 \end{bmatrix}\begin{bmatrix} x_1 \\ x_2 \\ x_3 \end{bmatrix} + \begin{bmatrix} u_1 \\ u_2 \\ u_3 \end{bmatrix}.$$

  a. Draw a SFG for this system.
  b. Draw an analog computer circuit for this system.

5-21. Given the following state equations. Draw a SFG for each, and from these draw the analog computer programs,

$$\frac{d}{dt}\begin{bmatrix} x_1 \\ x_2 \end{bmatrix} = \begin{bmatrix} 2 & -1 \\ 3 & 3 \end{bmatrix}\begin{bmatrix} x_1 \\ x_2 \end{bmatrix} + \begin{bmatrix} 1 & -1 \\ 0 & 1 \end{bmatrix}\begin{bmatrix} u_1 \\ u_2 \end{bmatrix},$$

$$\frac{d}{dt}\begin{bmatrix} x_1 \\ x_2 \\ x_3 \end{bmatrix} = \begin{bmatrix} 0 & 2 & 1 \\ -1 & 0 & 0 \\ 3 & -3 & 0 \end{bmatrix}\begin{bmatrix} x_1 \\ x_2 \\ x_3 \end{bmatrix} + \begin{bmatrix} 1 & 0 \\ -1 & 1 \\ 1 & 1 \end{bmatrix}\begin{bmatrix} u_1 \\ u_2 \end{bmatrix}.$$

242 System Response

5-22. Given the following differential equations. Write these in normal form, draw appropriate SFG-s, and then prepare computer programs for each.

$$(3p^2 + 2p - 1)y = 0 \qquad y(0) = 0, \quad \dot{y}(0) = 1$$
$$(2p^3 - 3p^2 + p + 5)y = u \qquad y(0) = \dot{y}(0) = 0, \quad \ddot{y}(0) = 1$$
$$(p^2 - 3p)y = 4u \qquad y(0) = 2, \quad \dot{y}(0) = 5.$$

5-23. Draw the SFG and from this prepare analog computer programs for each of the following. From these deduce the state matrixes, A, B, P, Q.

$$(p^3 + 3p^2 + 7p + 3)y = (2p + 3)x,$$
$$\left.\begin{array}{l}(p^2 + 2p + 3)y_1 + (2p + 1)y_2 = x_1 \\ -(2p + 1)y_1 + (3p^2 + p + 2)y_2 = x_2\end{array}\right\}.$$

5-24. Specify the system equations in differential equation and state form for which the computer program is that given.

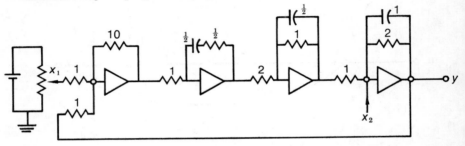

5-25. Develop the program graphs and the analog computer programs for the equations given, subject to the specified initial conditions,

$$(p^2 + 3p + 2)y = (2p + 5)u \qquad \dot{y}(0) = 2, \quad y(0) = 1, \quad \dot{u}(0) = -3, \quad u = f(t)$$
$$(p^3 + 3p^2 + 2p - 5)y = 7 \qquad y(0) = -1, \quad \dot{y}(0) = 3, \quad \ddot{y}(0) = 0.$$

5-26. a. Develop an analog computer program for the set of equations,

$$(p^2 + 2p - 3)y = x,$$
$$(p^2 + 3)x = (p + 1)y,$$

subject to: $\dot{y}(0) = 1, y(0), \dot{x}(0) = 0, x(0) = 5$.

b. Write the state description from the analog computer program.

6

General Time Domain Considerations

Chapter 5 has been concerned with developing techniques for studying the behavior of physical systems by direct solution of the equations that describe the dynamics of these systems. The mathematical techniques discussed in Section 5-a are limited, for the most part, to relatively simple systems which are subjected to relatively simple excitation functions. Even for such simple systems, the techniques are not particularly well matched to hand computing. Furthermore, the solution of the describing equations for nonlinear systems are almost impossible by these methods. The advent of the digital computer adds a new dimension to system analysis, since with it, the numerical solution can be carried out for nonlinear as well as for linear problems. As a result, it is no longer necessary to limit oneself to linear problems, nor is it necessary to linearize nonlinear problems. Consequently, the introduction of linear transformation techniques as a step in the solution of system problems is less important than it was in the past. Transformation methods remain very important techniques of system analysis, and we shall study these in some detail in later chapters. This chapter will consider some general time domain methods that allow the numerical solution of complicated system problems. We shall study the problems and complications that exist in these time domain methods.

6-1 SINGULARITY FUNCTIONS

We have already introduced the unit step as an excitation function in Section 5-7, and also, we referred to impulse functions in Section 5-6 in connection with initial and switching conditions. We shall have need for these and other so-called *singularity functions*, and we wish, therefore, to examine these somewhat more carefully.

The unit step function $u_{-1}(t)$ is defined as,

$$u_{-1}(t) = \begin{cases} 1 & \text{for } t > 0 \\ 0 & \text{for } t < 0 \end{cases}. \tag{6-1}$$

We shall later have occasion to consider the derivative of the step function. Here we encounter mathematical difficulties because of the discontinuity at $t = 0$, since our classical definition of a derivative assumes a continuous function. We can employ a subterfuge to avoid this difficulty — by replacing the step function by the better-behaved function shown in Fig. 6-1. In this figure, T may be very small but not zero. We may differentiate this function to give the form shown in Fig. 6-1b. As we let

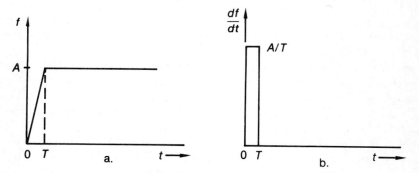

Fig. 6-1. Approximations to the step function and the impulse function.

the width T get smaller, the height of the pulse A/T gets larger although the area under the curve remains constant ($= A$). The pulse, in the limit as $T \to 0$, is defined as the impulse of amplitude A, and is written,

$$\lim_{T \to 0} \frac{df}{dt} = A u_0(t),$$

where $u_0(t)$ denotes the unit impulse. For simplicity we write this,

$$u_0(t) = \frac{d}{dt} u_{-1}(t). \tag{6-2}$$

It follows from these results that,

$$u_0(t) = 0 \quad \text{if} \quad t \neq 0,$$
$$u_0(t) = \infty \quad \text{if} \quad t = 0, \qquad (6\text{-}3)$$

and,

$$\int_{-\infty}^{\infty} u_0(t)\, dt = \int_{-\epsilon}^{\epsilon} u_0(t)\, dt = 1 \qquad \epsilon > 0. \qquad (6\text{-}4)$$

Equations (6-3) specify, as they must, that $u_0(t)$ is zero everywhere except at $t = 0$, where it becomes infinite in such a way that Eq. (6-4) is satisfied.

Another approach to the introduction of the impulse function is possible. In this approach $u_0(t)$ is defined in the sense of a so-called *generalized* (or symbolic) *function*. Now one proceeds by introducing the continuous function $\psi(t)$ (which is called the *testing* function) which vanishes identically outside of some finite interval. The impulse function $u_0(t)$ is defined as a symbolic function by the relation,

$$\int_{-\infty}^{\infty} u_0(t)\psi(t)\, dt = \psi(0). \qquad (6\text{-}5)$$

This expression has no meaning as an ordinary integral. Instead, the integral and the function $u_0(t)$ are defined by the number $\psi(0)$ and assigned to the function $\psi(t)$.

The unit step and the impulse functions are only two of a whole class of singularity functions, the more important of which are given in Table 6-1. The singularity functions are related to adjacent functions in Table

TABLE 6-1. Singularity Functions

Function name	Designation
.	.
.	.
.	.
Unit ramp	$u_{-2}(t) = t$
Unit step	$u_{-1}(t) = 1$
Unit impulse	$u_0(t)$
Unit doublet	$u_1(t)$
Unit triplet	$u_2(t)$
.	.
.	.
.	.

6-1 by an extension of Eq. (6-2). These are such that,

$$\frac{d}{dt}u_{-n}(t) = u_{-n+1}(t),$$

$$\int_{-\infty}^{\infty} u_{-n+1}(t)\,dt = u_{-n}(t). \tag{6-6}$$

As a practical matter, generating the unit ramp and unit step functions may be accomplished easily with electronic circuits. The unit impulse may be approximated by a narrow pulse whose time duration is small compared with any significant time constants of the circuit. The higher order singularity functions are more difficult to generate, and in fact, are of lesser theoretical interest.

Because of the relationship between the step function and the impulse function, and in view of the discussion in Section 2-9, it follows that we can determine the impulse response of a linear system by differentiating the step function response. In fact, by appropriate differentiation or integration of the step function response, we can determine the response of a linear system to any singularity function.

6-2 SUPERPOSITION INTEGRAL

A feature of the superposition integral is that it makes possible the time response of a physical system to a general excitation function from a knowledge of the system response to a unit step excitation. In this respect, it occupies a somewhat parallel position to that occupied by the method of variation of parameters discussed in Section 5-5. From a practical viewpoint, finding the response of a system to a step function may in itself be a complicated problem, and then carrying out the subsequent steps required by the superposition integral can be quite involved mathematically. The digital computer makes the mechanics of the solution reasonably tractable, and so the superposition integral and the related convolution integral (to be discussed later) have numerical value as well as analytic interest.

We consider a causal linear system to which a unit step excitation is applied. For the initially relaxed system, the response function, which is obtained in principle in the manner discussed in Section 5-a, is termed the *indicial* response, and is written $W(t)$. Suppose that we refer to Fig. 6-2 which shows an assumed indicial response. As shown, we designate the indicial response of this same system relative to a delayed time reference as $W(t-\lambda)u_{-1}(t-\lambda)$.

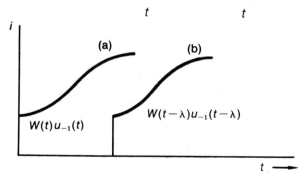

Fig. 6-2. Indicial response of a network: (a) unit step applied at $t = 0$ (b) unit step applied at $t = \lambda$.

We wish to find the system response when the excitation is a general time function, knowing the indicial response to the unit step function. The procedure is to approximate the given excitation function by a stepwise approximation, with time intervals $\Delta\lambda$. The manner of the approximation is shown in Fig. 6-3. Clearly, the given waveform is approximated by a number of incremental step functions which are delayed in accordance with the sampling time intervals. The system response will then be the superposition of the response of the system to each incremental step function, with due account of the time delay associated with each step. We can formalize this process mathematically.

The components of current corresponding to the successive voltage steps are the following,

$$i_0 = v(0)\, W(t),$$

$$i_1 = \left.\frac{dv}{d\lambda}\right|_{\Delta\lambda} \Delta\lambda\, W(t-\Delta\lambda),$$

$$i_2 = \left.\frac{dv}{d\lambda}\right|_{2\Delta\lambda} \Delta\lambda\, W(t-2\Delta\lambda), \tag{6-7}$$

$$i_3 = \left.\frac{dv}{d\lambda}\right|_{3\Delta\lambda} \Delta\lambda\, W(t-3\Delta\lambda),$$

and so on.

The response at any time t, say the interval between $n\Delta\lambda$ and $(n+1)\Delta\lambda$ is given approximately by the expression,

$$i(t) = v(0)\, W(t) + \left[\left.\frac{dv}{d\lambda}\right|_{\Delta\lambda} W(t-\Delta\lambda) + \cdots + \left.\frac{dv}{d\lambda}\right|_{n\Delta\lambda} W(t-n\Delta\lambda)\right]\Delta\lambda,$$

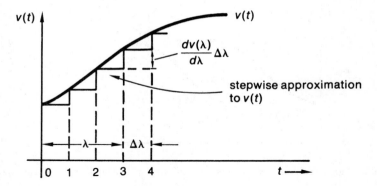

Fig. 6-3. Stepwise approximation to a given waveform.

which is,

$$i(t) = v(0)\,W(t) + \sum_{n=1}^{N} \left.\frac{dv}{d\lambda}\right|_{n\Delta\lambda} W(t - n\Delta\lambda)\,\Delta\lambda. \tag{6-8}$$

This form is suitable for numerical calculations using the digital computer.

To improve the approximation, the increment $\Delta\lambda$ is made progressively smaller. In the limit as n becomes infinite, $\Delta\lambda \to d\lambda$ and assuming that the product $n\Delta\lambda \to \lambda$, then the sum becomes an integral. Equation (6-8) then leads to,

$$i(t) = v(0)\,W(t) + \int_0^t W(t-\lambda)\frac{dv}{d\lambda}\,d\lambda. \tag{6-9}$$

This is one of many forms of the superposition (Duhamel) integral that relates the response of the system to a general waveform $v(t)$ from a knowledge of the indicial response to the unit step.

Another form of interest is obtained by carrying out the indicated integration in Eq. (6-9). Let,

$$u' = W(t-\lambda) \qquad du' = -\frac{dW(t-\lambda)}{d(t-\lambda)}\,d\lambda,$$

$$dv' = \frac{dv}{d\lambda}\,d\lambda \qquad v' = v(\lambda).$$

Equation (6-9) can be written,

$$i(t) = v(0)\,W(t) + W(t-\lambda)v(\lambda)\,\bigg|_0^t + \int_0^t v(\lambda)\frac{dW(t-\lambda)}{d(t-\lambda)}\,d\lambda,$$

which becomes,

$$i(t) = v(t) W(0) + \int_0^t v(\lambda) \frac{dW(t-\lambda)}{d(t-\lambda)} d\lambda. \qquad (6\text{-}10)$$

This form of the superposition integral warrants special attention. Recall that $W(t)$ is the response of the initially relaxed network to a unit step. Then $dW(t)/dt$, in accordance with the discussion in Section 6-1, is the response of the same initially relaxed network to the unit impulse. We interpret this to show that the superposition integral in the form given in Eq. (6-10) resolves the excitation function $v(t)$ into a set of successive impulse functions with amplitudes $v(\lambda)d\lambda$ at each instant λ, and the impulse response at each instant is appropriately summed. The impulse response of a system is often referred to as the *Green's function* of the system.

A third form of the superposition integral follows directly from Eq. (6-10) by a simple change of variable. If we write $t - \lambda = \lambda'$, then the integral in Eq. (6-10) becomes,

$$\int_0^t v(\lambda) \frac{dW(t-\lambda)}{d(t-\lambda)} d\lambda = -\int_t^0 v(t-\lambda') \frac{dW(\lambda')}{d\lambda'} d\lambda',$$

$$= \int_0^t v(t-\lambda) \frac{dW(\lambda)}{d\lambda} d\lambda.$$

The final step is valid because the definite integral has a value that is not dependent on the variable that is used. Hence Eq. (6-8) becomes,

$$i(t) = v(t) W(0) + \int_0^t v(t-\lambda) \frac{dW(\lambda)}{d\lambda} d\lambda. \qquad (6\text{-}11)$$

Other forms are possible (See Problems 6-8).

6-3 CONVOLUTION INTEGRAL

The superposition integral in the forms given in Eqs. (6-10) and (6-11) are frequently written in forms which reflect special properties of the system or input. First, let us consider the term $v(t)W(0)$ which, as is evident from Fig. 6-3, is a measure of the response of the system due to the existence of an initial excitation at time $t = 0$. We might represent this response as being due to an excitation that had been applied at negative time, with the value of the response being precisely $v(t)W(0)$ at $t = 0$. Thus we write,

$$v(t)W(0) = \int_{-\infty}^{0} v(\lambda) \frac{dW(t-\lambda)}{d(t-\lambda)} d\lambda = \int_{-\infty}^{0} v(t-\lambda) \frac{dW(\lambda)}{d\lambda} d\lambda. \quad (6\text{-}12)$$

If we combine this expression with Eq. (6-11), then we write,

$$i(t) = \int_{-\infty}^{t} v(\lambda) \frac{dW(t-\lambda)}{d(t-\lambda)} d\lambda. \quad (6\text{-}13)$$

Clearly, therefore, this equation contains the initial response, but it is hidden in the form of the integral.

For convenience, we now designate the impulse response of the system,

$$h(t-\lambda) = \frac{dW(t-\lambda)}{d(t-\lambda)}. \quad (6\text{-}14)$$

Using this notation, Eq. (6.13) assumes the form,

$$i(t) = \int_{-\infty}^{t} v(\lambda) h(t-\lambda) d\lambda. \quad (6\text{-}15)$$

Correspondingly, if we proceed from Eq. (6-11) the equivalent form will be,

$$i(t) = \int_{-\infty}^{t} v(t-\lambda) h(\lambda) d\lambda. \quad (6\text{-}16)$$

In fact, since it is customarily assumed that the input starts at $t = 0$; that is, $h(t) = 0$ for $t < 0$, then we may write,

$$i(t) = \int_{0}^{t} v(\lambda) h(t-\lambda) \, d\lambda = \int_{0}^{t} v(t-\lambda) h(\lambda) \, d\lambda. \quad (6\text{-}17)$$

Both of these integrals are known as *convolution* integrals and specify the convolution of $v(t)$ and $h(t)$.

A reasonable interpretation of convolution can be given. Refer to the first equation in Eq. (6.17). Suppose that we specify $v(t)$ as an infinite sum of weighted impulses according to the expression,

$$v(t) = \int_{-\infty}^{\infty} v(\lambda) u_0(t-\lambda) \, d\lambda. \quad (6\text{-}18)$$

But we have already noted that the response of a linear system to a unit impulse is given as $h(t)$. It then follows that the response to $v(\lambda) u_0(t-\lambda)$

will be $v(\lambda)h(t-\lambda)$. By superposition, the response to $v(t)$ will be,

$$i(t) = \int_{-\infty}^{\infty} v(\lambda)h(t-\lambda)\,d\lambda. \tag{6-19}$$

For the usual causal (physical) systems, since no response exists before the application of an excitation, then,

$$v(t) = 0 \quad (t < 0),$$

and

$$h(t-\lambda) = 0 \quad (\lambda > t).$$

Because of these, the time limits are changed to yield,

$$i(t) = \int_0^t v(\lambda)h(t-\lambda)\,d\lambda,$$

which is Eq. (6-17).

We observe, based on our development, since $h(t-\lambda)$ is the response to the unit impulse, that the first expression on the right represents the sum of the responses at the observation time t to impulses of strength $v(\lambda)\,d\lambda$ which occur at time λ, with λ varying from the time of application of the input until the observation time t. To interpret the second form of Eq. (6-17), suppose that $v(t)$ denotes an arbitrary voltage waveform that is applied to the network. Refer to Fig. 6-4. Observe that when $t = 0$, then $v(t-\lambda) = v(-\lambda)$; also when $t = \lambda$, $v(t-\lambda) = v(0)$. It is clear that λ is a quantity that is measured backwards from any t, and is, therefore, a time interval measured negatively from some specified reference time t, as illustrated. The quantity λ can vary from t to 0, the limits of integration, and as λ varies from 0 to t, then $v(t-\lambda)$ ranges through all past values of the input.

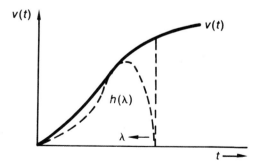

Fig. 6-4. For discussing the convolution integral.

Now consider the quantity $h(t)$ and so $h(\lambda)$. As defined, $h(t)$ is the transient response to the unit impulse (the steady state response is zero), and is given explicitly by Eq. (6-14), which is the complementary function of the system response for the impulse excitation, and this depends explicitly on the system differential equation. Suppose that $h(t)$ has the form illustrated in Fig. 6-5. A plot of $h(\lambda)$ can be superposed on Fig. 6-4 (in the backward direction, as shown). Equation (6-17) requires that the

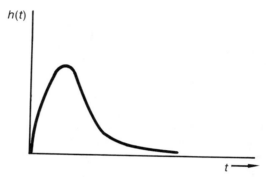

Fig. 6-5. Assumed impulse response of a general system.

product of $h(\lambda)$ and $v(t-\lambda)$ must be integrated from 0 to t to give the output response. This process might be considered to be a weighting of all past values of the input by the unit impulse response. But for any system with damping $h(\lambda)$ will be small for large λ; consequently the output at any time, found by integration, is influenced primarily by recent values of the input. This means that old values of the input have little effect on the present output. Strictly speaking, the present output is determined by all past history of the input, weighted by the unit impulse response. Involved in determining the present output is the process of folding (to obtain $h(-\lambda)$), shifting, multiplying, and integrating. The detailed process is better understood by considering a specific example, the solution of which is carried out by numerical approximation.

Example 6-3.1

Determine by numerical approximation methods the response at time $t = 4$ units of a system having a known impulse response function (and this might be determined analytically or experimentally) to a specified excitation function. The impulse response and the excitation function are given in the accompanying figures.

Convolution Integral 253

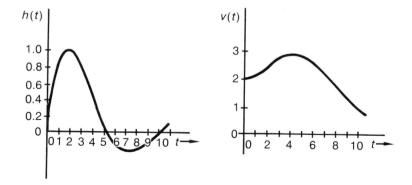

Fig. 6-3.1-1

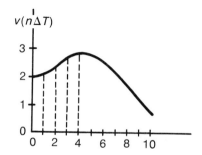

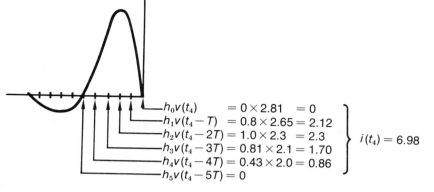

$h_0 v(t_4)\ \ \ \ \ = 0 \times 2.81 = 0$
$h_1 v(t_4 - T) = 0.8 \times 2.65 = 2.12$
$h_2 v(t_4 - 2T) = 1.0 \times 2.3\ \ = 2.3$ $\Big\}\ i(t_4) = 6.98$
$h_3 v(t_4 - 3T) = 0.81 \times 2.1 = 1.70$
$h_4 v(t_4 - 4T) = 0.43 \times 2.0 = 0.86$
$h_5 v(t_4 - 5T) = 0$

Fig. 6-3.1-2

254 General Time Domain Considerations

Solution. From the given figures we measure the ordinates at each time interval. These are:

$t = n\Delta T$	h_n	v_n
0	0	2.0
1	0.8	2.1
2	1.0	2.3
3	0.81	2.65
4	0.43	2.81
5	0.0	2.81
6	−0.14	2.65
7	−0.21	2.3
8	−0.20	1.8
9	−0.12	1.35
10	−0	0.9

The system response is expressed approximately by the convolution summation which is obtained from the second part of Eq. (6-17),

$$i(t) = \sum_{n=0}^{N} h_n\, v(t - n\Delta T), \qquad (6\text{-}20)$$

where N is the number of intervals of ΔT to characterize completely the impulse response. h_n is the ordinate value of the impulse response at the time point characterized by n.

At the time specified by $4\Delta T$, the response is made up in the manner shown in Fig. 6-3.1-2. It is clear that this process can be mechanized by plotting $v(t)$ and $h(\lambda)$ on transparent tapes and then sliding the tapes as required. The appropriate ordinates can then be read from the tapes.

6-4 CONVOLUTION SUMMATION

For analysis in the discrete time domain the foregoing approach must be modified somewhat. To study this, consider a linear discrete time system that is originally relaxed at time $t_n < 0$, which we write simply as $n < 0$. An input $u(0) = 1$ results in an output $y(n)$ over the interval $1 \leq n \leq \infty$, which is written,

$$y(n) = h(n) \qquad (n = 1, 2, \ldots, \infty), \qquad (6\text{-}21)$$

where $h(n)$ specify real numbers. Correspondingly for a linear system for a $u(0)$ of any magnitude, the output is,

$$y(n) = h(n)\, u(0) \qquad (n = 1, 2, \ldots, \infty). \qquad (6\text{-}22)$$

If instead of an input $u(0)$ at $t=0$ we apply $u(1)$ at $t=1$, then the output at time $t=n$ is,

$$y(n) = h(n-1)\,u(1) \qquad (n=2,3,\ldots,\infty). \tag{6-23}$$

In general $u(i)$ at $t=i$ will yield,

$$y(n) = h(n-i)\,u(i) \qquad (n=i+1, i+2, \ldots, \infty). \tag{6-24}$$

For a sequence of inputs $\{u(i)\}$ with $0 \le i \le \infty$ there will be the contributions from each input of the sequence. The composite output is the linear summation of the weighted contributions from each $u(i)$ for $i < n$, and is,

$$y(n) = \sum_{i=0}^{n-1} h(n-i)\,u(i) \qquad (n=1,2,\ldots,\infty). \tag{6-25}$$

Here $h(n-i)$ specifies the weight with which the input values $u(i)$, $i=1, 2, \ldots, n-1$ contribute to the output $y(n)$ at $t=n$. The *weighting sequence* $h(n)$, $n=1, 2, \ldots, \infty$ is the weighting sequence of the system.

Equation (6-25) is one form of the convolution summation of the system, and is the discrete time counterpart of Eq. (6-17). If we let $i = n-j$ in Eq. (6-25), we get the alternate form;

$$y(n) = \sum_{j=1}^{n} h(j)\,u(n-j), \tag{6-26}$$

which corresponds to the second form in Eq. (6-17), and Eq. (6-20).

6-5 STATE EQUATIONS

In a formal way, the time domain solutions to the state equations are readily obtained. When numerical answers are required the numerical integration of the differential equations is burdensome for nonlinear networks, and is extremely slow when the network has a wide spread of eigenvalues. Nevertheless from a computational point of view the procedure is direct and avoids the problems incident to any transform method. Furthermore, many of the time domain methods may be extended to nonlinear systems.

We begin by considering the force-free state equation,

$$\dot{x} = Ax, \tag{6-27}$$

with known initial conditions $x(0)$. In this connection, we examine the function,

$$\Phi(t) = e^{At} = I + At + \frac{A^2 t^2}{2!} + \cdots = \sum_{k=0}^{\infty} \frac{A^k t^k}{k!}, \tag{6-28}$$

which is called the *fundamental matrix* (sometimes referred to as the state transition matrix) of the system. It can be shown by a term-by-term differentiation that,

$$\frac{d}{dt}(e^{At}) = Ae^{At} = e^{At}A. \tag{6-29}$$

Consequently, by direct substitution, the solution to Eq. (6-27) is,

$$x(t) = e^{At} x(0), \tag{6-30}$$

or expressed more generally as,

$$x(t) = \Phi(t) x(0). \tag{6-31}$$

This result may be generalized to initial time $t = t_0$ instead of $t = 0$. In this case the solution is,

$$x(t) = e^{A(t-t_0)} x(t_0), \tag{6-32}$$

or more generally,

$$x(t) = \Phi(t - t_0) x(t_0). \tag{6-33}$$

As defined, the fundamental matrix relates the state at time t with that at time t_0.

A physical interpretation of Eq. (6-31) is possible which provides some insight into the meaning of this expression. In this, we observe that for the linear case the all-integrator programmed analog computer appropriate to any given system allows a direct evaluation of each term in the solution. This follows from the fact that since the integrator outputs may be chosen as a set of state variables, then an initial condition of unity at any integrator output corresponds to a unit impulse input to that integrator at time $t = 0$. In this process, care must be taken to insure that all other initial conditions remain at zero. Specifically, let us consider a second order system which is written explicitly in the form,

$$\begin{bmatrix} x_1(t) \\ x_2(t) \end{bmatrix} = \begin{bmatrix} \phi_{11}(t) & \phi_{12}(t) \\ \phi_{21}(t) & \phi_{22}(t) \end{bmatrix} \begin{bmatrix} x_1(0) \\ x_2(0) \end{bmatrix}.$$

It is seen that to find $\phi_{11}(t)$ we must set $x_1(0) = 1$ and $x_2(0) = 0$, in which case $x_1(t) = \phi_{11}(t)$. In general, $\phi_{ij}(t)$ is the response at the output of the ith integrator to an impulse at the input of the jth integrator when the initial conditions to all other integrators are zero.

Example 6-5.1

A system is specified by the matrix,

$$A = \begin{bmatrix} 0 & 1 \\ -\frac{1}{2} & -1 \end{bmatrix}.$$

Determine the response of the system to specified initial conditions $x(0)$.

Solution. From the known value of A we may determine the expressions for the various powers of A. Thus,

$$A^2 = \begin{bmatrix} -\frac{1}{2} & -1 \\ \frac{1}{2} & \frac{1}{2} \end{bmatrix} \quad A^3 = \begin{bmatrix} \frac{1}{2} & \frac{1}{2} \\ -\frac{1}{4} & 0 \end{bmatrix}.$$

Hence we have that,

$$\Phi(t) = e^{At} = \begin{bmatrix} 1 + 0t - \frac{1}{2}\frac{t^2}{2!} + \frac{1}{2}\frac{t^3}{3!} + \cdots & 0 + t - \frac{t^2}{2!} + \frac{1}{2}\frac{t^3}{3!} + \cdots \\ 0 - t + \frac{1}{2}\frac{t^2}{2!} - \frac{1}{4}\frac{t^3}{3!} + \cdots & 1 - t + \frac{1}{2}\frac{t^2}{2!} + 0 + \cdots \end{bmatrix}.$$

It is not possible to recognize these series, but it is possible in the present case to find a closed form solution. We carry this out in a round-about way. Let us reduce the state equations to a higher order differential equation. That is, we start with the state equations specified by the A matrix,

$$\dot{x}_1 = x_2,$$
$$\dot{x}_2 = -\frac{x_1}{2} - x_2.$$

The first of these is differentiated and the result is combined with the second. This yields,

$$\ddot{x}_1 = -\frac{x_1}{2} - \dot{x}_1,$$

so that,

$$\ddot{x}_1 + \dot{x}_1 + \frac{x_1}{2} = 0.$$

From this differential equation we find that,

$$s^2 + s + \tfrac{1}{2} = 0,$$

with the roots,

$$s = -\tfrac{1}{2} \pm j\tfrac{1}{2}.$$

Thus the solution is of the form,

$$x_1 = e^{-(1/2)t}\left(a \cos \frac{t}{2} + b \sin \frac{t}{2}\right).$$

From this and the state equations,

$$x_2 = e^{-(1/2)t}\left(-\frac{a}{2} \sin \frac{t}{2} + \frac{b}{2} \cos \frac{t}{2}\right) - e^{-(1/2)t}\left(a \cos \frac{t}{2} + b \sin \frac{t}{2}\right),$$

$$= \frac{e^{-(1/2)t}}{2}\left[-(a+b) \sin \frac{t}{2} - (a-b) \cos \frac{t}{2}\right].$$

From these we can write,

$$x_1(0) = a,$$

$$x_2(0) = -\frac{(a-b)}{2},$$

and so,

$$a = x_1(0),$$
$$b = 2x_2(0) + x_1(0).$$

Combine these with the expressions for x_1 and x_2,

$$x_1 = e^{-(1/2)t} x_1(0) \cos \frac{t}{2} + 2x_2(0) \sin \frac{t}{2} + x_1(0) \sin \frac{t}{2},$$

$$x_2 = e^{-(1/2)t} - x_1(0) + x_2(0) \sin \frac{t}{2} + x_2(0) \cos \frac{t}{2},$$

from which we write

$$\begin{bmatrix} x_1(t) \\ x_2(t) \end{bmatrix} = \begin{bmatrix} e^{-(1/2)t}\left(\cos \frac{t}{2} + \sin \frac{t}{2}\right) & 2e^{-(1/2)t} \sin \frac{t}{2} \\ -e^{-(1/2)t} \sin \frac{t}{2} & e^{-(1/2)t}\left(\cos \frac{t}{2} - \sin \frac{t}{2}\right) \end{bmatrix} \begin{bmatrix} x_1(0) \\ x_2(0) \end{bmatrix}.$$

This specifies $\Phi(t)$ or e^{At}, which can be equated to the form above. Considerable attention has been directed to techniques for evaluating the transition matrix (*see* references).

We now seek a solution to the general state equation,

$$\dot{x}(t) = Ax(t) + Bu(t). \tag{6-34}$$

To do this, each term in this equation is multiplied by the factor e^{-At}. This gives,

$$e^{-At}\dot{x} = e^{-At}Ax + e^{-At}Bu.$$

This expression is rearranged to the form,

$$e^{-At}\dot{x} - e^{-At}Ax = e^{-At}Bu,$$

or

$$\frac{d}{dt}(e^{-At}x) = e^{-At}Bu. \tag{6-35}$$

Now multiply by dt and integrate over the time interval t_0 to t. Also, to avoid confusion, a change of variable is made. The result is,

$$\int_{t_0}^{t} \frac{d}{dt}(e^{-A\tau}x)\,d\tau = \int_{t_0}^{t} e^{-A\tau}Bu\,d\tau,$$

which is,

$$e^{-At}x(t) - e^{-At_0}x(t_0) = \int_{t_0}^{t} e^{-A\tau}Bu\,d\tau.$$

Now all terms in this equation are premultiplied by e^{At}, which gives,

$$x(t) = e^{A(t-t_0)}x(t_0) + \int_{t_0}^{t} e^{+A(t-\tau)}Bu(\tau)\,d\tau. \tag{6-36}$$

When written in terms of the fundamental matrix, the expression is,

$$x(t) = \Phi(t-t_0)x(t_0) + \int_{t_0}^{t} \Phi(t-\tau)Bu(\tau)\,d\tau. \tag{6-37}$$

The first term on the right represents the contribution to the total response that arises from the free or natural response of the system due to initial conditions. The second term is the contribution due to the particular applied drivers or forcing functions, and appears in a matrix convolution integral.

Particular note is called to the fact that Eq. (6-37), which was obtained directly from the state equations, is in closed form, although obtaining a solution is usually a complicated task. Equation (6-36) is sometimes changed in form in an endeavor to obtain a solution to this equation. Suppose that we write $t + h$ for t in this equation. The result is,

$$x(t+h) = e^{Ah}e^{A(t-t_0)}x(t_0) + \int_{t_0}^{t+h} e^{+A(h+t-\tau)}Bu(\tau)\,d\tau,$$

which is written,

$$x(t+h) = e^{Ah}\left[e^{A(t-t_0)}x(t_0) + \int_{t_0}^{t} e^{A(t-\tau)}Bu(\tau)\,d\tau\right] + \int_{t}^{t+h} e^{A(h+t-\tau)}Bu(\tau)\,d\tau. \tag{6-38}$$

This expression is combined with Eq. (6-36) with the result,

$$x(t+h) = e^{Ah}x(t) + \int_t^{t+h} e^{+A(h+t-\tau)} Bu(\tau)d\tau. \tag{6-39}$$

By a change of variable, writing $\tau' = \tau - t$,

$$x(t+h) = e^{Ah}x(t) + \int_0^h e^{A(h-\tau')} Bu(t+\tau')d\tau'. \tag{6-40}$$

This expression is exact, but its utility depends on the computation of e^{Ah}. Liou et al. (7) have addressed themselves to this equation. It is proposed, since the infinite series in Eq. (6-28) is absolutely convergent for all finite h, that one truncate the series, and use a finite number of terms in the series expansion of e^{Ah}. This requires a criterion for the selection of the number of terms in the series. When the matrix A has a large eigenvalue, the integration step size h must be kept small in order to permit the rapid convergence of the series expansion for e^{Ah}. The question of numerical integration step size has received some attention in Section 5-13.

A quasi-analytic approach to the solution of Eq. (6-34), the general state equation, has been reported by Cohen and Flatt (8) and by Certaine (9). Here the diagonal terms and the off-diagonal terms of A are separated, with the off-diagonal terms being lumped together with $Bu(t)$ as part of the driving function. That is, Eq. (6-34) is written,

$$\dot{x} = Ax + Bu = Dx + (Nx + Bu), \tag{6-41}$$

where $A = D + N$, with D containing all diagonal terms, and N containing all off-diagonal terms of A. Here $(Nx + Bu)$ serves as an equivalent driving function. The analytic solution to this equation can be obtained by modifying Eq. (6-40) to read

$$x(t+h) = e^{Dh}x(t) + \int_0^h e^{D(h-\tau)}[Nx(t+\tau) + Bu(t+\tau)]d\tau. \tag{6-42}$$

The quantity e^{Dh} is easy to compute since D is a diagonal matrix. As to the integral, it can be carried by using a polynomial approximation to any desired degree for $(Nx + Bu)$.

When the matrix A is dominantly diagonal, that is, the diagonal term exceeds the sum of the magnitudes of the off-diagonal terms in the same row or column, much larger integration step size h may be used in solving Eq. (6-42) than that allowed in the usual methods. Unfortunately, network problems do not often have the property of diagonal dominance of the A matrix, and so Eq. (6-42) is not, therefore, a major break-through in the integration problem.

6-6 NUMERICAL SOLUTION OF CONTINUOUS TIME SYSTEMS

Another approach to the solution of the state equation formulation of a system is a numerical one, with the procedure for the numerical solution of systems of continuous equations being a logical extension of the methods previously discussed in Section 5-b. This involves writing difference equations for each of the differential equations making up the state vector and evaluating these at successive sampling times. This leads to a straightforward recursive procedure, but it can involve a large number of calculations. Specifically, suppose that we consider the state equations,

$$\begin{aligned}\dot{x}_1 &= -4x_1 + x_2, \\ \dot{x}_2 &= -3x_1 + x_3 + 3, \\ \dot{x}_3 &= -x_1 + 2u, \\ y &= x_1.\end{aligned} \quad (6\text{-}43)$$

These may be replaced by the set of difference equations at $t = nT$, by writing,

$$\begin{aligned}\frac{x_1(n+1)T - x_1(nT)}{T} &= -4x_1(nT) + x_2(nT), \\ \frac{x_2(n+1)T - x_2(nT)}{T} &= -3x_1(nT) + x_3(nT) + 3u(nT), \\ \frac{x_3(n+1)T - x_3(nT)}{T} &= x_1(nT) + 2u(nT), \\ y(nT) &= x_1(nT).\end{aligned} \quad (6\text{-}44)$$

These equations are now solved for $x_1(n+1)T$, $x_2(n+1)T$, $x_3(n+1)T$, and $y(nT)$ to yield the following set of equations,

$$\begin{aligned}x_1(n+1)T &= (1-4T)x_1(nT) + Tx_2(nT), \\ x_2(n+1)T &= -3Tx_1(nT) + x_2(nT) + Tx_3(nT) + 3Tu(nT), \\ x_3(n+1)T &= -Tx_1(nT) + x_3(nT) + 2Tu(nT), \\ y(nT) &= x_1(nT).\end{aligned} \quad (6\text{-}45)$$

These are the recursive formulas for the state equations. From these, for known x_1, x_2, x_3 at any time t, we can find y at this same time, and also x_1, x_2, x_3 one sample period later. This process can be repeated step-by-step, thus generating $y(0)$, $y(T)$, $y(2T)$, ..., for as long as desired, for a specified initial state $x_1(0)$, $x_2(0)$, $x_3(0)$.

6-7 DISCRETE TIME SYSTEMS

The governing set of equations for the discrete time case is in a natural form that allows a direct solution by recursion. Refer to Eq. (2-52) which describes a discrete time system, namely,

$$x(n+1) = Ax(n) + Bu(n),$$
$$y(n) = Px(n) + Qu(n). \tag{6-46}$$

The solution may be built up by writing,

$$x(1) = Ax(0) + Bu(0),$$
$$x(2) = Ax(1) + Bu(1) = A^2 x(0) + ABu(0) + Bu(1),$$
$$\vdots \tag{6-47}$$
$$x(n) = A^n x(0) + \sum_{k=0}^{n-1} A^{n-k-1} Bu(k).$$

This equation shows that the fundamental matrix relating the state vector $x(n)$ to its initial value $x(0)$ is,

$$\Phi(n) = A^n. \tag{6-48}$$

In terms of the fundamental matrix, the solution becomes,

$$x(n) = \Phi(n)x(0) + \sum_{k=0}^{n-1} \Phi(n-k-1) Bu(k). \tag{6-49}$$

This equation gives the state vector at time $n > 0$ as the sum of two major terms, one representing the contribution due to the initial state $x(0)$, and the other, the contribution due to the input u over the interval $(0, k-1)$. The system output vector is written, in a similar way,

$$y(n) = P\Phi(n)x(0) + \sum_{k=0}^{n-1} P\Phi(n-k-1) Bu(k) + Qu(n). \tag{6-50}$$

Observe that Eq. (6-49) is the discrete time analog of Eq. (6-37).

6-8 CONTINUOUS TIME SYSTEMS WITH SAMPLED INPUTS

A continuous time system may, for appropriate inputs, be treated as a discrete time system. That is, if a linear, time invariant system is excited by piecewise constant inputs in which the transitions take place at multiples of a fundamental period T, the state of the system will be described by a linear difference equation. If the input is a sampled signal which is completely specified by its ordinates at the sampling instants, the response is given by methods already discussed. Now the response at the sampling instants is given with less labor than would be necessary for finding the continuous response function.

We proceed to show this by considering Eq. (6-27), the general linear state equation in the absence of driving sources, and its solution, which is given by Eq. (6-30). Specifically, we consider Eq. (6-30) at two successive time intervals. We write, with $t = nT$,

$$x(nT) = e^{AnT}x(0), \qquad (6\text{-}51)$$

and for $t = (n+1)T$,

$$x(n+1)T = e^{A(n+1)T}x(0). \qquad (6\text{-}52)$$

From these we write the following relationship,

$$x(n+1)T = e^{AT}x(nT). \qquad (6\text{-}53)$$

This is the recursive formula for the solution of Eq. (6-27).

We proceed in a similar way to consider the more general case, which is given by Eq. (6-34). Thus for $t = nT$,

$$x(nT) = e^{AnT}x(0) + e^{AnT}\int_0^{nT} e^{-A\tau}Bu(\tau)\,d\tau, \qquad (6\text{-}54)$$

and for $t = (n+1)T$,

$$x(n+1)T - e^{A(n+1)T}x(0) + e^{A(n+1)T}\int_0^{(n+1)T} e^{-A\tau}Bu(\tau)\,d\tau. \qquad (6\text{-}55)$$

Multiply Eq. (6.54) by e^{AT} and subtract from Eq. (6-55). The result is,

$$x(n+1)T = e^{AT}x(nT) + e^{A(n+1)T}\int_{nT}^{(n+1)T} e^{-A\tau}Bu(\tau)\,d\tau. \qquad (6\text{-}56)$$

Recursive formulas can be derived from this expression for specified driving source function $u(t)$. Eq. (6-56) should be compared with Eq. (6-39).

Suppose that $u(t)$ is stepwise constant during the interval $nT \leq t \leq (n+1)T$, the case when one employs a sample-and-hold (quantizier) circuit for the driving source. The integration can be carried out under these conditions. The result is, after some algebraic manipulation,

$$x[(n+1)T] = e^{AT}x(nT) + (e^{AT} - I)A^{-1}Bu(nT), \qquad (6\text{-}57)$$

which can be written,

$$x[(n+1)T] = Fx(nT) + Gu(nT), \qquad (6\text{-}58)$$

where

$$F = e^{AT},$$
$$G = (e^{AT} - I)A^{-1}B.$$

This is the recursive formula when the forcing function is constant during each time interval. It is observed that the matrixes F and G are constants and are evaluated at the beginning of a computation. This development shows that a system that is described by Eq. (6-56) with piecewise constant (quantized) inputs at regular sampling intervals is equivalent, at the sampling instants, to the system that is governed by the difference equation given by Eq. (6-58). Furthermore, a comparison of Eq. (6-58) and Eq. (6-46) for a discrete time system shows the equations to be identical in form, thus allowing the continuous time system to be treated as a discrete time system, as already noted.

The next higher approximation (10) in the solution of Eq. (6-56) is possible if it is supposed that $u(t)$ can be approximated over the interval $nT \leq t \leq (n+1)T$ by an expression of the form,

$$u(nT + \tau) = u(nT) + \tau \frac{du}{dt}(nT). \qquad (6\text{-}59)$$

Now the integration can be carried out, the result being,

$$e^{AT}\int_0^T e^{-A\tau}B\left[u(nT) + \tau \frac{du}{dt}(nT)\right]d\tau = (e^{AT} - I)A^{-1}Bu(nT)$$

$$+ [(e^{AT} - I)A^{-2} - TA^{-1}]B\frac{du}{dt}(nT). \qquad (6\text{-}60)$$

The complete solution is then,

$$x(n+1)T = e^{AT}x(nT) + (e^{AT} - I)A^{-1}Bu(nT)$$

$$+ [(e^{AT} - I)A^{-2} - TA^{-1}]B\frac{du}{dt}(nT). \qquad (6\text{-}61)$$

This expression is to be contrasted with Eq. (6-57) for the case when a simple sample and hold circuit is used.

6-9 STEADY-STATE OUTPUT TO PERIODIC INPUTS

Here we consider a very special case of Eq. (6-36) for periodic inputs. We now suppose that $u(t)$ is periodic with period T, so that,

$$u(t) = u(t+T).$$

We choose the interval $(0, T)$ and write Eq. (6-36) as,

$$x(t) = e^{At} x(0) + \int_0^t e^{A(t-\tau)} Bu(\tau) d\tau. \tag{6-62}$$

But for the periodic case, we require that $x(0)$ at $t = 0$ be equal to the forced value of the function $x(t)|_{t=T} = x(T)$. This requires that,

$$x(T) = e^{AT} x(0) + \int_0^T e^{A(T-\tau)} Bu(\tau) d\tau = x(0).$$

This expression yields for $x(0)$,

$$x(0) = [I - e^{AT}]^{-1} \int_0^T e^{A(T-\tau)} Bu(\tau) d\tau. \tag{6-63}$$

When this is combined with Eq. (6-62) we have that,

$$x(t) = e^{At}[I - e^{AT}]^{-1} \int_0^T e^{A(T-\tau)} Bu(\tau) d\tau + \int_0^t e^{A(t-\tau)} Bu(\tau) d\tau,$$

which is written,

$$x(t) = e^{At} \left\{ [I - e^{AT}]^{-1} e^{AT} \int_0^T e^{-A\tau} Bu(\tau) d\tau + \int_0^t e^{-A\tau} Bu(\tau) d\tau \right\}. \tag{6-64}$$

Some attention has been directed to the solution of this equation (11).

PROBLEMS

6-1. a. Sketch the function, $f(t) = \frac{1}{2} + (1/\pi) \tan^{-1} at$ for $|\tan^{-1} at| \leq (\pi/2), a > 0$.
  b. Show that $f(t)$ approaches $u_{-1}(t)$ as $a \to \infty$.
  c. Show that df/dt approaches $u_0(t)$ as $a \to \infty$.

6-2. a. Sketch the functions

$$e^{-at^2}; \quad \frac{1}{\pi} \frac{a}{1+a^2 t^2}.$$

  b. Show that each has $u_0(t)$ as its limit as $a \to \infty$.

6-3. Sketch and dimension each of the following functions:
 a. $u_{-1}(t) - u_{-1}(t-4)$,
 b. $u_{-1}(t)e^{-t}$,
 c. $u_{-1}(t)(t-1)e^{-(t-1)}$,
 d. $u_{-2}(t) - 2u_{-1}(t-5)$,
 e. $u_{-1}(t-2)\sin 3t$.

6-4. Express the waves illustrated by functions which explicitly involve singularity functions.

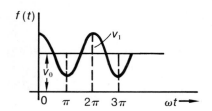
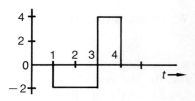
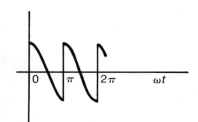
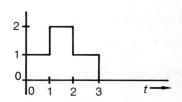

6-5. a. Write the function shown in terms of singularity functions.
 b. Find the derivative of the function and sketch the results.

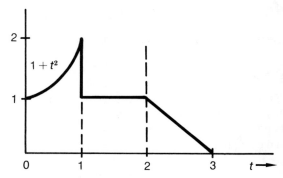

6-6. Consider a simple RC circuit; and the response is the voltage across the capacitor.

a. The input is the pulse shown. Find the response of the initially relaxed circuit.

b. Now let the amplitude of the pulse be V/T. Find the response of the circuit as $T \to 0$. This gives the impulse response.

c. Compare the results under b. with the result obtained by finding the system response to the unit step $u_{-1}(t)$ and then differentiating the result.

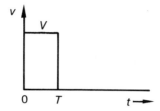

6-7. a. Repeat Problem 6-6 for the current $i(t)$ when the input is the pulse shown. Specifically, find $i(t)$ during the intervals $(0, T/2)$, $(T/2, T)$. Sketch the results.

b. Show the results as $T \to 0$. This approximates the doublet response.

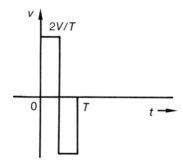

6-8. Prove the following two forms of the superposition integral,

$$i(t) = \frac{d}{dt} \int_0^t W(t-\lambda) \, v(\lambda) \, d\lambda,$$

$$i(t) = \frac{d}{dt} \int_0^t W(\lambda) \, v(t-\lambda) \, d\lambda.$$

6-9. The indicial response of a system is $3e^{-t}$. Find the response to the excitation shown.

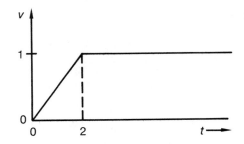

6-10. Consider the circuit shown. Find the output $v_2(t)$ employing the superposition integral.

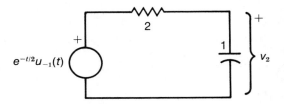

6-11. The impulse response of a network is given in a. of the accompanying figure. Find the response of the system to the excitation function shown in b.
NOTE: No simple circuit would have the impulse response characteristic shown.

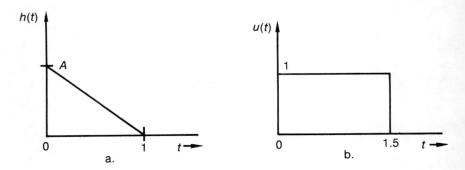

6-12. The step function response of a linear system is $te^{-t/\tau}u_{-1}(t)$. Find the response of the system to the function illustrated.

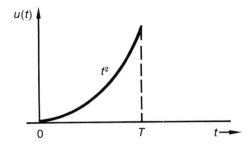

6-13. The impulse response of a system is given by $h(t) = (e^{-2t} + e^{-3t})u_{-1}(t)$.
  a. Find the step response of the system.
  b. Find the response to an input $u(t) = e^{-2t}u_{-1}(t)$ by using the appropriate form of the superposition integral.

6.14. Given the following A matrixes. Calculate the fundamental matrix e^{At} from Eq. (6-24), also e^{-At},

$$\begin{bmatrix} 1 & 1 \\ 1 & 1 \end{bmatrix} \quad \begin{bmatrix} 1 & -1 \\ 1 & -1 \end{bmatrix} \quad \begin{bmatrix} 1 & -1 & -1 \\ 1 & -1 & -1 \\ 0 & 0 & 0 \end{bmatrix}.$$

6-15. Calculate the fundamental matrix of the networks shown.

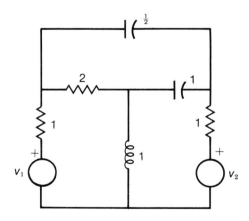

6-16. Find solutions for the following state equations,

$$\frac{d}{dt}\begin{bmatrix} x_1 \\ x_2 \end{bmatrix} = \begin{bmatrix} 0 & 5 \\ -1 & -3 \end{bmatrix} \begin{bmatrix} x_1 \\ x_2 \end{bmatrix} \quad x(0) = \begin{bmatrix} 1 \\ 0 \end{bmatrix},$$

$$\frac{d}{dt}\begin{bmatrix} x_1 \\ x_2 \end{bmatrix} = \begin{bmatrix} 0 & 1 \\ -1 & 2 \end{bmatrix} \begin{bmatrix} x_1 \\ x_2 \end{bmatrix} \quad x(0) = \begin{bmatrix} 1 \\ 0 \end{bmatrix}.$$

$$\frac{d}{dt}\begin{bmatrix} x_1 \\ x_2 \end{bmatrix} = \begin{bmatrix} 0 & 2 \\ -3 & 0 \end{bmatrix}\begin{bmatrix} x_1 \\ x_2 \end{bmatrix} \quad x(0) = \begin{bmatrix} 2 \\ 0 \end{bmatrix}.$$

6-17. Refer to Section 5-10 which contains a discussion of the series *RLC* circuit and its dual.

   a. Show that Eq. (5-100) can be written in the form,

$$\frac{d}{dt}\begin{bmatrix} x_1 \\ x_2 \end{bmatrix} = \begin{bmatrix} 0 & 1 \\ -\omega_n^2 & -2\zeta\omega_n \end{bmatrix}\begin{bmatrix} x_1 \\ x_2 \end{bmatrix} + \begin{bmatrix} 0 \\ 1 \end{bmatrix} u; \quad \begin{bmatrix} x_1(0) \\ x_2(0) \end{bmatrix} = \begin{bmatrix} 0 \\ 0 \end{bmatrix}.$$

Define all parameters.

   b. Obtain the analytical expression for x_1 and x_2 in terms of $u(t)$.
   c. Discuss the solution x_1 when $u(t)$ is a unit step function.
   d. Sketch x_1 for $\zeta = 0, 0.1, 1, 2$.
   e. Sketch the trajectory of x in the x_1, x_2 plane for $\zeta = 0, 0.1, 1, 2$.

6-18. In the following integral, A is a nonsingular real matrix and B is a constant matrix. Prove that,

$$\int_0^\infty e^{-At} B \, dt = A^{-1} B.$$

6-19. The state equation of a system is given in standard form,

$$\dot{x} = Ax + Bu,$$

where

$$A = \begin{bmatrix} 2 & -4 \\ 3 & -6 \end{bmatrix} \quad B = \begin{bmatrix} 1 & 0 \\ 0 & 1 \end{bmatrix} \quad u = \begin{bmatrix} 1 \\ 1 \end{bmatrix} \quad x(0) = \begin{bmatrix} 1 \\ 2 \end{bmatrix}.$$

Determine the solution, in accordance with Eq. (6-36).

6-20. A system is described by the differential equation,

$$\frac{d^2 y}{dt^2} + 3\frac{dy}{dt} + 5y = u(t).$$

   a. Reduce this equation to normal form.
   b. Determine the fundamental matrix $\Phi(t)$.
   c. Write the solution of the state equations.

6-21. A system is characterized by the state equations,

$$\dot{x}_1 = x_2,$$
$$\dot{x}_2 = -3x_2 - 5x_1 + u,$$
$$y = 3x_1 + 2x_2.$$

   a. Determine the impulse response of the system.
   b. Determine the step function response of the system.
   c. Verify the results in a. from the results in b.

6-22. Refer to the SFG-s shown.

   a. Find the differential equation of each, and from these find matrix A of the normal form of the differential equations.

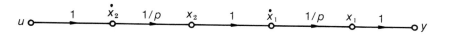

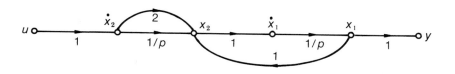

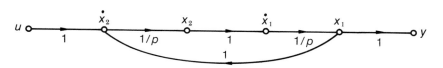

b. Find $\Phi(t) = e^{At}$ using the series form of the exponential function.
c. Find $\Phi(t)$ by finding its components $\Phi_{ij}(t)$ as impulse responses.
d. Find the impulse response of each system.

6-23. Refer to the SFG given.
  a. Write the controlling differential equation in normal form.
  b. Find matrix A and the fundamental matrix $\Phi(t)$ from A.
  c. Find the elements $\Phi_{ij}(t)$ of the fundamental matrix as the response at the appropriate point to assumed initial conditions.

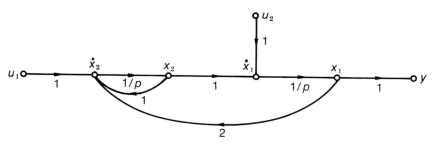

6-24. Suppose that the systems of Problem 6-22 have staircase inputs with sampling times of 1 sec. Write the difference equation for each system, as discussed in Section 6-6.

6-25. Show that the input-output relation for a discrete time system, given the step response, is,

$$y(n) = u(n_0) T u_1(n, k) + \sum_{k=n_0+1}^{\infty} T u_1(n, k) [u(k) - u(k-1)],$$

where
$$u(n) = 0 \quad (n < n_0).$$

6-26. The state equations of a given system are contained in the expression,

$$\frac{d}{dt}\begin{bmatrix}x_1\\x_2\end{bmatrix} = \begin{bmatrix}0 & -5\\2 & -3\end{bmatrix}\begin{bmatrix}x_1\\x_2\end{bmatrix} + \begin{bmatrix}u(t)\\0\end{bmatrix}.$$

Take $t = nT$ and derive a discrete time state equation for this system.

6-27. Refer to the discrete time SFG shown.

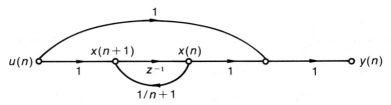

a. Determine the fundamental matrix and the unit response of the system.
b. Find the transmission matrix of the system.

6-28. Determine a numerical solution of the state equations given for discrete time values $t = nT$ for n from 1 to 10. Show these values on a sketch.

$$\frac{d}{dt}\begin{bmatrix}x_1\\x_2\\x_3\end{bmatrix} = \begin{bmatrix}1 & -1 & -1\\1 & -1 & -1\\0 & 0 & 0\end{bmatrix}\begin{bmatrix}x_1\\x_2\\x_3\end{bmatrix} \quad x(0) = \begin{bmatrix}1\\1\\1\end{bmatrix}.$$

6-29. The step function response of a discrete time system is given in the sketch. Find the response to the input shown.

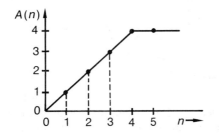 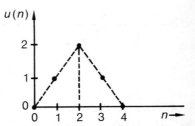

6-30. Suppose that the system specified in Problem 6-20 is subject to a staircase input

$$u(t) = 2^k \quad kT < t < KT + T \quad (k = 0, 1, 2, \ldots, \infty).$$

Write the recurrence equation solution for this system.

Part III System Response

b. Frequency Domain

7
The Laplace Transform

Chapters 5 and 6 were concerned with the time domain solution of system equations. Time domain techniques, except for the very simple cases, are not particularly convenient for hand solutions, even for linear problems, and are almost impossible for nonlinear problems. With the general availability of the digital computer, the burdensome calculations are transferred to the machine, and time domain methods are of basic importance.

For linear time invariant systems, which constitute a very important part of systems analysis, and especially where hand solutions are required, it has become customary to effect a mathematical transformation of the integrodifferential equations that describe the dynamics of the system from the time domain to a convenient algebraic domain. Such a transformation technique makes possible many algebraic manipulations as part of the solution process. Ultimately, of course, the time response is desired, since the time domain is the domain of the real world. Thus it is essential that an inverse transformation be effected back to the time domain from the algebraic domain of the transformation. Clearly, with increasing use of the digital computer, such transformation techniques are of lesser importance in providing time solutions to linear systems problems. However, such transformation techniques remain very important techniques of analysis, and we shall devote considerable time to this study.

The Laplace transform which is now to be studied permits, in a formal way, the transformation of the integrodifferential equation from the time domain to the s-domain, a complex frequency domain. There are many subtle aspects that concern the convergence of the Laplace transform, a study of which requires an understanding of the theory of functions of a complex variable. Because of this, we shall limit ourselves to the simpler aspects of the Laplace transform methods. These are usually sufficient to carry out analyses of network behavior.

7-1 THE LAPLACE TRANSFORM

This is one of a family of integral transforms; the Fourier integral which we shall later study, is another of the family. The general class of integral transforms here being considered is defined by the integral,

$$F(s) = \int_a^b f(x) \, K(s,x) \, dx, \qquad (7\text{-}1)$$

where $F(s)$ is the integral transform of $f(x)$, with $f(x)$ being any transformable function of the real variable x. $K(s, x)$ is known as the kernel of the transform. Here s is a complex variable ($s = \sigma + j\omega$). In particular, we distinguish (12) between the one-sided Laplace transform which is written,

$$F(s) = \int_0^\infty f(x) \, e^{-sx} \, dx, \qquad (7\text{-}2)$$

and the two-sided Laplace transform,

$$F(s) = \int_{-\infty}^\infty f(x) \, e^{-sx} \, dx.$$

For our purposes the one-sided Laplace transform is adequate, but in certain communication theory applications, one must use the two-sided transform. The Fourier transform that will be studied in some detail later,

$$F(j\omega) = \int_{-\infty}^\infty f(x) \, e^{-j\omega x} \, dx, \qquad (7\text{-}3)$$

is seen to be a two-sided transform. It is noted that the Fourier transform of $f(x) \, e^{-\sigma x}$ is identical with the two-sided Laplace transform of $f(x)$.

Since we normally operate primarily in the time domain in dynamic systems problems, our basic functions are time functions, and the Laplace transform, which is generally abbreviated as the $\mathscr{L}$-transform of $f(t)$, is written,

$$\mathscr{L}[f(t)] = \int_0^\infty f(t) \, e^{-st} \, dt = F(s). \qquad (7\text{-}4)$$

Laplace Transforms of Elementary Functions

For Laplace transformability, $f(t)$ must satisfy the Dirichlet conditions, which are sufficient although not necessary conditions. These are:

1. $f(t)$ must be piecewise continuous, i.e., it must be single-valued but can have a finite number of finite isolated discontinuities for $t \geq 0$.
2. $f(t)$ must be of exponential order; i.e., $f(t)$ must remain less than $Me^{a_0 t}$ as t approaches ∞, where M is a positive constant, and a_0 is a real number.

For example, such functions as $\tan \beta t$, $\cot \beta t$, e^{t^2}, do not possess Laplace transforms.

There is the inherent assumption in Eq. (7-4) that $f(t)$ must remain as defined for all of time. From a practical viewpoint, one may have some skepticism about meeting this requirement, although hypothetically this causes no difficulty.

7-2 LAPLACE TRANSFORMS OF ELEMENTARY FUNCTIONS

Laplace transforms do exist for most functions that are of interest in our studies. We shall evaluate a number of such functions.

Example 7-2.1

Find the Laplace transform of the unit step function $f(t) = u_{-1}(t)$.
Solution. By Eq. (7-4),

$$\mathscr{L}[u_{-1}(t)] = \int_0^\infty u_{-1}(t) e^{-st}\, dt = \int_0^\infty e^{-st}\, dt.$$

This integrates to,

$$\mathscr{L}[u_{-1}(t)] = -\frac{1}{s} e^{-st} \Big|_0^\infty = \frac{1}{s}.$$

This result shows the basis for designating the unit step as $u_{-1}(t)$.

Example 7-2.2

Find the Laplace transform of the unit impulse function $f(t) = u_0(t)$.
Solution. This evaluation requires care since $u_0(t)$ does not satisfy the Dirichlet conditions. Since $u_0(t)$ is nonzero only at the origin $t = 0$, then in general for any function $\psi(t)$ as shown in Eq. (6-5),

$$\int_{-\infty}^\infty \psi(t) u_0(t)\, dt = \psi(0).$$

278   The Laplace Transform

If we choose $\psi = e^{-st}$, which equals unity for $t = 0$, then for $t \geq 0$,

$$\int_{-\infty}^{\infty} u_0(t) e^{-st} \, dt = \int_{0}^{\infty} u_0 e^{-st} \, dt = 1,$$

since, by definition, the area under the unit impulse is unity.

Example 7-2.3

Find the Laplace transform of the exponential function e^{-at} for any real a.

Solution. We write directly,

$$\mathscr{L}[e^{-at}] = \int_{0}^{\infty} e^{-at} e^{-st} \, dt = \frac{e^{-(s+a)t}}{-(s+a)} \bigg|_{0}^{\infty}$$

$$= \frac{e^{-(\sigma+a)t} e^{-j\omega t}}{-(s+a)} \bigg|_{0}^{\infty}.$$

Observe that the numerator remains finite (and goes to zero) as $t \to \infty$ only if $\sigma + a > 0$; thus,

$$\mathscr{L}[e^{-at}] = \frac{1}{s+a} \qquad (\sigma > -a).$$

The restriction is necessary since it demands that the product $e^{-at} e^{-st}$ decrease with increasing t; for $\sigma \leq -a$ the integral diverges. In the case that a is complex, the restriction becomes $\sigma > -\operatorname{Re} a$.

Example 7-2.4

Find the Laplace transforms of $\sin \omega t$ and $\cos \omega t$.

Solution. By direct application of the results of Example 7-2.3 we see, by writing,

$$\sin \omega t = \frac{e^{j\omega t} - e^{-j\omega t}}{2j},$$

$$\cos \omega t = \frac{e^{j\omega t} + e^{-j\omega t}}{2},$$

and noting that,

$$\mathscr{L}[e^{j\omega t}] = \frac{1}{s - j\omega} \qquad \mathscr{L}[e^{-j\omega t}] = \frac{1}{s + j\omega},$$

then,

$$\mathscr{L}[\sin \omega t] = \frac{1}{2j} \left(\frac{1}{s - j\omega} - \frac{1}{s + j\omega} \right) = \frac{\omega}{s^2 + \omega^2},$$

$$\mathscr{L}[\cos \omega t] = \frac{1}{2}\left(\frac{1}{s-j\omega} - \frac{1}{s+j\omega}\right) = \frac{s}{s^2+\omega^2}.$$

One can proceed in this way to develop a table of Laplace transform pairs. Such a table can be extended using the properties of the Laplace transform described in the next section. Table 7-1 contains some very useful transform pairs. Extensive tables of Laplace transforms are available (13).

TABLE 7-1. Elementary Laplace Transform Pairs

$f(t)$	$F(s)$
1. $u_0(t)$	1
2. $u_{-1}(t)$	$\dfrac{1}{s}$
3. $t^n \quad (n > 0)$	$\dfrac{n!}{s^{n+1}}$
4. e^{-at}	$\dfrac{1}{s+a}$
5. te^{-at}	$\dfrac{1}{(s+a)^2}$
6. $(-1)^n \dfrac{t^{n-1}e^{-at}}{(n-1)!}$	$\dfrac{1}{(s+a)^n}$
7. $\dfrac{1}{b-a}(e^{-at} - e^{-bt})$	$\dfrac{1}{(s+a)(s+b)}$
8. $-\dfrac{1}{b-a}(ae^{-at} - be^{-bt})$	$\dfrac{s}{(s+a)(s+b)}$
9. $\sin \omega t$	$\dfrac{\omega}{s^2+\omega^2}$
10. $\cos \omega t$	$\dfrac{s}{s^2+\omega^2}$
11. $e^{-at} \sin \omega t$	$\dfrac{\omega}{(s+a)^2+\omega^2}$
12. $e^{-at} \cos \omega t$	$\dfrac{s+a}{(s+a)^2+\omega^2}$
13. $\sinh \omega t$	$\dfrac{\omega}{s^2-\omega^2}$
14. $\cosh \omega t$	$\dfrac{s}{s^2-\omega^2}$

7-3 PROPERTIES OF THE LAPLACE TRANSFORM

We wish now to examine a number of important properties of the Laplace transform.

The Laplace Transform

a. Linearity. If the functions $f_1(t)$ and $f_2(t)$ are Laplace transformable and with K_1 and K_2 being constants, then,

$$\mathscr{L}[K_1 f_1(t) + K_2 f_2(t)] = K_1 F_1(s) + K_2 F_2(s). \tag{7-5}$$

This result follows directly from Eq. (7-4), since,

$$\mathscr{L}[K_1 f_1(t) + K_2 f_2(t)] = \int_0^\infty [K_1 f_1(t) + K_2 f_2(t)] e^{-st} dt,$$

$$= \int_0^\infty K_1 f_1(t) e^{-st} dt + \int_0^\infty K_2 f_2(t) e^{-st} dt,$$

$$= K_1 F_1(s) + K_2 F_2(s).$$

Note that this property was used in carrying out Example 7-2.4. Actually the same results would have been obtained by carrying out the integrations using $\sin \omega t$ and $\cos \omega t$ as the functions $f(t)$.

b. Differentiation. The transform of the derivative of a time function is important in the solution of differential equations. If the function $f(t)$ and its derivative $df(t)/dt$ are Laplace transformable then, from the basic definition of the Laplace transform, we can write,

$$\mathscr{L}\left[\frac{d}{dt} f(t)\right] = \int_0^\infty \frac{df}{dt} e^{-st} dt. \tag{7-6}$$

Integrating by parts, by writing,

$$u = e^{-st} \qquad du = -se^{-st} dt,$$
$$dv = \frac{df}{dt} dt \qquad v = f.$$

then,

$$\mathscr{L}\left[\frac{df}{dt}\right] = fe^{-st} \Big|_0^\infty + s \int_0^\infty f e^{-st} dt.$$

The limit of $f(t)e^{-st}$ as $t \to \infty$ must be zero, otherwise the transform would not exist (this result follows from L'Hospital's rule, provided that $f(t)$ and all of its derivatives are not infinite at $t \to \infty$). Thus,

$$\mathscr{L}\left[\frac{df}{dt}\right] = s\mathscr{L}[f(t)] - f(0+),$$

so that,

$$\mathscr{L}\left[\frac{df}{dt}\right] = sF(s) - f(0+). \tag{7-7}$$

Properties of the Laplace Transform

In this expression $f(0+)$ denotes the value of the function $f(t)$ at the time $t = 0$. But by entry 1 in Table 7-1, this respresents an impulse of strength $f(0+)$. This is a very important result in network problems since it shows that initial conditions associated with derivative functions are automatically included as series impulse functions in the network description. In particular, if $f(t)$ denotes the current through an inductor, then terms of the form $v = L\,di/dt$ are included in the network equations. In this case, when such a term is Laplace transformed, there will be terms of the form,

$$\mathscr{L}\left[L\frac{di}{dt}\right] = LsI(s) - Li(0+),$$

in the transformed equation, where $i(0+)$ denotes the initial current through the inductor. This result means that an inductor with an initial current can be regarded in subsequent calculations as the equivalent of an initially relaxed inductor plus an impulse current source at initial time $t = 0$.

To find the transform of the second time derivative, we use Eq. (7-7) plus induction. We note that,

$$\frac{d^2f}{dt^2} = \frac{d}{dt}\left(\frac{df}{dt}\right),$$

then,

$$\mathscr{L}\left[\frac{d^2f}{dt^2}\right] = \mathscr{L}\left[\frac{d}{dt}\left(\frac{df}{dt}\right)\right] = s\mathscr{L}\left[\frac{df}{dt}\right] - \frac{df}{dt}(0+).$$

Again using Eq. (7-7) then,

$$\mathscr{L}\left[\frac{d^2f}{dt^2}\right] = s^2F(s) - sf(0+) - f^{(1)}(0+), \tag{7-8}$$

where, for convenience in writing, $f^{(1)}(0+)$ denotes df/dt at $t = 0+$.

The Laplace transform of the nth time derivative follows as a direct extension of the foregoing development. The result is found to be,

$$\mathscr{L}\left[\frac{d^nf}{dt^n}\right] = s^nF(s) - s^{n-1}f(0+) - s^{n-2}f^{(1)}(0+) - \cdots - f^{(n-1)}(0+), \tag{7-9}$$

where $f^{(k)}(0+)$ denotes the kth derivative $d^kf(0+)/dt^k$ at $t = 0+$. If all initial values are zero, this expression reduces to the form,

$$\mathscr{L}\left[\frac{d^nf}{dt^n}\right] = s^nF(s). \tag{7-10}$$

Example 7-3.1

Find the Laplace transform of the function $f(t) = t^n$ using Eq. (7-10) for the Laplace transformed form for $\mathscr{L}[t^n]$.

Solution. From Eq. (7-10),

$$\mathscr{L}\left[\frac{d^n f(t)}{dt^n}\right] = s^n F(s),$$

Furthermore, when $f = t^n$,

$$\frac{d^n t^n}{dt^n} = n(n-1)\ldots 3\cdot 2\cdot 1\cdot u_{-1}(t) = n!\, u_{-1}(t),$$

and so,

$$\mathscr{L}\left[\frac{d^n t^n}{dt^n}\right] = n!\,\mathscr{L}[u_{-1}(t)] = n!\,\frac{1}{s} = s^n F(s).$$

It follows from this that,

$$F(s) = \frac{n!}{s^{n+1}} = \mathscr{L}[t^n].$$

c. Integration. The transform of a time integral proceeds in a straightforward manner from the definition of the Laplace transform. If the function $f(t)$ is Laplace transformable, then its integral is written,

$$\mathscr{L}\left[\int f(t)\,dt\right] = \int_0^\infty \left[\int f(t)\,dt\right] e^{-st}\,dt.$$

This is integrated by parts by setting,

$$u = \int f(t)\,dt \qquad du = f(t)\,dt,$$

$$dv = e^{-st}\,dt \qquad v = -\frac{1}{s}e^{-st}.$$

Then,

$$\mathscr{L}\left[\int f(t)\,dt\right] = \left[-\frac{1}{s}e^{-st}\int f(t)\,dt\right]_0^\infty + \frac{1}{s}\int_0^\infty f(t)e^{-st}\,dt,$$

from which,

$$\mathscr{L}\left[\int f(t)\,dt\right] = \frac{1}{s}F(s) + \frac{1}{s}f^{(-1)}(0+), \qquad (7\text{-}11)$$

where $(f^{(-1)}(0+)/s)$ is the initial value of the integral of $f(t)$ at $t = 0+$. Note by entry 2 in Table 7-1 that this term denotes a step function of amplitude $f^{(-1)}(0+)$. This is a very important result in network problems since it shows that initial conditions associated with integral functions are automatically included as step functions in the Laplace transform development. In particular, if $f(t)$ denotes a current $i(t)$ through a capacitor, then the voltage across the capacitor is expressed by the relation,

$$v(t) = \frac{q(t)}{C} = \frac{1}{C}\int i\,dt.$$

The Laplace transform of such a term then becomes,

$$\mathscr{L}\left[\frac{q}{C}\right] = \mathscr{L}\left[\frac{1}{C}\int i\,dt\right] = \frac{I(s)}{Cs} + \frac{q(0+)}{Cs},$$

where $q(0+)$ is the charge on the capacitor at initial time $t = 0+$. This result means that an initially charged capacitor can be regarded, in so far as the subsequent action in a circuit is concerned, as an initially relaxed capacitor plus a series step function voltage source, or equivalently, a shunting step-function current source. Recall that this was precisely the manner in which initial conditions were established in the analog computer.

By Eq. (7-11) and mathematical induction we can readily show that,

$$\mathscr{L}\left[\iint f(t)\,dt\,dt\right] = \frac{F(s)}{s^2} + \frac{f^{(-1)}}{s^2}(0+) + \frac{f^{(-2)}}{s}(0+), \qquad (7\text{-}12)$$

where $f^{(-1)}(0+)$ denotes the value of the first integral, and $f^{(-2)}(0+)$ denotes the value of the second integral at $t = 0+$. In the cases when the initial values of the integrals vanish, then,

$$\mathscr{L}\left[\int f(t)\,dt\right] = \frac{F(s)}{s},$$

$$\mathscr{L}\left[\iint f(t)\,dt\,dt\right] = \frac{F(s)}{s^2}. \qquad (7\text{-}13)$$

d. Time Translation. We shall find occasion when it is necessary to express the transform of a function that has been translated along the time axis. This is the situation illustrated in Fig. 7-1, which shows the translation of a function $f(t)$ to the right by λ units of time.

Example 7-3.2

Find an expression for the Laplace transform of the delayed unit impulse function.

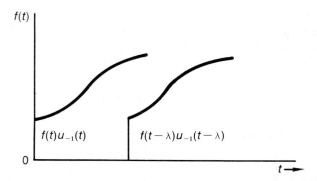

Fig. 7-1. A function $f(t)u_{-1}(t)$ and the same function delayed by a time $t = \lambda$.

Solution. We consider this problem from the point of view of $(d/dt)u_{-1}(t-\lambda)$. Thus,

$$\mathscr{L}\left[\frac{d}{dt}u_{-1}(t-\lambda)\right] = s\,\mathscr{L}[u_{-1}(t-\lambda)] - u_{-1}(-\lambda).$$

Observe the presence of the term $u_{-1}(-\lambda)$, the value of the step when the argument is $-\lambda$, instead of $u_{-1}(0+)$ (this is required to avoid an incorrect specification at $t = 0$). Note further that $u_{-1}(-\lambda) = 0$, and since $u_{-1}(0-) = 0$, we could have written the quantity $u_{-1}(-\lambda) = u_{-1}(0-)$ for the form to conform to accepted terminology.

We observe additionally that if $f(t) = 0$ for $t < 0$, the translated function is $f(t-\lambda)u_{-1}(t-\lambda)$. This is precisely the situation illustrated in Fig. 6-2. For such a time shifted function,

$$\mathscr{L}[f(t-\lambda)u_{-1}(t-\lambda)] = \int_0^\infty f(t-\lambda)u_{-1}(t-\lambda)e^{-st}\,dt = \int_\lambda^\infty f(t-\lambda)e^{-st}\,dt,$$

which specifies that $f(t-\lambda) = 0$ for $t < \lambda$. With a change of variable, $\tau = t - \lambda$ this expression may be written,

$$\mathscr{L}[f(t-\lambda)u_{-1}(t-\lambda)] = \int_0^\infty f(\tau)e^{-s\tau}e^{-s\lambda}\,d\tau,$$

or

$$\mathscr{L}[f(t-\lambda)u_{-1}(t-\lambda)] = e^{-s\lambda}F(s). \tag{7-14}$$

This expression shows that translation to the right in the time domain by λ units corresponds to multiplication by $e^{-s\lambda}$. This result will be found of particular importance when considering a recurrent waveform, and also the Z-transform. Retaining the factor $u_{-1}(t-\lambda)$ in Eq. (7-14) is desirable since it automatically calls attention to the fact that values for $f(t-\lambda)$ with negative arguments must be zero.

In an entirely similar way, a translation to the left in time with $\lambda > 0$, leads to,

$$\mathscr{L}[f(t+\lambda)] = \int_0^\infty f(t+\lambda)e^{-st}\,dt = \int_\lambda^\infty f(\tau)e^{-s\tau}e^{s\lambda}\,d\tau,$$

$$= \int_0^\infty f(\tau)u_{-1}(\tau-\lambda)e^{-s\tau}e^{s\lambda}\,d\tau.$$

This may be written,

$$\mathscr{L}[f(t+\lambda)] = e^{s\lambda}G(s), \qquad (7\text{-}15)$$

where

$$G(s) = \mathscr{L}[f(t)u_{-1}(t-\lambda)].$$

This shows that the Laplace transform of a function that is translated to the left may be expressed in terms of its Laplace transform before translation, but with provision that the function must vanish over a positive time range adjacent to the origin $(0 < t < \lambda)$ and equal in length to the translation interval.

Example 7-3.3

Refer to the accompanying figure which shows a recurrent square wave. Deduce the Laplace transform of the first pulse, the first two pulses, and the total recurring waveform.

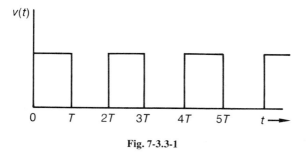

Fig. 7-3.3-1

Solution. Consider initially a single pulse. Clearly in the time domain this may be represented by a step function beginning at time $t=0$ followed in time by an equal step of opposite sense beginning at time $t=T$ as illustrated in Fig. 7-3.3-2. Clearly, the single pulse can be

286 The Laplace Transform

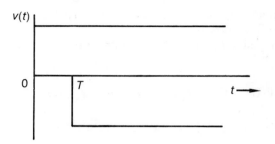

Fig. 7-3.3-2

represented analytically by the expression,

$$v_1(t) = V u_{-1}(t) - V u_{-1}(t-T).$$

The Laplace transform of this expression is then,

$$V_1(s) = \frac{V}{s} - \frac{V}{s} e^{-sT} = \frac{V}{s}(1 - e^{-sT}).$$

For the case of two pulses, we could proceed by adding appropriately delayed step functions, one at time $2T$ and the opposite step function at time $3T$. The same result can be accomplished by noting that the second pulse can be obtained from the first pulse by effecting a translation in time by an amount $2T$ to the right. That is,

$$v_2(t) = v_1(t - 2T) \qquad (t \geq 2T).$$

By the time translation theorem given by Eq. (7-14) we see that we can write,

$$V_2(s) = V_1(s) e^{-2sT}.$$

Thus the Laplace transform function for the two pulse group is,

$$V(s) = V_1(s) + V_2(s) = V_1(s)(1 + e^{-2sT}) = \frac{V}{s}(1 - e^{-sT})(1 + e^{-2sT}).$$

For the recurring pulse waveform we need to include the contribution by each pulse. By a direct extension of the foregoing results we write,

$$V(s) = V_1(s) + V_1(s) e^{-2sT} + V_1(s) e^{-4sT} + \cdots,$$

which is,

$$V(s) = V_1(s)(1 + e^{-2sT} + e^{-4sT} + \cdots).$$

This expression may be written as,

$$V(s) = \frac{V_1(s)}{1 - e^{-2sT}}.$$

This is combined with the known form for $V_1(s)$. The result is,

$$V(s) = \frac{V}{s}\left(\frac{1 - e^{-sT}}{1 - e^{-2sT}}\right),$$

or finally,

$$V(s) = \frac{V}{s}\left(\frac{1}{1 + e^{-sT}}\right).$$

e. Initial Value. Evidently for a Laplace transformable $f(t)$ one can deduce the appropriate $F(s)$. Hence for a specified $F(s)$ one can, by inversion, find the appropriate $f(t)$, and from this calculate the initial value $f(0)$. We shall show here that the initial value property permits $f(0)$ to be calculated directly from $F(s)$ without the need for inversion. To accomplish the result, we begin with Eq. (7-7) and consider the expression as $s \to \infty$. That is, we examine the expression,

$$\lim_{s \to \infty} \int \frac{df}{dt} e^{-st} dt = \lim_{s \to \infty} [sF(s) - f(0+)].$$

It is here assumed, of course, that $f(t)$ and its first derivative are Laplace transformable, and that the limit of $sF(s)$ as s approaches infinity exists. But the integral vanishes for $s \to \infty$, and further, $f(0+)$ is independent of s, so that,

$$\lim_{s \to \infty} [sF(s) - f(0+)] = 0.$$

Furthermore, $f(0+) = \lim_{t \to 0+} f(t)$, so that,

$$\lim_{s \to \infty} sF(s) = \lim_{t \to 0+} f(t). \quad (7\text{-}16)$$

If $f(t)$ has a discontinuity at the origin, this expression specifies the value of the impulse $f(0+)$. If $f(t)$ contains an impulse term, then the left hand side does not exist, and the initial value property does not hold.

Example 7-3.4

Consider the two functions given. Apply the initial value theorem to these.

a. $F_1(s) = \dfrac{s}{s^2+3}$.

b. $F_2(s) = \dfrac{s^2+s+3}{s^2+3}$.

Solution. When the initial value theorem is applied to function a.

$$\lim_{s\to\infty} sF(s) = \lim_{s\to\infty} \frac{s^2}{s^2+3} = 1.$$

When applied to function b. the initial value of $f(t)$ cannot be found. The reason is easily discovered by writing $F_2(s)$ in the form,

$$F_2(s) = 1 + \frac{s}{s^2+3},$$

which shows the existence of an impulse in the function.

f. Final Value. The value of $f(t)$ for large values of t can also be found directly from its transform. We again begin with Eq. (7-7), but now we let s approach zero. Thus we examine,

$$\lim_{s\to 0} \int_0^\infty \frac{df}{dt} e^{-st}\,dt = \lim_{s\to 0}\,[sF(s) - f(0+)].$$

Consider the quantity on the left. Since s and t are independent, and since $e^{-st} \to 1$ as $s \to 0$, then the integral on the left becomes, in the limit,

$$\int_0^\infty \frac{df}{dt}\,dt = \lim_{t\to\infty} f(t) - f(0+).$$

We combine the latter two equations to get,

$$\lim_{t\to\infty} f(t) - f(0+) = \lim_{s\to 0} sF(s) - f(0+),$$

from which it follows that the final value of $f(t)$ is given by,

$$\lim_{t\to\infty} f(t) = \lim_{s\to 0} sF(s). \qquad (7\text{-}17)$$

This result applies if $F(s)$ possesses a simple pole at the origin, but it does not apply if $F(s)$ has imaginary axis poles, poles in the right half plane, or higher order poles at the origin.

Example 7-3.5

Apply the final value theorem to the following two functions:

$$F_1(s) = \frac{s+a}{(s+a)^2+b^2}, \qquad F_2(s) = \frac{s}{s^2+b^2}.$$

Solution. For the first function,

$$\lim_{s \to 0} \frac{s(s+a)}{(s+a)^2 + b^2} = 0.$$

The second function yields for $sF(s) = s^2/(s^2 + b^2)$, which has singularities on the imaginary axis at $s = \pm jb$, and the theorem does not apply.

g. Convolution in the s-Plane. We shall find, when we apply the Laplace transform to the equations of linear systems, that products of Laplace transforms occur, say of the form $V(s)H(s)$, each of which is the Laplace transform of a time function. Often one of these, say $H(s)$ describes the features of the system, and the second, $V(s)$, is the Laplace transform of the driving function. In other problems, it may happen that each of the factors gives a system description, and the combined result, which we now denote $H_1(s)H_2(s)$, implies the cascade of two systems which are to be so combined that the second system does not load the first system. This situation arises approximately in electronic circuits where a transistor or tube may be connected so as to isolate the circuit elements making up successive stages.

If a function $V(s)H(s)$ is deduced for a system for which the response is desired, any of a number of methods to be discussed may be used. We are actually less concerned here with solutions to specific problems than with the relation between the product $V(s)H(s)$ and the individual transforms $V(s) = \mathscr{L}[v(t)]$ and $H(s) = \mathscr{L}[h(t)]$. We can deduce such a relation using the convolution integral that was discussed in Chapter 6. Particularly, using the notation of Section 6-3, we shall show that the Laplace transform of the convolution integral relates the impulse response of time domain analysis to the product of the Laplace transforms of the time functions that are being convolved. That is, we shall prove that the Laplace transform of the function,

$$\mathscr{L}[i(t)] = \mathscr{L}\left[\int_0^t v(\lambda)h(t-\lambda)\,d\lambda\right] = V(s)H(s), \qquad (7\text{-}18)$$

where,

$$V(s) = \mathscr{L}[v(t)] \qquad H(s) = \mathscr{L}[h(t)].$$

To prove result, we begin with Eq. (7-18) which is written,

$$I(s) = \int_0^\infty i(t)\,e^{-st}\,dt = \int_0^\infty e^{-st}\,dt \int_0^t v(\lambda)h(t-\lambda)\,d\lambda. \qquad (7\text{-}19)$$

Observe that the upper limit on the right hand integral may be changed to infinity since $h(t-\lambda) = 0$ for $\lambda > t$. Upon inverting the order of integration†

$$I(s) = \int_0^\infty v(\lambda)\,d\lambda \int_0^\infty h(t-\lambda)e^{-st}\,dt. \qquad (7\text{-}20)$$

We now effect a change of variable by writing $\tau = t - \lambda$. The expression becomes,

$$I(s) = \int_0^\infty v(\lambda)\,d\lambda \int_{-\lambda}^\infty h(\tau)e^{-s\tau}e^{-s\lambda}\,d\tau.$$

But the lower limit on the right hand integral may be changed from $-\lambda$ to zero because $h(\tau) = 0$ for $\tau < 0$, so that,

$$I(s) = \int_0^\infty v(\lambda)e^{-s\lambda}\,d\lambda \int_0^\infty h(\tau)e^{-s\tau}\,d\tau = V(s)H(s), \qquad (7\text{-}21)$$

which proves the result.

Example 7-3.6

Find the Laplace transform of $\int_0^t (t-\lambda)\cos\lambda\,d\lambda$.

Solution. This expression is precisely in the form of Eq. (7-18) so that,

$$\mathscr{L}\left[\int_0^t (t-\lambda)\cos\lambda\,d\lambda\right] = \mathscr{L}[t]\mathscr{L}[\cos t].$$

But we have that,

$$\mathscr{L}[t] = \frac{1}{s} \qquad \mathscr{L}[\cos t] = \frac{s}{s^2+1}.$$

Hence it follows that,

$$\mathscr{L}\left[\int_0^t (t-\lambda)\cos\lambda\,d\lambda\right] = \frac{1}{s(s^2+1)}.$$

7-4 INVERSE LAPLACE TRANSFORM

The operation that changes $F(s)$ back to its equivalent $f(t)$ is the inverse Laplace transformation. A rigorous proof of the inverse Laplace transform is rather involved, and is beyond the scope of this book. However, a proof that is based on an intuitive extension of the Fourier integral inversion is readily possible, and this will be given in Chapter 11. Here we simply note that the Laplace inversion integral is written,

$$\mathscr{L}^{-1}[F(s)] = f(t) = \frac{1}{2\pi j}\int_{\sigma-j\infty}^{\sigma+j\infty} F(s)\,e^{st}\,ds. \qquad (7\text{-}22)$$

†This assumes that $H(s)$ is absolutely convergent.

This requires that the integration be performed in the s-plane along a line $s = \sigma$, where σ is a constant damping factor which will ensure the convergence of the integral,

$$\int_{-\infty}^{\infty} |f(t)| e^{-\sigma t} \, dt = \int_{-\infty}^{\infty} |f(t)| e^{-\sigma t} \, dt < \infty. \tag{7-23}$$

This is actually a necessary condition for $f(t)e^{-\sigma t}$ to be Fourier integral transformable, as is discussed in Section 11-7.

Integration in the s-plane is not generally required for linear problems of the usual type with which we are concerned. This follows from the uniqueness property of the Laplace transform, namely, that corresponding to a function $f(t)$, the transform $F(s)$ is unique, and vice versa. This means that there is a one-to-one correspondence between the direct and the inverse transforms, as expressed by the pair of operations,

$$F(s) = \mathscr{L}[f(t)],$$
and $\tag{7-24}$
$$f(t) = \mathscr{L}^{-1}[F(s)].$$

Consequently one needs merely to refer to Table 7-1 (a table of Laplace transforms), or one of the more extensive tables that are readily available, to write the $f(t)$ appropriate to a given $\mathscr{L}^{-1}[F(s)]$. In the event that the available tables do not include a given $F(s)$ and if the $F(s)$ cannot be reduced to available forms, then one might be faced with carrying out the inversion integral of Eq. (7-22).

7-5 PROBLEM SOLVING BY LAPLACE TRANSFORMS

We shall consider several examples which will show the manner of employing the Laplace transform in the solution of network problems, and also the relationship of the results to those using the time domain methods already discussed.

Example 7-5.1

Find an expression for the current in the simple RL series circuit shown. Assume that an initial current $i(0-)$ existed in the inductor when the switch S is closed.

Solution. The system differential equation is seen to be,

$$L\frac{di}{dt} + Ri = V u_{-1}(t).$$

By taking the Laplace transform of each term in this differential equation,

$$L[sI(s) - i(0+)] + RI(s) = \frac{V}{s}.$$

This is rearranged to,

$$(Ls + R)I(s) = \frac{V}{s} + Li(0+).$$

We solve for $I(s)$,

$$I(s) = \frac{V}{L}\frac{1}{s(s+R/L)} + \frac{i(0+)}{(s+R/L)}.$$

We write this as,

$$I(s) = \frac{V}{R}\left(\frac{1}{s} + \frac{1}{s+R/L}\right) + i(0+)\frac{1}{s+R/L}.$$

We observe that the expression for $I(s)$ is made up of three terms, one being due to the applied excitation or forcing function (this leads to the particular solution) $Vu_{-1}(t)$, which appears in its transformed form V/s; the second arises from the network owing to the excitation being applied suddenly (the transient or complementary function); the third term is also a transient term that is due to the initial current, and which appears as an impulse excitation of strength $i(0+)$. We observe also that the nature of the network appears in the function $1/(s + R/L)$, which is known as the system function of the network for input voltage and output current. More will be said about the system functions in Chapter 8.

As to the initial condition, conservation of flux linkages requires that $i(0+) = i(0-)$, and $i(0+)$ is assumed to be known. Using appropriate entries in Table 7-1, the solution, which is the inverse transform of the foregoing equation, is,

$$i(t) = \frac{V}{R}(1 - e^{-Rt/L}) + i(0+)e^{-Rt/L}.$$

This method of problem solution should be compared with that in Section 5-7.

As discussed in Section 5-4 a quantity of importance in systems analysis is the so-called "steady-state" value, the result when the transient terms have died away. This is the particular solution of the differential equation for the response appropriate to the excitation function. Clearly, this value can be found from the complete solution by setting $t \to \infty$, although this is the long way to this result if only this quantity is required. It can also be found using the methods discussed in Section 5-4. Often, it can be established by inspection of the system diagram. In this example, the steady-state value of the current is $i(t)|_{t \to \infty} = V/R$.

Example 7-5.2

Find the transformed expressions for the currents $i_1(t)$ and $i_2(t)$ for the coupled circuit shown, assuming that it is initially relaxed.

Solution. We first write the integrodifferential equations for the network using the KVL. The result is, using p to denote d/dt,

$$10 = (2+p)i_1 - pi_2,$$

$$0 = -pi_1 + \left(p + 3 + \frac{1}{p}\right)i_2.$$

Laplace transform these equations to find,

$$\frac{10}{s} = (2+s)I_1(s) - sI_2(s),$$

$$0 = -sI_1(s) + \left(s + 3 + \frac{1}{s}\right)I_2(s).$$

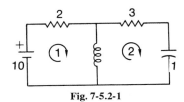

Fig. 7-5.2-1

The expressions for $I_1(s)$ and $I_2(s)$ are found by Cramer's rule,

$$I_1(s) = \frac{\begin{vmatrix} \dfrac{10}{s} & -s \\ 0 & s+3+\dfrac{1}{s} \end{vmatrix}}{\begin{vmatrix} 2+s & -s \\ -s & s+3+\dfrac{1}{s} \end{vmatrix}} = \frac{\dfrac{10}{s}\left(s+3+\dfrac{1}{s}\right)}{(2+s)\left(s+3+\dfrac{1}{s}\right)-s^2} = \frac{10(s^2+3s+1)}{s(5s^2+7s+2)}$$

$$I_2(s) = \frac{\begin{vmatrix} 2+s & \dfrac{10}{s} \\ -s & 0 \end{vmatrix}}{\Delta} = \frac{10s}{5s^2+7s+2}$$

Unless one has a rather extensive table of Laplace transform pairs, the likelihood of handling these equations directly is slight. Hence we wish to consider methods for handling such equations.

7-6 EXPANSION THEOREM

Example 7-5.2 has led to expressions for the transforms of the currents $I_1(s)$ and $I_2(s)$ which may not be readily available in the usual table of Laplace transform pairs. This suggests the need for a method that will permit such forms to be developed into forms that are more amenable to simple tabular look-up. Fortunately, it is possible to develop a procedure

The Laplace Transform

for accomplishing this end. It is observed that both $I_1(s)$ and $I_2(s)$ are rational fractions, i.e., the ratio of two polynomials. Furthermore, each is a proper fraction since the degree of the denominator polynomial is higher than the degree of the numerator polynomial.

To provide for the generalizations that we wish to make, let us suppose that the functions $I(s)$ have numerator polynomials that are of higher degree than the denominator polynomial. Now we would divide the numerator polynomial by the denominator polynomial, carrying out the long division until the numerator polynomial is of degree one less than the denominator polynomial. This results in a number of power terms plus a proper fraction. The proper fraction may be expanded into a partial fraction expansion. The result of these operations is the expression,

$$I(s) = B_0 + B_1 s + \cdots + \frac{A_1}{s - s_1} + \frac{A_2}{s - s_2} + \cdots + \frac{A_{p1}}{s - s_p}$$
$$+ \frac{A_{p2}}{(s - s_p)^2} + \cdots + \frac{A_{pr}}{(s - s_p)^r}. \quad (7\text{-}25)$$

This expression has been written in a form to show three types of terms, (a) polynomial, (b) simple partial fraction including all terms with distinct roots, and (c) partial fraction appropriate to multiple roots. The quantities $s_1, s_2, \ldots$, in the denominator of Eq. (7-25) are called the *poles* of the function $I(s)$.

To find the constants $A_1, A_2, \ldots$, the polynomial terms are removed thereby leaving the proper fraction,

$$I(s) - (B_0 + B_1 s + \cdots) = N(s), \quad (7\text{-}26)$$

where,

$$N(s) = \frac{A_1}{s - s_1} + \frac{A_2}{s - s_2} + \cdots + \frac{A_k}{s - s_k} + \cdots + \frac{A_{pl}}{(s - s_p)} + \cdots + \frac{A_{pr}}{(s - s_p)^r}.$$

To find the constants A_k, which in complex variable terminology are the residues of the function $N(s)$ at the simple poles s_k, it is only necessary to note that as $s \to s_k$, the term $A_k/(s - s_k)$ will become large compared with all other terms. In the limit,

$$A_k = \lim_{s \to s_k} [(s - s_k) N(s)]. \quad (7\text{-}27)$$

For each simple pole, the table of Laplace transform pairs show that upon taking the inverse transform, each will be of a simple exponential form,

$$\mathcal{L}^{-1}\left[\frac{A_k}{s - s_k}\right] = A_k e^{s_k t}. \quad (7\text{-}28)$$

Note also that since $N(s)$ contains only real coefficients, then if s_k is a complex pole with residue A_k, there will also be a pole s_k^* with residue A_k^*, where the asterisk (*) denotes the conjugate complex quantity. For such complex poles,

$$\mathscr{L}^{-1}\left[\frac{A_k}{s-s_k}+\frac{A_k^*}{s-s_k^*}\right]=A_k e^{s_k t}+A_k^* e^{s_k^* t},$$

which may be combined in the following way,

$$\begin{aligned}\text{response}&=(a+jb)e^{(\sigma_k+j\omega_k)t}+(a-jb)e^{(\sigma_k-j\omega_k)t},\\&=e^{\sigma_k t}[(a+jb)(\cos\omega_k t+j\sin\omega_k t)+(a-jb)(\cos\omega_k t-j\sin\omega_k t)],\\&=2e^{\sigma_k t}(a\cos\omega_k t-b\sin\omega_k t).\end{aligned}$$

This may be written,

$$\begin{aligned}\text{response}&=2\sqrt{a^2+b^2}\,e^{\sigma_k t}\cos(\omega_k t+\theta_k),\\&=2|A_k|e^{\sigma_k t}\cos(\omega_k t+\theta_k),\end{aligned}\qquad(7\text{-}29)$$

where,

$$\theta_k=\tan^{-1}b/a.$$

When the proper fraction contains a multiple pole of order r, the coefficients in the partial fraction, $A_{p1}, A_{p2}, \ldots, A_{pr}$, which are involved in the terms,

$$\frac{A_{p1}}{s-s_p}+\frac{A_{p2}}{(s-s_p)^2}+\cdots+\frac{A_{pk}}{(s-s_p)^k}+\cdots+\frac{A_{pr}}{(s-s_p)^r},\qquad(7\text{-}30)$$

must be evaluated. A simple application of Eq. (7-27) is not suitable. In the procedure required, both sides of Eq. (7-27) are multiplied by $(s-s_p)^r$. This gives,

$$(s-s_p)^r N(s)=(s-s_p)^r\left[\frac{A_{p1}}{s-s_1}+\cdots+\frac{A_{p-1}}{s-s_{p-1}}\right]\\+A_{p1}(s-s_p)^{r-1}+\cdots+A_{pr}\qquad(7\text{-}31)$$

In the limit as $s=s_p$ all terms on the right vanish, with the exception of A_{pr}. Suppose now that this equation is differentiated once with respect to s. The constant A_{pr} will disappear in the differentiation, but $A_{p(r-1)}$ will be determined by setting $s=s_p$. This procedure may be continued to find each of the coefficients A_{pk}. Specifically, we see that,

$$A_{pk}=\frac{1}{(r-k)!}\left\{\frac{d^{r-k}}{ds^{r-k}}[N(s)(s-s_p)^r]\right\}_{s=s_p}.\qquad(7\text{-}32)$$

Example 7-6.1

Expand the following function into a partial fraction expansion,

$$F(s) = \frac{s^2+2s+3}{s^2(s+1)}.$$

Solution. We write the following expression,

$$F(s) = \frac{s^2+2s+3}{s^2(s+1)} = \frac{A_{11}}{s} + \frac{A_{12}}{s^2} + \frac{A_2}{s+1}.$$

The coefficients are now specified. These are,

$$A_2 = (s+1)F(s)|_{s\to -1} = \frac{s^2+2s+3}{s^2}\bigg|_{s\to -1} = 2,$$

$$A_{12} = s^2 F(s)|_{s\to 0} = \frac{s^2+2s+3}{s+1}\bigg|_{s\to 0} = 3.$$

By Eq. (7-32),

$$A_{11} = \frac{d}{ds}[s^2 F(s)]_{s\to 0} = \frac{d}{ds}\left(\frac{s^2+2s+3}{s+1}\right)\bigg|_{s\to 0}$$

$$= -\frac{s^2+2s+3}{(s+1)^2} + \frac{2s+2}{s+1}\bigg|_{s\to 0} = -1.$$

The solution is,

$$F(s) = \frac{s^2+2s+3}{s^2(s+1)} = -\frac{1}{s} + \frac{3}{s^2} + \frac{2}{s+1}.$$

The form that arises when the inverse transform of a multiple order pole exists is readily found. Consider the case for a second order pole. This requires the evaluation of the expression,

$$\mathscr{L}^{-1}\left[\frac{A_{p2}}{(s-s_p)^2}\right] = -A_{p2}\mathscr{L}^{-1}\left[\frac{d}{ds}\frac{1}{(s-s_p)}\right]. \tag{7-33}$$

The resulting expression in the time domain is,

$$f(t) = A_2 t e^{s_p t}. \tag{7-34}$$

For the general term, as an extension of the foregoing,

$$\mathscr{L}^{-1}\left[\frac{A_{pk}}{(s-s_p)^k}\right] = A_{pk}\mathscr{L}^{-1}\left[\frac{1}{(k-1)!}\frac{d^{k-1}}{ds^{k-1}}\frac{1}{(s-s_p)}\right],$$

with the result that,

$$f(t) = (-1)^{k-1}\frac{A_{pk}}{(k-1)!} t^{k-1} e^{s_p t}. \tag{7-35}$$

Insofar as the polynomial terms in Eq. (7-25) are concerned, the B_0 term introduces an impulse function, $B_1 s$ introduces a doublet of strength B_1, and higher order terms introduce higher order singularity functions.

Attention is specifically called to the need, in carrying out the above determinations for the residues, for finding the roots of the denominator polynomial. As already discussed, this can be a very unpleasant task, especially if the polynomial is of a higher order than quadratic. This is particularly true if any of the roots are closely spaced. Digital computer routines have been developed which will carry out root finding (*see* Section 5-11 for the Newton-Raphson method for finding real roots), hence some of the drudgery of such an operation can be alleviated. Most computer routines cannot separate the very closely spaced roots, if any exist.

7-7 LINEAR STATE EQUATIONS

It is of some analytic interest to examine the solution of the state equations for linear systems by Laplace transform techniques. The procedure is direct, but due account must be taken of the matrix nature of the equations.

We begin with the general state equations,

$$\dot{x} = Ax + Bu,$$
$$y = Px + Qu. \qquad (7\text{-}36)$$

Consider the first of this set. Upon Laplace transforming this equation, there results,

$$sX(s) - x(0+) = AX(s) + BU(s), \qquad (7\text{-}37)$$

where $x(0+)$ is the initial state vector, a quantity to be discussed in the next section. Equation (7-37) is rearranged to the form,

$$[sI - A]X(s) = x(0+) + BU(s), \qquad (7\text{-}38)$$

with I the unit matrix. We now premultiply this expression by $[sI - A]^{-1}$. The result is,

$$X(s) = [sI - A]^{-1} x(0+) + [sI - A]^{-1} BU(s). \qquad (7\text{-}39)$$

This is conveniently written,

$$X(s) = \Phi(s) x(0+) + \Phi(s) BU(s), \qquad (7\text{-}40)$$

where,

$$\Phi(s) = [sI - A]^{-1}.$$

$\Phi(s)$ is called the *characteristic or state transition matrix* of the system.

To find $x(t)$ from Eq. (7-40), the inverse transform is taken. First consider the case when the sources are zero. In this case Eq. (7-40) becomes,

$$X(s) = \Phi(s)x(0+), \qquad (7-41)$$

and the inverse transform may be written,

$$x(t) = \mathscr{L}[\Phi(s)]x(0+). \qquad (7-42)$$

But this corresponds exactly to the case discussed in Section 6-5, whence it is clear that we may write from Eq. (6-27),

$$\mathscr{L}^{-1}[\Phi(s)] = \Phi(t) = e^{At}, \qquad (7-43)$$

so that,

$$x(t) = \Phi(t)x(0+) = e^{At}x(0+). \qquad (7-44)$$

Example 7-7.1

A system is specified by the matrix,

$$A = \begin{bmatrix} 0 & 1 \\ -\frac{1}{2} & -1 \end{bmatrix}.$$

Determine the response of the system. Note that this is Example 6-5.1, but done by Laplace methods.

Solution. The procedure involves the evaluation of the quantity,

$$[sI - A]^{-1} = \left\{ \begin{bmatrix} s & 0 \\ 0 & s \end{bmatrix} - \begin{bmatrix} 0 & 1 \\ -\frac{1}{2} & -1 \end{bmatrix} \right\}^{-1} = \begin{bmatrix} s & -1 \\ \frac{1}{2} & s+1 \end{bmatrix}^{-1},$$

which is,

$$[sI - A]^{-1} = \frac{\begin{bmatrix} s+1 & 1 \\ -\frac{1}{2} & s \end{bmatrix}}{s(s+1) + \frac{1}{2}} = \begin{bmatrix} \dfrac{s+1}{s^2+s+\frac{1}{2}} & \dfrac{1}{s^2+s+\frac{1}{2}} \\ \dfrac{-\frac{1}{2}}{s^2+s+\frac{1}{2}} & \dfrac{s}{s^2+s+\frac{1}{2}} \end{bmatrix}. \qquad (7-45)$$

The inverse transform of each term in this resulting matrix is evaluated. The terms are,

$$\mathscr{L}^{-1}\left[\frac{s+1}{s^2+s+\frac{1}{2}}\right] = \mathscr{L}^{-1}\left[\frac{s+1}{s^2+s+\frac{1}{4}+(\frac{1}{2})^2}\right] = \mathscr{L}^{-1}\left[\frac{s+1}{(s+\frac{1}{2})^2+(\frac{1}{2})^2}\right],$$

$$= \mathscr{L}^{-1}\left[\frac{(s+\frac{1}{2})+\frac{1}{2}}{(s+\frac{1}{2})^2+(\frac{1}{2})^2}\right] = e^{-(1/2)t}\cos\frac{t}{2} + e^{-(1/2)t}\sin\frac{t}{2},$$

$$= \sqrt{2}\,e^{-(1/2)t}\cos\left(\frac{t}{2}-45°\right),$$

$$\mathcal{L}^{-1}\left[\frac{1}{s^2+s+\frac{1}{2}}\right] = 2\,e^{-(1/2)t}\sin\frac{t}{2},$$

$$\mathcal{L}^{-1}\left[\frac{-\frac{1}{2}}{s^2+s+\frac{1}{2}}\right] = -e^{-(1/2)t}\sin\frac{t}{2},$$

$$\mathcal{L}^{-1}\left[\frac{s}{s^2+s+\frac{1}{2}}\right] = e^{-(1/2)t}\cos\frac{t}{2} - e^{-(1/2)t}\sin\frac{t}{2} = \sqrt{2}\,e^{-(1/2)t}\cos\left(\frac{t}{2}+45°\right).$$

Hence,

$$\Phi(t) = \mathcal{L}^{-1}[s\mathbf{I}-\mathbf{A}]^{-1} = \begin{bmatrix} \sqrt{2}\,e^{-(1/2)t}\cos\left(\frac{t}{2}-\frac{\pi}{4}\right) & 2\,e^{-(1/2)t}\sin\frac{t}{2} \\ -e^{-(1/2)t}\sin\frac{t}{2} & \sqrt{2}\,e^{-(1/2)t}\cos\left(\frac{t}{2}+\frac{\pi}{4}\right) \end{bmatrix}.$$

The solution to the complete expression for $X(s)$ in Eq. (7-40) can be written by Eq. (7-42) and a generalization of the convolution integral, Eq. (7-18). Thus we have,

$$x(t) = \mathcal{L}^{-1}[\Phi(s)x(0+)+\Phi(s)BU(s)], \qquad (7\text{-}46)$$

which is given by the time-domain expression,

$$x(t) = \Phi(t)x(0+) + \int_0^t \Phi(t-\tau)BU(\tau)\,d\tau. \qquad (7\text{-}47)$$

We have here used the fact that,

$$\mathcal{L}^{-1}[\Phi(s)] = \Phi(t),$$

as specified by Eq. (7-43), plus the fact that,

$$\mathcal{L}^{-1}[\Phi(s)BU(s)] = \int_0^t \Phi(t-\tau)BU(\tau)\,d\tau,$$

which is in agreement with the theorem for the Laplace transformation of the convolution of two functions (*see* Eq. (7-18)). Equation (7-47) is also in agreement with Eq. (6-37).

Now refer to the output vector, which is obtained by Laplace transforming the second of Eq. (7-36). This gives,

$$Y(s) = PX(s) + QU(s), \qquad (7\text{-}48)$$

By Eq. (7-40) this may be written,

$$Y(s) = P\Phi(s)x(0+) + [P\Phi(s)B+Q]U(s). \qquad (7\text{-}49)$$

This expression shows that the output is the superposition of the response due to the initial state vector (initial conditions) and that due to the drivers. In the time domain, which is deduced by taking the inverse transform of this expression, we have, using the same procedure as in going from Eq. (7-46) to Eq. (7-47),

$$y(t) = P\Phi(t)x(0+) + \int_0^t [P\Phi(t-\tau)B + Q]u(\tau)\, d\tau. \qquad (7\text{-}50)$$

The real problems in carrying out the solution of the integrals in Eq. (7-47) and Eq. (7-50) have been discussed in some detail in Section 6-5.

An approach to a closed form solution for linear systems is possible, in principle, and we wish to examine this. We first observe that,

$$\Phi(s) = [sI - A]^{-1} = \frac{\text{adjoint } [sI - A]}{\Delta(s)} = \frac{\Delta_a(s)}{\Delta(s)}. \qquad (7\text{-}51)$$

Here $\Delta_a(s)$, the adjoint of the characteristic matrix, is $\Delta_a(s) = [|sI-A|_{ji}] = [|sI-A|_{ij}]^T$ which is the transpose of the cofactor of the ijth element of $|sI - A|$. $\Delta(s) = |sI - A|$ is the determinant of $[sI - A]$, and is called the *characteristic polynomial* of matrix A. The roots of $\Delta(s)$ are the eigenvalues of the system, that is, they are the values of s_j for which,

$$|sI - A| = 0. \qquad (7\text{-}52)$$

This equation is called the *characteristic equation*. In a system of order n, A is an $n \times n$ matrix and the determinant $|sI - A|$ is an nth degree polynomial which may be written,

$$\begin{aligned}|sI - A| &= s^n + a_{n-1}s^{n-1} + a_{n-2}s^{n-2} + \cdots + a_0, \\ &= (s - s_1)^{m_1}(s - s_2)^{m_2}, \ldots, (s - s_m)^{m_m},\end{aligned} \qquad (7\text{-}53)$$

with $m \leq n$, depending on whether repeated roots exist. We note that computing the cofactors of $[sI - A]$ is quite laborious for $n > 3$, owing to the number of calculations involved in evaluating the determinants in both $\Delta_a(s)$ and $\Delta(s)$, a matter to be discussed in Chapter 8. The situation is eased somewhat if we express $\Phi(s)$ explicitly as,

$$\Phi(s) = \frac{\Delta_a(s)}{\Delta(s)} = \frac{B_0 s^{n-1} + B_1 s^{n-2} + \cdots + B_{n-1}}{s^n + a_1 s^{n-1} + \cdots + a_n}, \qquad (7\text{-}54)$$

where $B_0, B_1, B_2, \ldots$, are $n \times n$ matrixes. (These B-s are not related to the B in Eq. (7-36).) This expression arises from the fact that each element of the adjoint matrix in the numerator of Eq. (7-51) is a polynomial of degree

$n-1$. The matrixes B_k and the coefficients a_k can be found by means of the algorithm (14)

$$a_k = -\frac{1}{k} Tr(AB_{k-1}) \quad (k = 1, 2, \ldots, n),$$
$$B_k = AB_{k-1} + a_k I \quad (k = 1, 2, \ldots, (n-1)). \tag{7-55}$$

where Tr denotes the trace (the sum of the diagonal elements) of the matrix. The last relation is B_{n-1} hence for $k = n$, $B_n = 0$, and so $B_n = AB_{n-1} + a_n I = 0$.

A method proposed by Branin (15) possesses some computational advantages. This method is accomplished by using the eigenvalues and the eigenvectors of the matrix A, although, as already discussed, finding these is often a difficult problem, but for computer-size problems, say those requiring a frequency response calculation, the method is very effective. The method is based on finding a linear transformation matrix W such that the eigenvalue matrix is diagonal. The columns of W are the corresponding eigenvectors. This procedure is based on finding a linear matrix transformation,

$$x = WX, \tag{7-56}$$

such that W transforms the state variable X into the state variable x. The derivative of this expression gives,

$$\dot{x} = W\dot{X}. \tag{7-57}$$

These expressions are combined with the state equation Eq. (7-36) to give,

$$W\dot{X} = AWX + Bu. \tag{7-58}$$

Premultiply by W^{-1} to find,

$$\dot{X} = W^{-1}AWX + W^{-1}Bu. \tag{7-59}$$

Note that since this differential equation represents the same equation as that of Eq. (7-36), the characteristic equations must be equal. This may be shown by considering the determinant of Eq. (7-59),

$$\det|W^{-1}AW - sI| = \det|W^{-1}(A - sI)W|,$$
$$= \frac{1}{\det|W|} \det|A - sI| \det|W|, \tag{7-60}$$
$$= \det|A - sI| = 0.$$

This means that the eigenvalues and the eigenvectors of the original differential equation and the transformed differential equation are the same.

A particularly simple and useful state space is one for which the matrix $W^{-1}AW$ contains only diagonal elements†. This space is sometimes called the normal coordinate space, and W is called the normal or diagonalization transformation. In this case,

$$\Lambda = W^{-1}AW, \qquad (7\text{-}61)$$

defines a *similarity transformation*, and the diagonal matrix Λ is called the Jordan matrix. Thus the Jordan matrix formulation uncouples the response due to each eigenvector. This same transformation also diagonalizes $[sI - A]$ giving,

$$W^{-1}[sI - A]W = sI - \Lambda. \qquad (7\text{-}62)$$

This expression is solved for $[sI - A]$ and then inverted. The result is,

$$[sI - A]^{-1} = W[sI - \Lambda]^{-1}W^{-1}.$$

Equation (7-46) now becomes,

$$x(t) = \mathscr{L}^{-1}[W[sI - \Lambda]^{-1}W^{-1}x(0+) + W[sI - \Lambda]^{-1}W^{-1}BU(s)]. \qquad (7\text{-}63)$$

Observe that the problem has now been converted to one which requires the inversion of the diagonal matrix $[sI - \Lambda]$ at each frequency for the initial value terms.

Let us consider the solution of Eq. (7-49) in the light of the foregoing discussion. In the manner of Section 7-6, Eq. (7-49) for $Y(s)$ may be expanded into partial fraction form. This gives,

$$Y(s) = \frac{Y(s)_1}{s - s_1} + \frac{Y(s)_2}{s - s_2} + \cdots,$$

where,

$$Y(s)_i = X_i + U_i,$$

with

$$X_i = \lim_{s \to s_i} [(s - s_i)P\Phi(s)x(0+)] \qquad (i = 1, 2, \ldots, n),$$

$$U_i = \lim_{s \to s_i} [(s - s_i)[P\Phi(s)B + Q]U(s)] \qquad (i = a, b, \ldots, k).$$

†Every $n \times n$ matrix that is real and symmetric or that has n distinct eigenvalues can be diagonalized. Conversely, a matrix that is not real and symmetric and that does not have distinct eigenvalues cannot, in general, be reduced to diagonal form.

The resultant solution will then have the general form,

$$y(t) = \sum_{i=1}^{n} X_i e^{s_i t} + \sum_{j=1}^{k} U_j e^{s_j t}, \qquad (7\text{-}64)$$

where the X_i and U_j are the appropriate residue matrixes for the two factors in the expression for $Y(s)$. This is not a particularly convenient form for numerical solution.

7-8 INITIAL CONDITIONS AND INITIAL STATE VECTORS

A knowledge of the initial state vector is of fundamental importance in the solution of the state equations. This is found from specified initial conditions, and hence a means is necessary for relating the initial conditions to the initial state vector. We discuss this matter in terms of a particular example.

We consider specifically a system which is described by the following state equations,

$$\begin{aligned}&\text{a. } \dot{x}_1 = -2x_1 + x_2, \\ &\text{b. } \dot{x}_2 = -3x_1 + x_3 + u, \\ &\text{c. } \dot{x}_3 = -5x_3 + 3u, \\ &\text{d. } y = x_1,\end{aligned} \qquad (7\text{-}65)$$

subject to a set of specified initial conditions $y(0), \dot{y}(0), \ddot{y}(0)$. These translate by Eq. (7-65d) into the equivalent conditions,

$$y(0) = x_1(0) \qquad \dot{y}(0) = \dot{x}_1(0) \qquad \ddot{y}(0) = \ddot{x}_1(0). \qquad (7\text{-}66)$$

When these are combined with Eq. (7-65a) we find that,

$$\begin{aligned} x_2(0) &= \dot{x}_1(0) + 2x_1(0), \\ &= \dot{y}(0) + 2y(0), \end{aligned} \qquad (7\text{-}67)$$

which thus leads to,

$$\dot{x}_2(0) = \ddot{y}(0) + 2\dot{y}(0). \qquad (7\text{-}68)$$

Now combine Eq. (7-68) and (7-66) with Eq. (7-65b) to get, omitting u, which is not involved,

$$x_3(0) = \dot{x}_2(0) + 3x_1(0) = \ddot{y}(0) + 2\dot{y}(0) + 3y(0). \qquad (7\text{-}69)$$

From these we write finally,

$$\begin{aligned} x_1(0) &= y(0), \\ x_2(0) &= \dot{y}(0) + 2y(0), \\ x_3(0) &= \ddot{y}(0) + 2\dot{y}(0) + 3y(0). \end{aligned} \qquad (7\text{-}70)$$

Another approach to establishing the initial state vector begins by inserting the initial conditions as step functions at each state variable node. This idea follows directly from the manner in which the initial conditions appear in the Laplace transform for the integral function. The essence of the method is illustrated in Fig. 7-2 which shows a SFG appropriate to the set of state equations,

$$\begin{aligned} \dot{x}_1 &= -x_1 + u, \\ \dot{x}_2 &= -2x_2 + u, \\ \dot{x}_3 &= -4x_3 + u, \\ y &= 3x_1 + 2x_2 - 5x_3. \end{aligned} \quad (7\text{-}71)$$

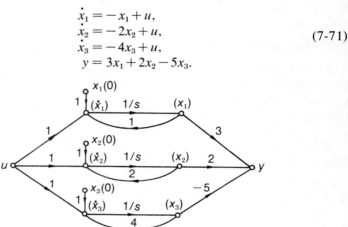

Fig. 7-2. A SFG representation of Eq. (7-71).

Observe that the initial conditions, which are step functions, are shown as being generated by the application of impulses of strength $x_j(0)$ through integrators to each node.

The output function corresponding to this set of state equations is,

$$Y(s) = \frac{3}{s+1} + \frac{2}{s+2} - \frac{5}{s+4} U(s) + \frac{3x_1(0)}{s+1} + \frac{2x_2(0)}{s+2} - \frac{5x_3(0)}{s+4}. \quad (7\text{-}72)$$

Now we observe that at time $t = 0 - U(s) = 0$, so that,

$$Y(s)|_{0-} = \frac{3x_1(0)}{s+1} + \frac{2x_2(0)}{s+2} - \frac{5x_3(0)}{s+4}. \quad (7\text{-}73)$$

Each term is expanded by long division, with the result that,

$$\begin{aligned} Y(s)|_{0-} = &\, 3\left(\frac{1}{s} - \frac{1}{s^2} + \frac{1}{s^3} - \cdots\right)x_1(0) + 2\left(\frac{2}{s} - \frac{4}{s^2} + \frac{8}{s^3} - \cdots\right)x_2(0) \\ &+ 5\left(-\frac{4}{s} + \frac{16}{s^2} - \frac{64}{s^3} + \cdots\right)x_3(0). \end{aligned}$$

Now we observe that the coefficients of $1/s$ determines $y(0)$, those of $1/s^2$ determines $\dot{y}(0)$, and those of $1/s^3$ determines $\ddot{y}(0)$. Thus we may write,

$$y(0) = 3x_1(0) + 4x_2(0) - 20x_3(0),$$
$$\dot{y}(0) = -3x_1(0) - 8x_2(0) + 80x_3(0), \qquad (7\text{-}74)$$
$$\ddot{y}(0) = 3x_1(0) + 16x_2(0) - 320x_3(0).$$

This is a set of algebraic equations in the quantities $x_1(0)$, $x_2(0)$ and $x_3(0)$. These may be inverted to give each $x_j(0)$ in terms of $y(0)$, $\dot{y}(0)$ and $\ddot{y}(0)$, thereby specifying the initial state vector $x(0)$.

PROBLEMS

7-1. Find the Laplace transform of the following:
Table 7-1, entries 5, 7, 8, 11, 13;

$$t^2 + 2t + 3; \qquad \frac{1 + \cos 4t}{2}; \qquad t \sin \omega t; \qquad \frac{\sin \omega t}{\omega t}.$$

7-2. Find the Laplace transforms of the functions illustrated.

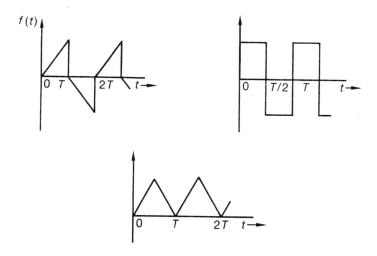

7-3. Prove that the residues (the constants in the partial fraction expansion) may be written,

$$A_k = \left[(s - s_k)\frac{P(s)}{Q(s)}\right]_{s=s_k} = \left[\frac{P(s)}{(dQ(s)/ds)}\right]_{s=s_k}.$$

7-4. Find the inverse Laplace transforms of the following functions:

$$\frac{1}{s(s+3)}; \quad \frac{s}{(s+2)(s+3)}; \quad \frac{s^2}{(s-1)^2}; \quad \frac{s+1}{(s+2)(s+3)^2}; \quad \frac{3s+2}{s(s^2+2s+5)};$$

$$\frac{2s+3}{s^2+5s+2}; \quad \frac{s}{(s^2-\alpha^2)}; \quad \frac{s}{s^2+\alpha^2}; \quad \frac{s^2+2s+3}{s^2(s+1)}; \quad \frac{s^2+2s+3}{s^2+3s+7}.$$

7-5. Solve the following differential equations by means of Laplace techniques:

$$\frac{d^2y}{dt^2} + 3\frac{dy}{dt} + 2y = t^2 + 2t \qquad y(0) = 2, \quad \frac{dy}{dt}(0+) = -15,$$

$$\frac{d^2y}{dt^2} + 3y = \sin t + e^{-t} \qquad \text{initial conditions zero,}$$

$$\frac{d^2y}{dt^2} + 2\frac{dy}{dt} + 3y = 0 \qquad y(0+) = 1, \quad \frac{dy}{dt}(0+) = 0.$$

7-6. Determine the impulse response of the following differential equations:

$$\frac{d^2y}{dt^2} + 3\frac{dy}{dt} + 2y = u_0(t) \qquad \text{initially relaxed,}$$

$$\frac{d^3y}{dt^3} + 2\frac{d^2y}{dt^2} + 2\frac{dy}{dt} + y = u_0(t) \qquad y(0+) = 1, \quad \frac{dy}{dt}(0+) = \frac{d^2y}{dt^2}(0+) = 0.$$

7-7. a. Find the solution to the differential equation,

$$\frac{d^2y}{dt^2} + 3\frac{dy}{dt} + 2y = x(t) = u_{-1}(t),$$

with the system initially relaxed.

b. Use the solution deduced in a. to find the solution to the following differential equations:

$$\frac{d^2y}{dt^2} + 3\frac{dy}{dt} + 2y = \frac{dx}{dt},$$

$$\frac{d^2y}{dt^2} + 3\frac{dy}{dt} + 2y = 2\frac{dx}{dt} + 3x.$$

7-8. a. Find the solution to the following differential equation, assuming the system to be initially relaxed,

$$\frac{dy}{dt} + 2y = 3.$$

b. Use the result of a. to find the solution to the differential equation,

$$\frac{d^2y}{dt^2} + 2\frac{dy}{dt} = 0.$$

c. Write b. in state equation form that uses the form in a. as one state equation. Write the solution to these state equations from a. and b.

7-9. Find the impulse response of the system that is specified by the SFG shown.

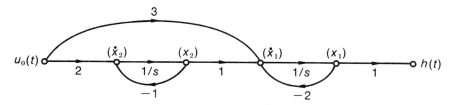

7-10. Determine the driving point current in the following circuits. Assume that these circuits are initially relaxed.

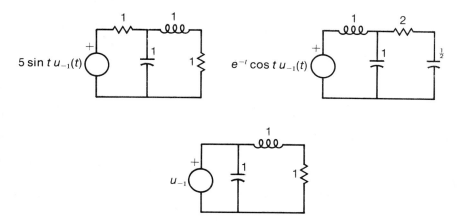

7-11. The switch S is initially closed, and the system is in its steady-state. Find $i(t)$ and $v(t)$ after the switch is opened.

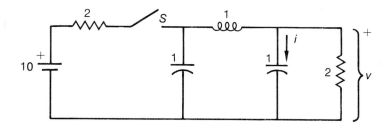

7-12. Determine the current $i(t)$ in the network shown, subject to the following switching sequence: S_1 closed at time $t = 0$, S_2 closed at time $t = 3$ sec.

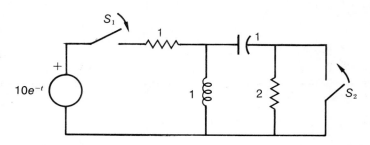

7-13. Refer to the accompanying network,

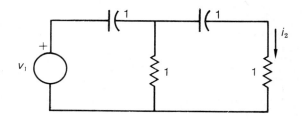

a. Find the transfer function $I_2(s)/V_1(s)$.

b. Determine the impulse response, if the network is initially relaxed.

c. Find an expression for $i_2(t)$ if the capacitors have initial charges $q_1(0)$ and $q_2(0)$.

7-14. Find the initial and the final values of the given functions. Check these results by reference to the time functions,

$$\frac{1}{s(s+a)}; \quad \frac{s}{(s+a)^2}; \quad \frac{1}{(s+a)^2+b^2}; \quad \frac{1}{s^2+a^2}.$$

7-15. Apply the initial and the final value theorems to determine the values $f(0)$ and $f(\infty)$ for the following Laplace functions,

$$\frac{1}{s(s^2+2\zeta\omega_n s+\omega_n^2)} \quad \frac{s^2+4s+2}{s(s^3+3s^2+3s+1)}.$$

7-16. Determine the inverse of the matrix $[sI - A]$, where $[A]$ is the matrix

$$\begin{bmatrix} 3 & 2 & 7 \\ 1 & 5 & 6 \\ -4 & -1 & -3 \end{bmatrix}.$$

a. By direct expansion.

b. By the use of the algorithm of Eq. (7-54).

7-17. Determine the eigenvalues of the systems described by the following A matrixes,

$$\begin{bmatrix} 0 & 1 \\ 1 & 0 \end{bmatrix} \quad \begin{bmatrix} 1 & -1 & 0 \\ -1 & 1 & -1 \\ 0 & -1 & 1 \end{bmatrix} \quad \begin{bmatrix} 1 & 2 & 1 \\ 0 & 3 & 7 \\ 4 & 1 & 5 \end{bmatrix}.$$

7-18. Given a system for which,

$$A = \begin{bmatrix} 0 & 1 \\ -2 & -3 \end{bmatrix}.$$

a. Find the eigenvalues of this system.

b. Find a transformation such that $W^{-1}AW$ is the diagonal matrix

$$\Lambda = \begin{bmatrix} -1 & 0 \\ 0 & -2 \end{bmatrix}.$$

8

s-Plane: Poles and Zeros

Our studies in Chapter 7 have shown that to apply the Laplace transform techniques to the solution of system equations requires that we Laplace transform each term in the describing equations. In this process initial value terms appear which account for initial conditions. In this chapter we shall be largely concerned with transformed equations for the driving function with all initial value terms set equal to zero. This is the case for the initially relaxed system. This process naturally leads to the *system function*, a quantity that we wish to study in some detail.

8-1 THE SYSTEM FUNCTION

To set the stage for a general discussion of the system function, we examine the case of simple circuits involving only a single electrical element. We write the instantaneous through-across relationships for these, and also show their Laplace transformed forms. These are the following,

$$\begin{aligned} v_R &= Ri & V_R(s) &= RI(s), \\ v_c &= \frac{1}{C}\int i\, dt & V_c(s) &= \frac{I(s)}{Cs}, \\ v_L &= L\frac{di}{dt} & V_L(s) &= LsI(s). \end{aligned} \quad (8\text{-}1)$$

We now define a *system function* $T(s)$ as denoting the Laplace transform of any output/input function. This may be a driving point or a transfer relationship. For the simple elements with the voltage as the excitation function and current as response, we now define the system functions of these simple circuits as,

$$T_R(s) = \frac{\text{response}}{\text{excitation}} = R,$$
$$T_c(s) = \frac{1}{Cs}, \qquad (8\text{-}2)$$
$$T_L(s) = Ls.$$

Now let us refer to Eq. (5-2), which is the general form of the differential equation relating response to excitation,

$$a_n \frac{d^n y}{dt^n} + a_{n-1} \frac{d^{n-1} y}{dt^{n-1}} + \cdots + a_0 y = b_m \frac{d^m u}{dt^m} + \cdots + b_0 u. \qquad (8\text{-}3)$$

The coefficients a_k and b_k are real constants, y denotes the response and u denotes the excitation, both of which are functions of time. When this equation is Laplace transformed, for zero initial conditions, we have that

$$(a_n s^n + a_{n-1} s^{n-1} + \cdots + a_0) Y(s) = (b_m s^m + \cdots + b) U(s). \qquad (8\text{-}4)$$

The system function is,

$$T(s) = \frac{\text{response}}{\text{excitation}} = \frac{b_m s^m + b_{m-1} s^{m-1} + \cdots + b_0}{a_n s^n + a_{n-1} s^{n-1} + \cdots + a_0} = \frac{P(s)}{Q(s)}. \qquad (8\text{-}5)$$

For a physical system $T(s)$ is the ratio of two rational polynomials in s. Attention is called to the fact that the response that is being measured may be at any one of a number of points in a general system; furthermore, the excitation may be applied at any one or more points of the system. Because of this, there are many $T(s)$ functions possible in a given system, as already noted. $T(s)$ in any particular case applies to the particular system and with specified ports for the excitation and the system response.

A formal feature of the system function, and this is particularly evident when Eq. (8-4) is compared with Eq. (8-3), and especially if Eq. (8-3) is written in terms of the operator $p = d/dt$. We see that in the Laplace transformation the operator p is replaced by the complex variable s, but otherwise the general form of the dynamic equation is retained in the transformation.

From an operational point of view $T(s)$ plays a key role in the solution of systems problems. That this is so follows from Eq. (8-4) which is written,

$$Y(s) = T(s) U(s), \qquad (8\text{-}6)$$

312 s-Plane: Poles and Zeros

and from which, by the inversion process,

$$y(t) = \mathscr{L}^{-1}[T(s)U(s)]. \tag{8-7}$$

As already discussed, carrying out this inversion process can be accomplished in general by:

a. A partial fraction expansion.
b. Convolution techniques.
c. Integration in the complex plane.

In our subsequent development we shall find that $T(s)$ is fundamental to the total system response; to the development of special theorems for the numerical solution of network behavior; and to the studies of the stability of a system.

8-2 IMPEDANCE AND ADMITTANCE FUNCTIONS

As already noted above, a variety of system functions exist in a given system, depending upon where the excitation is injected and the terminals which are to be designated as the output port. For example, for the series RLC circuit and its dual, as shown in Fig. 8-1, we write the *driving point* equations,

$$Ri + L\frac{di}{dt} + \frac{1}{C}\int i\,dt = v(t),$$

$$Gv + C\frac{dv}{dt} + \frac{1}{L}\int v\,dt = i(t). \tag{8-8}$$

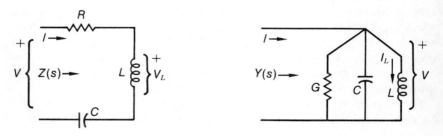

Fig. 8-1. The simple series RLC circuit and its dual.

Impedance and Admittance Functions

By Laplace transforming these equations and then solving, we write,

$$T(s) = Z(s) = R + Ls + \frac{1}{Cs}$$
$$T(s) = Y(s) = G + Cs + \frac{1}{Ls},$$
(8-9)

which are essentially the $T(s)$ functions for these cases, although for driving point functions it is customary to use the terms driving point impedance function or driving point admittance function, as appropriate.

It is not necessary, of course, that only driving point functions be considered. Refer to Fig. 8-2 which shows a one-port and a two-port system. As discussed above for the one-port network, two system functions are of interest

V/I driving point impedance.
I/V driving point admittance.

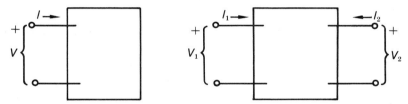

Fig. 8-2. General one-port and two-port systems.

For the two-port, a number of functions of possible interest exist, including,

V_1/I_1 driving point impedance at port 1.
I_1/V_1 driving point admittance at port 1.
V_2/I_2 driving point impedance at port 2.
I_2/V_2 driving point admittance at port 2.
V_1/I_2 transfer impedance from port 1 to port 2.
I_2/V_1 transfer admittance from port 2 to port 1.
V_2/I_1 transfer impedance from port 2 to port 1.
I_1/V_2 transfer admittance from port 1 to port 2.
V_2/V_1 voltage ratio from port 2 to port 1 (often called voltage gain).
V_1/V_2 voltage ratio from port 1 to port 2.
I_2/I_1 current ratio from port 2 to port 1 (often called current gain).
I_1/I_2 current ratio from port 1 to port 2.

314   s-Plane: Poles and Zeros

For an n-port system with voltages and currents defined at each port, a large variety of different system functions may be defined. In general, all will be of the general form given in Eq. (8-5).

Example 8-2.1

Each of the boxes shown are general combinations of electrical elements. Find the driving point characteristics of each.

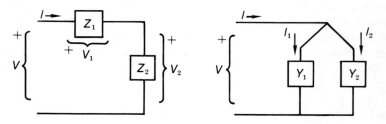

Fig. 8-2.1-1

Solution. By a simple extension of Eq. (8-9) we write, for the two circuits,

$$V = V_1 + V_2 \quad \text{series circuit,}$$
$$I = I_1 + I_2 \quad \text{parallel circuit.}$$

It follows from these that,

$$Z = V/I = Z_1 + Z_2 \quad \text{series circuit,}$$
$$Y = I/V = Y_1 + Y_2 \quad \text{parallel circuit.}$$

Example 8-2.2

Find the driving point impedance of the given network.

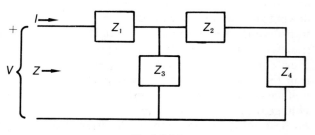

Fig. 8-2.2-1

Solution. From the simple combinatorial features of networks, as shown in Example 8-2.1, we write,

$$Z = Z_1 + \frac{1}{(1/Z_3) + (1/Z_2 + Z_4)},$$

from which,

$$Z = Z_1 + \frac{Z_3(Z_2 + Z_4)}{Z_2 + Z_3 + Z_4}.$$

Example 8-2.3

Find the voltage gain ratio V_2/V for the network shown. Express the results in the form of Eq. (8-6).

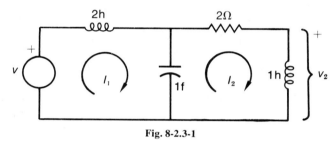

Fig. 8-2.3-1

Solution. An application of the KVL to the two loops leads to the equations in Laplace form,

$$V = \left(2s + \frac{1}{s}\right)I_1 - \frac{1}{s}I_2,$$

$$0 = -\frac{1}{s}I_1 + \left(s + 2 + \frac{1}{s}\right)I_2.$$

Further, since $V_2 = LsI_2$, then we find I_2 from these equations, and then find V_2. We find I_2 by an application of Cramer's rule for the solution of simultaneous equations. This is,

$$I_2 = \frac{\begin{vmatrix} \left(2s + \frac{1}{s}\right) & V \\ -\frac{1}{s} & 0 \end{vmatrix}}{\begin{vmatrix} \left(2s + \frac{1}{s}\right) & -\frac{1}{s} \\ -\frac{1}{s} & \left(s + 2 + \frac{1}{s}\right) \end{vmatrix}} = \frac{\frac{1}{s}V}{\left(2s + \frac{1}{s}\right)\left(s + 2 + \frac{1}{s}\right) - \frac{1}{s^2}}.$$

From this we write for V_2,

$$V_2 = LsI_2 = \frac{V}{\left(2s+\dfrac{1}{s}\right)\left(s+2+\dfrac{1}{s}\right) - \dfrac{1}{s^2}}.$$

By expansion, this expression becomes,

$$\frac{V_2}{V} = T(s) = \frac{s}{2s^3 + 4s^2 + 3s + 2}.$$

8-3 SYSTEM DETERMINANTS

Example 8-2.3 shows the general method for determining a specified system function. In general, the method employs the application of the KCL or the KVL, depending on which is appropriate to the network, writing the set of simultaneous equations that describes the dynamics of the system, and then carrying out solutions for the prescribed functions. The general equations are, for the loop network, upon applying the KVL as per Eq. (4-45) and then Laplace transforming the equations,

$$\begin{aligned} Z_{11}(s)I_1 + Z_{12}(s)I_2 + \cdots + Z_{1n}(s)I_n &= V_1 \\ Z_{21}(s)I_1 + Z_{22}(s)I_2 + \cdots + Z_{2n}(s)I_n &= V_2 \\ &\vdots \\ Z_{n1}(s)I_1 + Z_{n2}(s)I_2 + \cdots + Z_{nn}(s)I_n &= V_n. \end{aligned} \qquad (8\text{-}10)$$

Correspondingly for the node-pair network, upon applying the KCL, the equation set is, from Eq. (4-33),

$$\begin{aligned} Y_{11}(s)V_1 + Y_{12}(s)V_2 + \cdots + Y_{1n}(s)V_n &= I_1 \\ Y_{21}(s)V_1 + Y_{22}(s)V_2 + \cdots + Y_{2n}(s)V_n &= I_2 \\ &\vdots \\ Y_{n1}(s)V_1 + Y_{n2}(s)V_2 + \cdots + Y_{nn}(s)V_n &= I_n. \end{aligned} \qquad (8\text{-}11)$$

For our present purposes, we shall limit our discussion to the set Eq. (8-10).

To find the various currents $I_1, I_2, \ldots, I_n$ requires that the set of algebraic equations be solved simultaneously. One of the standard methods, and this was used in Example 8-2.3, is to use Cramer's rule, by which we can write for each I_k,

$$I_k = \frac{\Delta_k}{\Delta} = \frac{\begin{vmatrix} Z_{11} & Z_{12} & \cdots & V_1 & \cdots & Z_{1n} \\ Z_{21} & Z_{22} & \cdots & V_2 & \cdots & Z_{2n} \\ \cdot & \cdot & & \cdot & & \cdot \\ \cdot & \cdot & & \cdot & & \cdot \\ \cdot & \cdot & & \cdot & & \cdot \\ Z_{n1} & Z_{n2} & \cdots & V_n & \cdots & Z_{nn} \end{vmatrix}}{\begin{vmatrix} Z_{11} & Z_{12} & \cdots & Z_{1k} & \cdots & Z_{1n} \\ Z_{21} & Z_{22} & \cdots & Z_{2k} & \cdots & Z_{2n} \\ \cdot & \cdot & & \cdot & & \cdot \\ \cdot & \cdot & & \cdot & & \cdot \\ \cdot & \cdot & & \cdot & & \cdot \\ Z_{n1} & Z_{n2} & \cdots & Z_{nk} & \cdots & Z_{nn} \end{vmatrix}} \quad (8\text{-}12)$$

Observe that Δ is the determinant of the coefficients of the equations in Eq. (8-10), and Δ_k is the determinant with the kth row in Δ replaced by the terms on the right hand side of the equations, as shown. What we wish to examine at this time is the number of multiplications that are required in carrying out the expansion of the determinants specified in Eq. (8-12). We observe that the coefficients of each term may be complex, in general. But in the multiplications of two complex numbers of the form $(a+jb)(c+jd)$ four real multiplications are required. Hence we initially assume that all of the coefficients in Eq. (8-12) are real numbers.

Now consider the determinant Δ, which is of order $n \times n$. We consider the Laplace expansion method, and expand along the first column to get,

$$\Delta = Z_{11}\Delta_{11} - Z_{12}\Delta_{12} + Z_{13}\Delta_{13} + \cdots + Z_{1n}\Delta_{1n},$$

where Δ_{ij} is the cofactor of the jth row and kth column. This equation requires n multiplications. Now it is observed that each Δ_{jk} is one order less than Δ_k. Again expand the determinants in the manner just noted. This will reduce the order of the appropriate cofactors by one less than Δ_{jk}, and this process will have required $n-1$ multiplications. If this process is continued, we shall find that the total number of multiplications in expanding a determinant of order n by the Laplace method is,

$$n(n-1)(n-2)\ldots 2.1 = n!.$$

For the n currents I_k, we require the solution of n determinants Δ_k, plus Δ a total of $n+1$ determinants; this requires a total of $(n+1)n!$ multiplications. Further, each term in the $n!$ factors requires $(n-1)$ multiplications. Thus the total number of individual multiplications is $(n+1)n!(n-1) =$

$(n+1)!(n-1)$ if the coefficients of the set of equations are real, or up to 4 times this number if they are all complex. For a system of equations with $n = 10$, the number of multiplications for real coefficients is $11! \times 9 = 359{,}251{,}200$. This calls for an enormous amount of arithmetic, and one therefore seeks methods which will require fewer calculations. Fortunately, methods which require less arithmetic do exist.

Two classes of techniques are important in methods for solving sets of linear equations. One class of direct methods stems from the Gauss elimination technique. The standard triangular decomposition by Gaussian elimination requires on the order of $n^3/3$ multiplication and addition operations. The second class arises from iterative methods, and is of particular value when the coefficient matrixes of the system contain a number of zero elements—the situation that arises when one seeks to solve the finite difference approximation to Laplace's equation in field problems. These so-called sparse matrix decomposition methods require kn operations, where k depends on the nonzero structure of the circuit. Empirical results for typical 20–30 node integrated circuit amplifier designs vary from $4n$ to $16n$ operations.

The direct methods, while general in application, involve considerable arithmetic, especially for large order systems. The order of elimination of variables affects the amount of computation (particularly in a sparse matrix) and also the accuracy of computation. Further, the influence of round-off errors may be undesirably large, although this effect can be eliminated by using a pivotal strategy. We shall discuss this procedure.

For convenience, suppose that the set of simultaneous equations to be solved is written in matrix form,

$$[AX] = [B]. \qquad (8\text{-}13)$$

In the Gauss reduction process for the solution of the values of x, elementary row transformations are performed successively on the rows (adding a multiple of a row to some other row, interchanging two rows, multiplying a row by a constant, etc.). The aim of this procedure is to convert the series of equations given in Eq. (8-13) into an upper triangular form, as shown,

$$\begin{bmatrix} a'_{11} & a'_{12} & a'_{13} & \cdots & a'_{1n} \\ 0 & a'_{22} & a'_{23} & \cdots & a'_{2n} \\ 0 & 0 & a'_{33} & & a'_{3n} \\ \cdot & \cdot & \cdot & \cdot & \cdot \\ \cdot & \cdot & \cdot & & \cdot \\ \cdot & \cdot & \cdot & & \cdot \\ 0 & 0 & 0 & \cdots & a'_{nn} \end{bmatrix} \begin{bmatrix} x_1 \\ x_2 \\ x_3 \\ \cdot \\ \cdot \\ \cdot \\ x_n \end{bmatrix} = \begin{bmatrix} y'_1 \\ y'_2 \\ y'_3 \\ \cdot \\ \cdot \\ \cdot \\ y_n \end{bmatrix}, \qquad (8\text{-}14)$$

from which we immediately conclude, from the last equation, that $x_n = y'_n/a'_{nn}$. Back substitution permits us to find $x_{n-1}, x_{n-2}, \ldots, x_1$. A total of approximately $4n^3/3$ multiplications are involved in this process. The procedure including pivotal strategy, is best illustrated by an example.

Example 8-3.1†

Solve the following system of simultaneous equations:

$$x_1 + 4x_2 + 7x_3 = 2,$$
$$-3x_1 + 6x_2 + 2x_3 = 1,$$
$$2x_1 + 3x_2 + 4x_3 = 3.$$

Solution. Successive transformations are made, as indicated:
Stage I — First Column

$$\begin{vmatrix} 1 & 4 & 7 & 2 \\ \ominus 3 & 6 & 2 & 1 \\ 2 & 3 & 4 & 3 \end{vmatrix}$$
Coefficient matrix. First pivot is the coefficient with the largest absolute value in the first column

$$\begin{vmatrix} -3 & 6 & 2 & 1 \\ 1 & 4 & 7 & 2 \\ 2 & 3 & 4 & 3 \end{vmatrix}$$
Interchange first row with second (pivotal) row, so as to bring the pivotal element into diagonal position.

Stage II — Second Column

$$\begin{vmatrix} 1 & -2 & -\tfrac{2}{3} & -\tfrac{1}{3} \\ 0 & 6 & \tfrac{23}{3} & \tfrac{7}{3} \\ 0 & 7 & -\tfrac{16}{3} & \tfrac{11}{3} \end{vmatrix}$$
a. Divide pivotal row by pivot
b. Row 2 − (1 × modified Row 1)
c. Row 3 − (2 × modified Row 1)

$$\begin{vmatrix} 1 & -2 & -\tfrac{2}{3} & -\tfrac{1}{3} \\ 0 & \bigcirc{7} & \tfrac{16}{3} & \tfrac{11}{3} \\ 0 & 6 & \tfrac{23}{3} & \tfrac{7}{3} \end{vmatrix}$$
Interchange second row with third (pivotal) row.

Stage III — Third Column

$$\begin{vmatrix} 1 & 0 & \tfrac{6}{7} & \tfrac{5}{7} \\ 0 & 1 & \tfrac{16}{21} & \tfrac{11}{21} \\ 0 & 0 & \bigcirc{\tfrac{65}{21}} & -\tfrac{17}{21} \end{vmatrix}$$
a. Divide pivotal row by pivot
b. Row 1 − (−2 × modified Row 2)
c. Row 2 − (6 × modified Row 2)

$$\begin{vmatrix} 1 & 0 & 0 & \tfrac{61}{65} \\ 0 & 1 & 0 & \tfrac{47}{65} \\ 0 & 0 & 1 & -\tfrac{17}{65} \end{vmatrix}$$
a. Divide pivotal row by pivot
b. Row 1 − ($\tfrac{6}{7}$ × modified Row 3)
c. Row 2 − ($\tfrac{16}{21}$ × modified Row 3).

†IBM Student Text, "FORTRAN in the Physical Sciences", pg. 42, 1965.

The solution is,

$$x_1 = \tfrac{61}{65}, \qquad x_2 = \tfrac{47}{65}, \qquad x_3 = -\tfrac{17}{65}.$$

The Gauss procedure would seemingly be stopped if $a_{jj} = 0$ since this would require division by zero. The difficulty is avoided by simply permuting the jth column with one having a nonzero entry in the jth row. If the entire jth row is zero, then either, (a) the equations are inconsistent, if $y_j \neq 0$, or (b) if $y_j = 0$, the jth equation can be eliminated, since it is a $0 = 0$ identity, thereby resulting in $n-1$ equations in n unknowns, thus not leading to a unique solution.

The foregoing procedure may be modified slightly to indicate the effects of roundoff error when computer solutions are obtained. This consists in adding an extra column c in the computation of A, X, c, where the ith component of c is the negative of the algebraic sum of the elements in the ith row. Clearly, the row sums would be zero. Now the operations are carried out on the expanded matrix. The row sums, after completion of the final stage, are an indication of the influence of the roundoff error during the computation.

A meaningful estimate of the computational effort using the Gauss method is difficult when the entries of the matrix are rational polynomials in s. In the case of an RLC network with all three types of elements between all nodes, there will be $(3^{n-1})n!$ multiplications for each determinant $\Delta(s)$ when the Laplace expansion method is used, as against $4n^3/3$ multiplications for each complex Gauss reduction. For a frequency run involving new numbers, a new Gauss reduction process is required. The two calculations will be equal, for the calculation at N frequencies, when,

$$N \frac{(4n^3)}{3} = 3^{n-1} n!.$$

For the usual values of N (≤ 100), the repeated complex Gauss reduction is more economical in multiplications for $n \geq 5$. For two element kind networks (RL, RC, LC) the Gauss method is superior for $n \geq 7$.

8-4 THE s-PLANE

As already discussed, $T(s)$ is an explicit function of the complex variable s, and, as shown in Eq. (8-5), is the ratio of two polynomials in s. Consequently, for a specified value of s, $T(s)$ has a specific value. Before proceeding with a study of the properties of $T(s)$, we shall examine some of the features of complex numbers.

A complex number is one which is of the form,

$$s = a + jb, \qquad (8\text{-}15)$$

where both a and b are real numbers, and $j = \sqrt{-1}$. a is known as the real part of s, and b is the imaginary part of s, and these may be written,

$$a = Re\, s,$$
$$b = Im\, s.$$

When referred to a rectangular coordinate system with the X axis being the axis of real, and the Y axis being the axis of imaginaries, the complex number s has a unique representation as a point in this plane, as shown in Fig. 8-3. Conversely, we note that to every point of the complex plane, there corresponds a complex number.

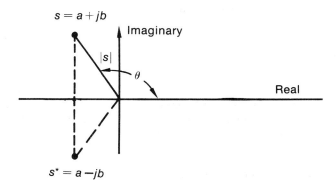

Fig. 8-3. The complex s-plane and the representation of a complex number.

The representation for s in Eq. (8-15) is called the algebraic form. A polar form is also possible, as illustrated in Fig. 8-3, which is written,

$$s = |s| e^{j\theta}, \qquad (8\text{-}16)$$

where $|s|$ is the modulus or magnitude of the complex number, and θ is the argument or angle of s (often written as arg s) measured in radians relative to the axis of reals.

We note that to every complex number $s = a + jb$ there is a complex number $s^* = a - jb$ (note that the sign of the imaginary term has been reversed). The number s^* is known as the conjugate complex (or complex conjugate) of the number s. Certain properties relating to s and s^* are evident from Fig. 8-3. These are seen to be,

$$|s| = |s^*| = \sqrt{a^2 + b^2},$$
$$\theta = \tan^{-1}\frac{b}{a} = \arg s = -\arg s^*, \qquad (8\text{-}17)$$
$$Re\, s = Re\, s^* = \tfrac{1}{2}(s + s^*),$$
$$Im\, s = -Im\, s^* = \tfrac{1}{2}(s - s^*).$$

From the foregoing properties, we add the following, which are evident:

a. The sum of complex numbers,

$$s_1 + s_2 = (a_1 + a_2) + j(b_1 + b_2). \qquad (8\text{-}18)$$

b. The distance between points s_1 and s_2,

$$|s_1 - s_2| = [(a_1 - a_2)^2 + (b_1 - b_2)^2]^{1/2}. \qquad (8\text{-}19)$$

c. The product of two complex numbers,

$$s_1 s_2 = (a_1 a_2 - b_1 b_2) + j(a_1 b_2 + a_2 b_1), \qquad (8\text{-}20)$$

and in polar form,

$$s_1 s_2 = s_1 e^{j\theta_1} s_2 e^{j\theta_2} = s_1 s_2 e^{j(\theta_1 + \theta_2)}.$$

A simple generalization of this is possible, namely, that the product of a number of complex numbers is the product of the magnitudes and the sum of the arguments.

8-5 T(s) AND ITS POLE-ZERO CONSTELLATION

As discussed in connection with Eq. (8-5) for $T(s)$, the system function is given as a rational algebraic function of the complex variable s, and consists of the ratio of two polynomials. Here m and n are integers and the coefficients a and b are real numbers, since they arise from the element parameters RLC, MKD, etc. It is possible to write each of the polynomials in factored form, with the numerator polynomial factored into its m roots and the denominator polynomial factored into its n roots. In the case of higher order polynomials, finding the roots may be a difficulty, although digital computer programs are available for carrying out the root-finding operation (see Section 5-12). When the polynomials are factored, $T(s)$ assumes the form,

$$T(s) = \frac{P(s)}{Q(s)} = H\frac{(s - s_1)(s - s_2)\cdots(s - s_m)}{(s - s_a)(s - s_b)\cdots(s - s_n)}. \qquad (8\text{-}21)$$

The factor $H = b_m/a_n$ is the scale factor. The roots of the numerator polynomial are numbered $s_1, s_2, s_3, \ldots, s_m$ and are the zeros of the system function. Correspondingly, the roots of the denominator polynomial are numbered $s_a, s_b, \ldots, s_n$ and are the poles of the system function. The poles and zeros of $T(s)$ are complex numbers in general. At each zero the numerator polynomial $P(s)$ is zero, and $T(s)$ is zero. Correspondingly, at each pole, the denominator polynomial $Q(s)$ is zero, and $T(s)$ becomes infinite.

The zeros and poles of $T(s)$ possess some important properties. As already noted, the coefficients a and b must be real, although the zeros and poles are complex, in general. This requires that for each complex zero, say s_z, there must be a corresponding conjugate complex zero s_z^*; similarly complex poles s_p must be accompanied by conjugate complex poles s_p^*. That is, complex roots must occur in complex conjugate pairs. In more precise form, Eq. (8-21) will show the complex conjugate roots explicitly, thus,

$$T(s) = H \frac{(s - s_1^*)(s - s_1)(s - s_2)(s - s_2^*) \cdots}{(s - s_a^*)(s - s_a)(s - s_b)(s - s_b^*) \cdots}. \tag{8-22}$$

$T(s)$ is often represented graphically. This is done by plotting the zeros and poles in the s-plane. It is customary to represent the zeros by small circles and the poles by small crosses. As a result, the p-z pattern of a given system function will appear as a group of circles and crosses at appropriate points in the s-plane, as illustrated in Fig. 8-4.

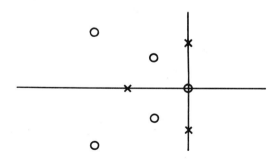

Fig. 8-4. A typical pole-zero constellation.

The numerical value of $T(s)$ corresponding to a particular value of s is obtainable from Eq. (8-5) or from Eqs. (8-21) or (8-22) which explicitly show the pole and zero forms. We now wish to show that this same information can be deduced graphically. To do this, refer to Eq. (8-21), and suppose that we wish to evaluate $T(s)$ for the particular value of $s = s_r$. This means that we wish to find the value,

$$T(s)|_{s_r} = T(s_r) = H \frac{(s_r - s_1)(s_r - s_2) \cdots (s_r - s_m)}{(s_r - s_a)(s_r - s_b) \cdots (s_r - s_n)}. \tag{8-23}$$

Actually, since s_r and all s_z and s_p are known, this is just an algebraic quantity, the numerical value of which can be calculated using the general rules of Section 8-4. If we consider s_r to be a point in the s-plane and $T(s_r)$ to be the corresponding value of $T(s)$ in the $T(s)$-plane, one says that the point s_r has been mapped on the $T(s)$ plane by the *mapping* function $T(s)$.

Let us consider a typical term of $T(s_r)$, say $(s_r - s_n)$. Refer to Fig. 8-5 which shows the graphical construction for finding $(s_r - s_n)$, which is just

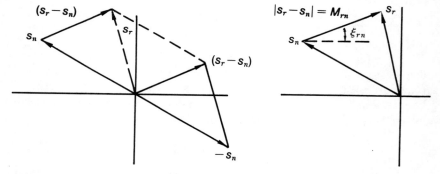

Fig. 8-5. (a) Construction for finding the magnitude and phase of $(s_r - s_n)$, (b) the simplified construction.

the distance from s_n to s_r, and with a phase angle ξ_{rn} between the line drawn from s_n to s_r and the positive real axis. Clearly, we can associate a magnitude and phase with each term in Eq. (8-23) and we write,

$$T(s_r) = H \frac{(M_{r1} e^{j\xi_{r1}})(M_{r2} e^{j\xi_{r2}}) \cdots (M_{rm} e^{j\xi_{rm}})}{(M_{ra} e^{j\eta_{ra}})(M_{rb} e^{j\eta_{rb}}) \cdots (M_{rn} e^{j\eta_{rn}})}. \tag{8-24}$$

It follows from this that,

$$\text{modulus of } T(s_r) = |T(s_r)| = H \frac{M_{r1} M_{r2} \cdots M_{rm}}{M_{ra} M_{rb} \cdots M_{rn}}, \tag{8-25}$$

$$\text{argument of } T(s_r) = \arg H + \sum_{k=1}^{m} \xi_{rk} - \sum_{k=1}^{n} \eta_{rk}.$$

For a real system H is a real quantity and the argument of H is zero.

Example 8-5.1

Show the pole-zero configuration, and find the value of the given $T(s)$ for $s = j2$.

$$T(s) = \frac{s-2}{(s+2+j1)(s+2-j1)}.$$

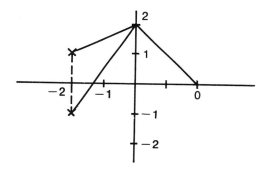

Fig. 8-5.1-1

Solution. The pole-zero configuration is illustrated in the figure. The directional vectors appropriate to each root are also shown in the figure. The modulus and argument of each factor is,

$$|j2 - 2| = 2\sqrt{2} \quad \arg = 135 \text{ deg},$$
$$|j2 + 2 + j1| = \sqrt{13} \quad \arg = 56.3 \text{ deg},$$
$$|j2 + 2 - j1| = \sqrt{5} \quad \arg = 26.5 \text{ deg}.$$

The resultant value of $T(j2)$ is,

$$T(j2) = \frac{2\sqrt{2} \,/\underline{135}}{\sqrt{13}\,/\underline{56.3}\,\sqrt{5}\,/\underline{26.5}} = 0.351\,/\underline{52.2}.$$

Of course, if these points are carefully drawn on the s-plane, the resultant lengths can be measured with a ruler, and the angles can be measured with the aid of a protractor.

Suppose that s_r denotes one point on a specified path or contour on the

326   s-Plane: Poles and Zeros

s-plane. We can then let s_r take on successive values appropriate to the path, with a construction appropriate to each s_r leading to a modulus and argument as in Eq. (8-25). In this way we produce a map or image of the path in the s-plane onto a corresponding map on the $T(s)$ plane. We can generalize this to say that any contour on the s-plane can be mapped onto the $T(s)$ plane for a specified mapping function $T(s)$. We shall find that a map of the imaginary or frequency axis onto the $T(j\omega)$-plane will be of subsequent interest, and will receive detailed attention in Chapter 9.

8-6 STEP AND IMPULSE RESPONSE

We wish to examine the response of a general system which is described by the system function $T(s)$ to unit step and impulse excitation functions in order to relate the procedure to the material of Section 7-6. Other important results will also appear in this study.

We initially consider the system to be excited by a step function of amplitude V. By Eq. (8-7) we write,

$$i(t) = \mathscr{L}^{-1}[T(s)U(s)] = \mathscr{L}^{-1}\left[T(s)\frac{V}{s}\right]. \qquad (8\text{-}26)$$

The procedure continues by expanding the right hand expression into partial fraction form and writing the values of the constants that arise in this expansion. When this is done in the manner prescribed in Section 7-6 the result is,

$$i(t) = \sum_{k=1}^{n}\left[(s-s_k)T(s)\frac{V}{s}\right]_{s=s_k}e^{s_k t} + VT(0)u_{-1}(t). \qquad (8\text{-}27)$$

In this form, it has been assumed that a $1/s$ term appears in the partial fraction expansion, and this gives rise to the second term on the right. The summation term on the right takes into account the remaining terms in the partial fraction expansion of $I(s) = T(s)U(s)$. As shown in this expression, it has been assumed that all poles are simple.

To determine the impulse response of the system, we again begin with Eq. (8-7), and write,

$$i(t) = \mathscr{L}^{-1}[T(s)V] = V\mathscr{L}^{-1}[T(s)]. \qquad (8\text{-}28)$$

By following a parallel procedure as that above that leads to Eq. (8-27), the result is,

$$i(t) = \sum_{k=1}^{m}[(s-s_k)T(s)V]_{s=s_k}e^{s_k t} + VT(0)u_0(t). \qquad (8\text{-}29)$$

An important conclusion can be drawn from this expression, namely, that $T(s)$ is a measure of the impulse response of a network. Vice versa, the result means also that given the impulse response of the network, one can deduce therefrom the system function $T(s)$.

A comparison of Eqs. (8-27) and (8-29) shows that the fundamental property of linear differential equations discussed in Section 3-15 is met, and that the impulse response of the system is the derivative of the step function response.

8-7 STEP AND IMPULSE RESPONSE OF A SYSTEM WITH ONE EXTERNAL POLE

Reference was made in Section 7-6 to the fact that a system response function $T(s)V(s)$ could have an external pole. This was said to be the case when the numerator polynomial in $T(s)$ was one degree higher than the denominator polynomial, with an applied step function. In this case $T(s)V(s)$ results in the ratio of two polynomials of equal degree. If the numerator polynomial is divided by the denominator polynomial, the result is of the form,

$$I(s) = B_0 + I_1(s), \qquad (8\text{-}30)$$

where B_0 is a constant, and where $I_1(s)$ is a proper fraction. As a result, the solution corresponding to Eq. (8-27) consists of terms of the form,

$$i(t) = \sum_{k=1}^{m} \left[(s-s_k) T(s) \frac{V}{s} \right]_{s=s_k} e^{s_k t} + VT(0) u_{-1}(t) + VB_0 u_0(t). \qquad (8\text{-}31)$$

The first terms are the regular transient terms; the second term is that appropriate to the step function excitation, and the last term is an impulse term that has arisen because of the appearance of the constant term in the expansion for $I(s)$.

In a parallel way, if the excitation is an impulse, then the numerator polynomial of the system response is one degree higher than that of the denominator, and upon carrying out the long division, the response function is,

$$I(s) = B_1 s + B_0 + I_1(s). \qquad (8\text{-}32)$$

The solution appropriate to this is the expression,

$$i(t) = \sum_{k=1}^{m} \left[(s-s_k) T(s) V \right]_{s=s_k} e^{s_k t} + VB_0 u_0(t) + VB_1 u_1(t). \qquad (8\text{-}33)$$

The first terms are the transient terms appropriate to the impulse excitation; the second term $B_0 u_0(t)$ is an impulse term; the third term is a doublet term specified by the doublet $u_1(t)$ of strength B_1.

8-8 STATE MODELS FROM SYSTEM FUNCTIONS

We have discussed the importance of the system function $T(s)$ in describing a system when expressed through a Kirchhoff formulation. In Eq. (7-49) we obtained an expression for the output function $Y(s)$ of the system when expressed in state form. We wish to show that given $T(s)$ we can deduce a state representation of the system, and vice versa. To begin this discussion, we examine the output function $Y(s)$ of Eq. (7-49) for the initially relaxed case, which is the expression,

$$Y(s) = [P\Phi(s)B + Q]U(s). \tag{8-34}$$

This is written as,

$$Y(s) = T(s)U(s), \tag{8-35}$$

where

$$T(s) = P\Phi(s)B + Q.$$

$T(s)$ is the system or transfer *matrix* of the network, and is the system function in equivalent matrix form. Clearly, $T(s)$ may be written explicitly as,

$$T(s) = \begin{bmatrix} T_{11}(s) & T_{12}(s) & \cdots & T_{1n}(s) \\ \cdot & \cdot & & \cdot \\ \cdot & \cdot & & \cdot \\ \cdot & \cdot & & \cdot \\ T_{n1}(s) & T_{n2}(s) & \cdots & T_{nn}(s) \end{bmatrix}, \tag{8-36}$$

and the ith component of $Y(s)$, which may be written $Y_i(s)$, is,

$$Y_i(s) = T_{i1}U_1 + T_{i2}U_2 + \cdots + T_{in}U_n. \tag{8-37}$$

It follows that $T_{ij}(s)$, which is the ijth element of $T(s)$, is the transfer function between $u_j(t)$ and $x_i(t)$. Based on the discussion in Section 8-6 we write,

$$T_{ij}(s) = \mathscr{L}[h_{ij}(t)], \tag{8-38}$$

where $h_{ij}(t)$ is the response at the ith output terminal to a unit impulse at the jth input terminal.

There are several reasons why one may wish to develop state equations from the system function. Of prime importance is the fact that the system function can be measured experimentally. This may be done by:

a. Impressing a sinusoidal excitation function at a number of frequencies and measuring the frequency response.

b. Using short (quasi impulse) pulses to approximate the impulse behavior of the network.

c. Measuring the system behavior to random input signals (a process that has not been discussed in this book).

From such experimental data a system function can be approximated. This assumes that no uncoupled modes exist, since these do not appear in the response, and such a system is not adequately described by the system function. Some discussion of this matter is given in Section 13-6.

We shall discuss three methods for obtaining state models from system functions:

a. Through the use of signal flow graphs.
b. By normal form decomposition.
c. By partial fraction decomposition.

The first of these methods was introduced in Section 4-5, and shall be given further consideration at this time.

a. SFG Decomposition. The features of this method are best discussed by considering a specific example. Suppose, therefore, that we begin with the system function,

$$T(s) = \frac{3s+2}{s^3+4s^2+3s+1}. \tag{8-39}$$

The numerator and denominator are divided by s^3 to get,

$$T(s) = \frac{(3/s^2)+(2/s^3)}{1+(4/s)+(3/s^2)+(1/s^3)}. \tag{8-40}$$

The system is of 3rd order, and we then identify 3 state variables (x_1, x_2, x_3) plus the input and output. Furthermore, each state equation involves the state variable and its first derivative; thus a total of 8 nodes are required:

$$u, sX_3(\dot{x}_3), X_3(x_3), sX_2(\dot{x}_2), X_2(x_2), sX_1(\dot{x}_1), X_1(x_1) \text{ and } y.$$

These can be identified graphically. In addition, three branches are directly identified, namely, the integral connection between each sX and its corresponding X. In addition, we expect the derivative of each state variable to be directly related to the previous state variable. Therefore, we arbitrarily connect each X_k to the adjacent node to its right by a unity transmittance. The result of this is the partial solution shown in Fig. 8-6. Attention is called to the fact that we show the designating nodes as time-domain

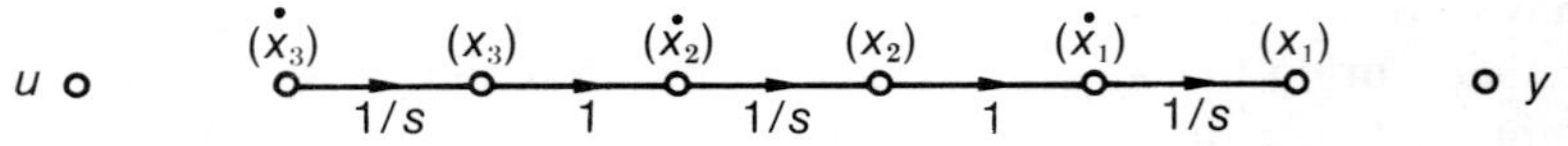

$u \circ \!\!-\!\!\underset{1/s}{\overset{(\dot{x}_3)}{\circ}}\!\!-\!\!\underset{1}{\overset{(x_3)}{\circ}}\!\!-\!\!\underset{1/s}{\overset{(\dot{x}_2)}{\circ}}\!\!-\!\!\underset{1}{\overset{(x_2)}{\circ}}\!\!-\!\!\underset{1/s}{\overset{(\dot{x}_1)}{\circ}}\!\!-\!\!\overset{(x_1)}{\circ}\!\!-\!\!\circ\, y$

Fig. 8-6. A step in the decomposition of the system function.

quantities. The parentheses around these node designations are a reminder that this is an s-domain SFG and each node carries with it a corresponding s-domain designation.

To continue with this SFG procedure, we call upon Eq. (3-18a), the Mason rule for the transmittance of a SFG. We observe from Eq. (3-18b) that if all feedback loops meet at a single node, then the graph determinant is 1−(sum of all different loop transmittances). The SFG of Example 3-5.1 is precisely of this form. This suggests that we seek a configuration with all loops through one node and such that,

$$\Sigma L = -\frac{4}{s} - \frac{3}{s^2} - \frac{1}{s^3}. \quad (8\text{-}41)$$

Of course, this is not the only path configuration, but it is the most direct. The specified constraints given by Eq. (8-41) can be satisfied by including 3 feedback paths from node x_1. The path to $\dot{x}_1$ must be chosen with path transmittance -4 so that the path gain is reduced by a term $-4/s$ in the denominator. The path from x_1 and $\dot{x}_2$ with transmittance -3 introduces the term $-3/s^2$. Finally, the path from x_1 to $\dot{x}_3$ with transmittance -1 introduces the term $-1/s^3$. The result of this procedure is shown in Fig. 8-7.

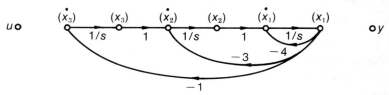

Fig. 8-7. The SFG that realizes the denominator of the specified $T(s)$.

To realize the numerator of Eq. (8-40), we again refer to Eq. (3-18). Now we make use of the numerator of the expression for the transmittance of the SFG, which states that the numerator specifies the weighted sum of all direct transmittances from u to y. Based on this, we proceed as follows: include a unity transmittance path from x_1 to y; include a path with forward gain 2 from u to $\dot{x}_3$ (weighting factor $\Delta_k = 1$) to realize the term $2/s^2$; include a path with forward gain 3 from u to $\dot{x}_2$ (weighting factor $\Delta_k = 1$) to realize the term $3/s^2$. The complete SFG is then that given in Fig. 8-8. Attention is called to the fact that this SFG is precisely of the form shown in Example 3-5.1.

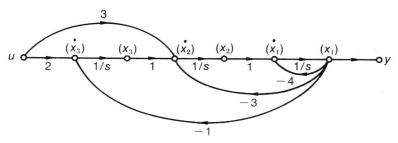

Fig. 8-8. The SFG realization of Eq. (8-39).

We use Fig. 8-8 to write the system equations, which are,

$$\dot{x}_1 = -4x_1 + x_2,$$
$$\dot{x}_2 = -3x_1 + x_3 + 3u,$$
$$\dot{x}_3 = -x_1 + 2u,$$
$$y = x_1,$$

(8-42)

which may be written in matrix form,

$$\frac{d}{dt}\begin{bmatrix} x_1 \\ x_2 \\ x_3 \end{bmatrix} = \begin{bmatrix} -4 & 1 & 0 \\ -3 & 0 & 1 \\ -1 & 0 & 0 \end{bmatrix}\begin{bmatrix} x_1 \\ x_2 \\ x_3 \end{bmatrix} + \begin{bmatrix} 0 \\ 3 \\ 2 \end{bmatrix}u,$$

$$y = \begin{bmatrix} 1 & 0 & 0 \end{bmatrix}\begin{bmatrix} x_1 \\ x_2 \\ x_3 \end{bmatrix}.$$

(8-43)

b. Normal Form Decomposition. This procedure in developing state equations from a specified $T(s)$ is to write the differential equation appropriate to $T(s)$ and then reduce the differential equation to normal form, in the manner discussed in Section 2-8. We continue to use the $T(s)$ of Eq. (8-39), and write from this the corresponding differential equation,

$$\frac{d^3y}{dt^3} + 4\frac{d^2y}{dt^2} + 3\frac{dy}{dt} + y = 3\frac{du}{dt} + 2u.$$

(8-44)

This differential equation is now replaced by the pair of equations

$$\frac{d^3x}{dt^3} + 4\frac{d^2x}{dt^2} + 3\frac{dx}{dt} + x = u,$$

$$y = 3\frac{dx}{dt} + 2x. \qquad (8\text{-}45)$$

As discussed in Section 3-5, we define the quantities,

$$\dot{x}_1 = x_2,$$
$$\dot{x}_2 = x_3, \qquad (8\text{-}46)$$

and when combined with Eq. (8-45) we have,

$$\dot{x}_3 = -4x_3 - 3x_2 - x_1 + u,$$
$$y = 3x_2 + 2x_1.$$

These equations, in matrix form, are,

$$\frac{d}{dt}\begin{bmatrix} x_1 \\ x_2 \\ x_3 \end{bmatrix} = \begin{bmatrix} 0 & 1 & 0 \\ 0 & 0 & 1 \\ -1 & -3 & -4 \end{bmatrix} \begin{bmatrix} x_1 \\ x_2 \\ x_3 \end{bmatrix} + \begin{bmatrix} 0 \\ 0 \\ u \end{bmatrix},$$

$$y = \begin{bmatrix} 2 & 3 & 0 \end{bmatrix} \begin{bmatrix} x_1 \\ x_2 \\ x_3 \end{bmatrix}. \qquad (8\text{-}47)$$

The SFG appropriate to this set of equations is given in Fig. 8-9.

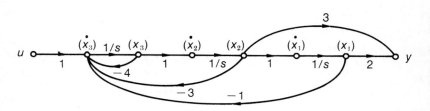

Fig. 8-9. The SFG realization of Eq. (8-46).

c. Partial Fraction Decomposition. This method is based on writing $T(s)$ in partial fraction form. This requires, of course, that the roots of the denominator polynomial of $T(s)$ be available. It is here assumed that these are known. Let us assume that $T(s)$ has been expanded into partial fraction form, using as representative, a system function with three different poles. This allows us to write,

$$T(s) = \frac{a}{s - s_1} + \frac{b}{s - s_2} + \frac{c}{s - s_3}, \quad (8\text{-}48)$$

where, in the manner discussed in Section 7-6, the constants a, b, c are,

$$\begin{aligned}
a &= (s - s_1)T(s)|_{s=s_1}, \\
b &= (s - s_2)T(s)|_{s=s_2}, \\
c &= (s - s_3)T(s)|_{s=s_3}.
\end{aligned} \quad (8\text{-}49)$$

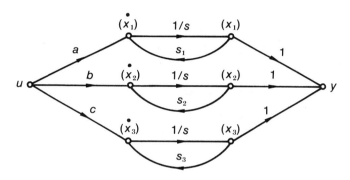

Fig. 8-10. The SFG realization of Eq. (8-48).

A SFG is readily drawn for Eq. (8-48), as given in Fig. 8-10. From this we write,

$$\begin{aligned}
\dot{x}_1 &= s_1 x_1 + au, \\
\dot{x}_2 &= s_2 x_2 + bu, \\
\dot{x}_3 &= s_3 x_3 + cu, \\
y &= x_1 + x_2 + x_3,
\end{aligned} \quad (8\text{-}50)$$

which are, when written in matrix form,

$$\frac{d}{dt}\begin{bmatrix} x_1 \\ x_2 \\ x_3 \end{bmatrix} = \begin{bmatrix} s_1 & 0 & 0 \\ 0 & s_2 & 0 \\ 0 & 0 & s_3 \end{bmatrix} \begin{bmatrix} x_1 \\ x_2 \\ x_3 \end{bmatrix} + \begin{bmatrix} a \\ b \\ c \end{bmatrix} u,$$

$$y = \begin{bmatrix} 1 & 1 & 1 \end{bmatrix} \begin{bmatrix} x_1 \\ x_2 \\ x_3 \end{bmatrix}. \quad (8\text{-}51)$$

334 s-Plane: Poles and Zeros

It is observed that in this form, the A matrix is diagonal, with the entries being the poles of the transfer function (the eigenvalues of A). This form of the state equation which separates or decouples the different modes is called the normal form, and is related to the Jordan matrix discussed in Section 7-7. It is a form that is often of particular value in analysis and design (16).

In the event that the poles of the denominator appear in complex conjugate pairs, a somewhat different procedure is adopted. This arises from the fact that the use of the decoupled form may be tedious, owing to the complex conjugate feedback paths in the SFG. In essence, instead of proceeding in terms of response components in the form $e^{s_1 t}$, $e^{-(\alpha - j\beta)t}$, $e^{-(\alpha + j\beta)t}$, it is preferable to use the real forms $e^{s_1 t}$, $e^{-\alpha t}\cos\beta t$, $e^{-\alpha t}\sin\beta t$. This is actually accomplished from $T(s)$ which, in partial fraction form, will be of the form,

$$T(s) = \frac{a}{s - s_1} + \frac{bs + c}{(s + \alpha)^2 + \beta^2}, \qquad (8\text{-}52)$$

which arises from Eq. (8-48) with the combination of the conjugate complex poles. This expression is written more conveniently in the form,

$$T(s) = \frac{a}{s - s_1} + \frac{b(s + \alpha)}{(s + \alpha)^2 + \beta^2} + \frac{((c - b\alpha)/\beta)\beta}{(s + \alpha)^2 + \beta^2}. \qquad (8\text{-}53)$$

To realize this in SFG form, we have only to write the two right hand terms in modified form,

$$T(s) = \frac{a}{s - s_1} + \frac{b/(s + \alpha)}{1 + (\beta^2/(s + \alpha)^2)} + \frac{((c - b\alpha)/\beta)(\beta/(s + \alpha)^2)}{1 + (\beta^2/(s + \alpha)^2)}. \qquad (8\text{-}54)$$

This equation has the SFG representation of Fig. 8-11. The state equa-

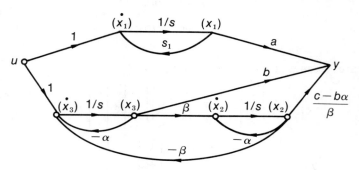

Fig. 8-11. A SFG representation of Eq. (8-54).

tions from this figure are,

$$\begin{aligned}
\dot{x}_1 &= s_1 x_1 + u, \\
\dot{x}_2 &= -\alpha x_2 + \beta x_3, \\
\dot{x}_3 &= -\beta x_2 - \alpha x_3 + u, \\
y &= x_1 + \frac{(c-b\alpha)}{\beta} x_2 + x_3,
\end{aligned} \quad (8\text{-}55)$$

which are, in vector form,

$$\frac{d}{dt}\begin{bmatrix} x_1 \\ x_2 \\ x_3 \end{bmatrix} = \begin{bmatrix} s_1 & 0 & 0 \\ 0 & -\alpha & \beta \\ 0 & -\beta & \alpha \end{bmatrix} \begin{bmatrix} x_1 \\ x_2 \\ x_3 \end{bmatrix} + \begin{bmatrix} 1 \\ 0 \\ 1 \end{bmatrix} u,$$

$$y = \begin{bmatrix} a & \dfrac{(c-b\alpha)}{\beta} & b \end{bmatrix} \begin{bmatrix} x_1 \\ x_2 \\ x_3 \end{bmatrix}. \quad (8\text{-}56)$$

These equations show that simple complex branch impedance can be avoided, but to do so requires that the decoupling between the states specified by the conjugate poles must be sacrificed.

It is noted that decoupling is also lost when a system function involves poles of higher order. To examine this, consider the particular system function,

$$T(s) = \frac{(s+\beta)}{(s+\alpha_1)(s+\alpha_2)^2}, \quad (8\text{-}57)$$

which may be expressed in partial fraction form,

$$T(s) = \frac{a}{s+\alpha_1} + \frac{b}{s+\alpha_2} + \frac{c}{(s+\alpha_2)^2}. \quad (8\text{-}58)$$

A SFG of this function will be of the form in Fig. 8-12. The state equa-

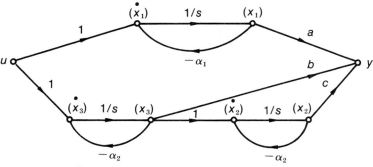

Fig. 8-12. A SFG representation of Eq. (8-58).

tions from this figure are,

$$\dot{x}_1 = -\alpha_1 x_1 + u,$$
$$\dot{x}_2 = -\alpha_2 x_2 + x_3,$$
$$\dot{x}_3 = -\alpha_2 x_3 + u,$$
$$y = ax_1 + cx_2 + bx_3.$$

(8-59)

In vector form, these are,

$$\frac{d}{dt}\begin{bmatrix} x_1 \\ x_2 \\ x_3 \end{bmatrix} = \begin{bmatrix} -\alpha_1 & 0 & 0 \\ 0 & -\alpha_2 & 1 \\ 0 & 0 & -\alpha_2 \end{bmatrix} \begin{bmatrix} x_1 \\ x_2 \\ x_3 \end{bmatrix} + \begin{bmatrix} 1 \\ 0 \\ 1 \end{bmatrix} u,$$

$$y = \begin{bmatrix} a & b & c \end{bmatrix} \begin{bmatrix} x_1 \\ x_2 \\ x_3 \end{bmatrix}.$$

(8-60)

Higher order poles would be handled in much the same way as above, with an appropriate cascade of $1/(s+\alpha)$ branches to realize the highest order pole.

8-9 SYSTEM FUNCTION REALIZATION USING OPERATIONAL AMPLIFIERS

Most of the content of this text is concerned with analysis; that is, given a specified system, to determine the response at some point in the system for specified excitation functions. The essence of the procedure using classical methods is to determine the system functions $T(s)$ for the given system appropriate to the output-input ports, and then employ classical or Laplace transform techniques to determine the response. The essence of the state variable method is rather similar—to describe the system by a set of state equations, and then by classical or Laplace transform methods, to determine the response characteristics of the system.

A second very broad class of system problems exist, called system synthesis, which requires specifying a system configuration that will provide a specified output response characteristic for a specified input function. In essence, therefore, what is required is a system configuration that will have a prescribed system function, $T(s)$. A well-organized study of network synthesis (17) exists which develops networks with specified $T(s)$ using passive elements.

System Function Realization Using Operational Amplifiers

More recently significant progress has been made in "active" network synthesis using amplifiers in the realization procedure (18). It is not our intention to develop a unified treatment of active network synthesis. We do wish to observe that much of Section 8-8 does actually provide the essence of network function realization using operational amplifiers. Moreover, since such operational amplifiers are readily available commercially in small modular or integrated circuit form, system function realization can be effected at modest cost. Such realization may be simpler than that using methods discussed by Hazony (17) and others that require only a single amplifier.

In particular, let us consider that the requisite system description is given by Eq. (8-4). Using the signal flow graph methods discussed in Section 8-8a, this equation is graphed as in Fig. 8-13. It is observed from

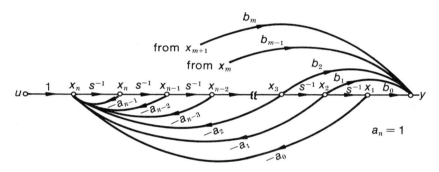

Fig. 8-13. SFG of a general system function $T(s)$.

this figure, and as pointed out by Steber and Krueger (19) that for the nth order equation, one requires n integrators plus four summing amplifiers, one for combining feedback paths with positive coefficients, and with negative coefficients at the $\dot{x}_n$ terminal, and similarly at the y terminal, for a total of $(n+4)$ operational amplifiers. Actually, the use of a summing integrator (*see* Fig. 5-22) at the $\dot{x}_n$ node will allow a reduction of one operational amplifier, to a minimum of $(n+3)$.

By proceeding in the manner discussed in Section 8-8c (20), a partial fraction decomposition of $T(s)$, when all poles are simple, leads to the SFG shown in Fig. 8-14, which describes the equation,

$$T(s) = \frac{A_1}{s-s_1} + \frac{A_2}{s-s_2} + \cdots + \frac{A_n}{s-s_n}. \qquad (8\text{-}61)$$

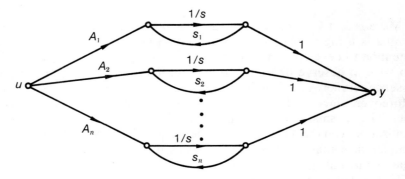

Fig. 8-14. SFG of $T(s)$ in partial fraction form (separate roots).

It is seen that each term in Eq. (8-61) is realized by an operational amplifier with a parallel RC feedback network, as in Fig. 5-24, with the R and C so chosen to yield the individual poles. Hence the total realization requires n integrators plus one summing amplifier at the output, for a total of $(n+1)$ operational amplifiers.

For the case when poles appear in conjugate pairs, the general form for $T(s)$ is like Eq. (8-53), with a SFG like Fig. 8-11. Realizing this requires a sign changer for each conjugate pair, plus an output summing amplifier. Hence the total number of operational amplifiers will depend on the number of complex pairs of roots, and in this configuration may exceed $(n+4)$. Similarly when there are higher order poles, as shown in Fig. 8-12 for a second order pole, a sign changer is required for each second order pole, plus the output summing amplifier. Clearly, the total number of operational amplifiers depends on the number of higher order poles, and may exceed $(n+4)$. In such cases, of course, if the criterion is the total number of operational amplifiers, the Steber and Krueger procedure may be preferred.

PROBLEMS

8-1. Write the differential equations of the systems which are specified by the following system functions,

$$\frac{s+1}{s^2+3s+5}, \quad \frac{s+3}{(s-j3)(s+j3)}, \quad \frac{s+2}{s(s+1)(s+3)}, \quad \frac{(s+2-j3)(s+2+j3)}{s(s^2+3s+5)}.$$

8-2. Determine the system function of each of the following:

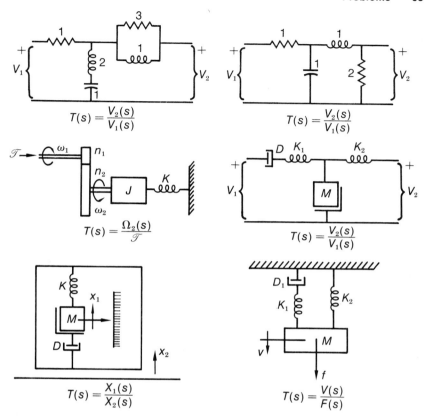

8-3. Show the $p-z$ configuration of the system functions of Problem 8-1.

8-4. Calculate the numerical value of the functions of Problem 8-1 for the following values of s: $-2, -j3$. Give the results in algebraic and in polar form.

8-5. Find the driving point functions of the networks shown, and plot the p-z constellation in the s-plane.

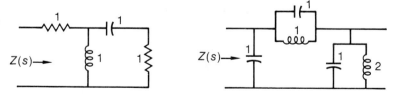

8-6. a. Deduce an expression for $Z(s)$ for the coupled network shown.
   b. Plot the $p-z$ pattern for the following values of M; 0; 0.1; 0.5; 1.0.

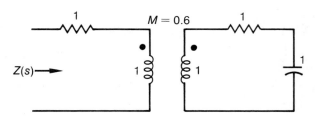

8-7. Systems are specified by the following system functions,

$$\frac{1}{s+1}, \quad \frac{s+1}{(s+2)(s+4)}, \quad \frac{s+1}{s^2+3s+5}, \quad \frac{s+2}{s(s^2+3s+5)}.$$

Determine the forced response, if the excitation function is $2e^{-2t}$.

8-8. Repeat Problem 8-7 when the excitation function is $2e^{-2t} + 3e^{-5t}$.

8-9. Refer to the accompanying figure.

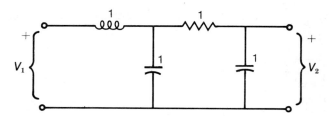

a. Determine $T(s) = V_2/V_1(s)$ and $Y(s) = I_1/V_1(s)$, and plot the p-z pattern.
b. For an applied $v_1 = V \cos t$, find the current i_1.

8-10. It is assumed that the circuits shown are initially relaxed. Find the quantities specified.

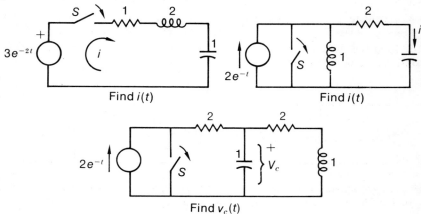

8-11. Refer to the circuit shown. Find the output voltage $v_2(t)$ when the excitation is; (a) the voltage v_1, and (b) the current i_1. In each case choose the excitation to be,

$$\text{excitation} = \cos t \quad (t > 0),$$
$$= 0 \quad (t < 0).$$

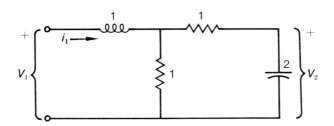

8-12. For a linear deflection of a magnetic cathode ray tube and for instantaneous retrace, the current is the sawtooth shown. The deflection coil circuit is closely approximated by a series RL circuit.

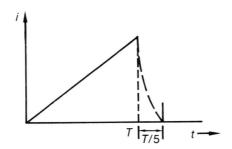

a. What must be the wave-shape v that is required in order to yield the specified current waveform?

b. If the retrace time need not be instantaneous, but may be a simple exponential that falls to $0.1\,I$ in $T/5$, what is the shape of v, and what circuit changes must be made, if any?

8-13. Consider a simple series RC circuit to which are applied the singularity functions illustrated. Sketch the current i_c for each case.

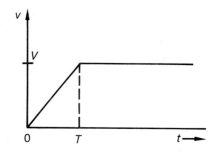
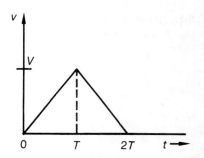

8-14. A series RL circuit, with $R = 1, L = 1$, is initially relaxed. Find the current when the excitation voltage is the finite wavetrain shown. Carry out the solution both analytically and numerically using a difference equation approximation. Sketch both results on the same curve sheet.

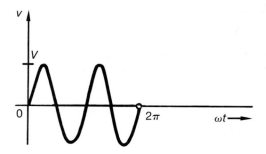

8-15. The input current to a two-port is $i_1 = u_{-1}(t)$; the current in the short-circuited output has the form shown. Now suppose that an impulse voltage is applied to the open-circuited port 2 terminals. What is the voltage v_1 across the open-circuited port 1 terminals?

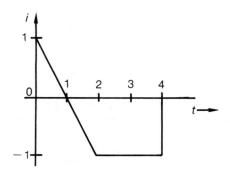

8-16. The transient response of a circuit to a unit step voltage is
$$i(t) = 2e^{-2t} + 3e^{-5t}.$$
Specify the poles and zeros of the system function.

8-17. The output response of a system to a unit impulse current at the input terminals is,
$$v(t) = e^{-t} + 3e^{-3t} + 2e^{-4t} \cos 2t.$$
a. Determine $T(s)$ for this system.
b. What is the response of the network to a unit step current.

8-18. The driving point current to a network for an applied impulse voltage is,
$$i(t) = e^{-t} + 3e^{-3t}.$$
Deduce a suitable network structure for the conditions specified.

8-19. Given the following system functions. Deduce for each, three different state models. Show the SFG of each.

$$\frac{s+1}{(s+2)(s+3)}, \quad \frac{2s+3}{s(s^2+3s+2)}, \quad \frac{s+2}{(s+3)(s+1)^2}.$$

8-20. Refer to the accompanying figure that gives a circuit for generating a complex pole pair:

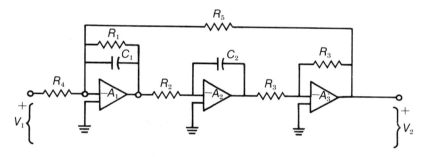

a. Deduce the $T(s)$ function for this circuit.
b. Draw the corresponding SFG.
c. Draw a SFG appropriate to the same $T(s)$ using the ideas in Eq. (8-54). Compare these.

9

System Response to Sinusoidal Functions

The properties of system response to sinusoidal excitation functions are of considerable importance in practical problems. There are several reasons for this: (a) sinusoidal sources and appropriate instrumentation are readily available over a wide range of frequencies, (b) many of our further studies will be carried out in the $j\omega$ plane, e.g., communications systems operate largely with sinusoidal waveforms. This chapter will be concerned with certain aspects of system response that are peculiar to sinusoidal excitation.

9-1 FEATURES OF SINUSOIDS

Ordinarily, the electric power supplied by the electric utilities is very closely sinusoidal in time. In fact, the alternator, the rotating machine that generates the power, has been carefully designed both mechanically and electrically to insure that the voltages that are produced are almost pure sinusoids. As a result, almost all electrical appliances and electrical instruments operate from sinusoidal power sources. Signal generators for engineering and scientific purposes often are basically sinusoidal sources with high purity of waveform over wide ranges of frequency, although not all signal sources are necessarily designed to produce sinusoidal signals.

We shall have occasion to use various properties of the sinusoid. Refer to Fig. 9-1 which illustrates the simple sinusoidal waveform. These are described analytically respectively as $I_m \sin \omega t$ and $I_m \sin(\omega t + \theta)$, where I_m denotes the peak value of the wave, and θ denotes the phase angle with respect to time $t = 0$. Among the important characteristics of the

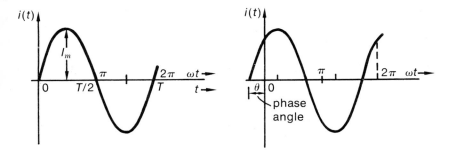

Fig. 9-1. The sinusoidal waveform.

sinusoid are the time average value (and the time duration must be indicated) and the so-called rms value. These are discussed in the following examples.

Example 9-1.1

Show that the full-cycle time average value of the sine wave is zero.

Solution. By definition, the full cycle time average is,

$$I_{\text{avg}} = \langle i(t) \rangle = \frac{1}{T} \int_0^T i\, dt = \frac{1}{2\pi} \int_0^{2\pi} i\, d(\omega t). \tag{9-1}$$

The current waveform is written explicitly,

$$i = I_m \sin \omega t = I_m \sin\left(\frac{2\pi}{T}\right)t.$$

Then,

$$i(t) = \frac{1}{2\pi} \int_0^{2\pi} I_m \sin \omega t\, d(\omega t) = \frac{1}{2\pi}[-I_m \cos \omega t]_0^{2\pi} = 0.$$

If it is recognized that the average value, as defined by Eq. (9-1), is a measure of the net area under the curve over one cycle, and since the sinusoid will have as much area above the zero axis as below it, then the result is obvious. In fact, the time average value of a sinusoid will be zero independent of any phase angle θ. For $\theta = 90$ deg. the wave will be a cosine function — but the net area under the curve over a complete cycle remains zero.

Consider now that a sinusoidal voltage source is applied across a resistor having a resistance R. We write directly, since,

$$v = iR,$$

346   System Response to Sinusoidal Functions

that
$$i = \frac{V_m}{R}\sin\omega t = I_m \sin\omega t,$$

where $I_m = V_m/R$. Furthermore, the instantaneous power to the resistor is,
$$p = vi = i^2 R.$$

The question is, what is the full-cycle time average value of the power? To answer this, we write, in accordance with definition, Eq. (9-1),
$$P = \langle p \rangle = \frac{1}{T}\int_0^T (i^2 R)\,dt = \frac{1}{2\pi}\int_0^{2\pi}(i^2 R)\,d(\omega t),$$

which we write,
$$P = \langle p \rangle = R\left[\frac{1}{T}\int_0^T i^2\,dt\right] = R\left[\frac{1}{2\pi}\int_0^{2\pi} i^2\,d(\omega t)\right]. \tag{9-2}$$

We now define the quantity I_{rms} from these expressions,
$$I_{\text{rms}} = \sqrt{\frac{1}{2\pi}\int_0^{2\pi} i^2\,d(\omega t)}. \tag{9-3}$$

Observe that this is actually the root-mean-square value of the current, which is deduced from the mean value of the current squared. Based on this definition, we see that we may write,
$$P = \langle p \rangle = I_{\text{rms}}^2 R. \tag{9-4}$$

Example 9-1.2

Deduce the value of I_{rms} in terms of I_m for the sinusoidal current wave.
Solution.   We proceed directly from Eq. (9-3), and write,
$$I_{\text{rms}} = \left[\frac{1}{2\pi}\int_0^{2\pi} I_m^2 \sin^2\omega t\,d(\omega t)\right]^{1/2}.$$

Carrying out the details of the integration,
$$I_{\text{rms}} = \left[\frac{I_m^2}{2\pi}\int_0^{2\pi}\frac{1-\cos 2\omega t}{2}d(\omega t)\right]^{1/2} = \left[\frac{I_m^2}{2\pi}\left(\frac{\omega t}{2}+\frac{\sin 2\omega t}{4}\right)\bigg|_0^{2\pi}\right]^{1/2},$$
$$= \left[\frac{I_m^2}{2\pi}\frac{2\pi}{2}\right]^{1/2} = \frac{I_m}{\sqrt{2}} = 0.707\,I_m.$$

It is observed from the foregoing examples that the full cycle average of a sinusoidal wave is zero, whereas the rms value is a constant value equal

Features of Sinusoids

to 0.707 of the peak value of the sinusoid. Moreover, these results are independent of any time phase displacement, with, say $i = I_m \sin(\omega t + \theta)$ replacing $I_m \sin \omega t$ in the foregoing examples.

An important feature of the sinusoid is that the sum of two sinusoids of the same period but of different amplitudes and different phase will also be a sine wave. This same idea is valid for any number of sine waves of equal period. Another important feature of the sinusoid is that the derivative and the integral of the function are also sinusoidal. This means that any linear operation on a sinusoid does not change the waveshape.

We examine the matter of combining sine waves of the same period. This will provide us with the opportunity to introduce some important notation for sinusoids. Suppose that we add the two sinusoids,

$$v_1 = V_1 \sin \omega t \qquad v_2 = V_2 \sin(\omega t + \theta). \tag{9-5}$$

We consider,

$$v = v_1 + v_2 = V_1 \sin \omega t + V_2 \sin(\omega t + \theta). \tag{9-6}$$

By direct trigonometric expansion we have that,

$$v = V_1 \sin \omega t + V_2 (\sin \omega t \cos \theta + \cos \omega t \sin \theta),$$

which may be combined as follows,

$$v = (V_1 + V_2 \cos \theta) \sin \omega t + V_2 \sin \theta \cos \omega t. \tag{9-7}$$

This is written in the form,

$$v = \sqrt{(V_1 + V_2 \cos \theta)^2 + V_2 \sin \theta)^2} \left[\frac{(V_1 + V_2 \cos \theta) \sin \omega t}{\sqrt{(V_1 + V_2 \cos \theta)^2 + (V_2 \sin \theta)^2}} \right.$$

$$\left. + \frac{V_2 \sin \theta \cos \omega t}{\sqrt{(V_1 + V_2 \cos \theta)^2 + (V_2 \sin \theta)^2}} \right]. \tag{9-8}$$

Define the quantities,

$$\cos \psi = \frac{V_1 + V_2 \cos \theta}{\sqrt{(V_1 + V_2 \cos \theta)^2 + (V_2 \sin \theta)^2}},$$

$$\sin \psi = \frac{V_2 \sin \theta}{\sqrt{(V_1 + V_2 \cos \theta)^2 + (V_2 \sin \theta)^2}}. \tag{9-9}$$

Equation (9-8) becomes, in the light of these,

$$v = \sqrt{(V_1 + V_2 \cos \theta)^2 + (V_2 \sin \theta)^2} (\sin \omega t \cos \psi + \cos \omega t \sin \psi),$$

which is,

$$v = \sqrt{V_1^2 + V_2^2 + 2V_1 V_2 \cos \theta} \sin(\omega t + \psi). \tag{9-10}$$

This development verifies that the resultant of the two sinusoids of the same period is also a sine wave. Clearly, because of the phase angle θ, the amplitude of the resultant may vary over wide limits.

Now let us reconsider the same problem but using the exponential forms for the sinusoids. Here we begin with the two functions,

$$v_1 = Im(V_1 e^{j\omega t}) \qquad v_2 = Im(V_2 e^{j(\omega t + \theta)}), \tag{9-11}$$

with Im denoting that the imaginary component in the expansion is being considered. The sum of these is,

$$\begin{aligned} v = v_1 + v_2 &= Im(V_1 e^{j\omega t}) + Im(V_2 e^{j(\omega t + \theta)}), \\ &= Im(V_1 e^{j\omega t} + V_2 e^{j(\omega t + \theta)}), \end{aligned} \tag{9-12}$$

or

$$v = Im(V_1 + V_2 e^{j\theta}) e^{j\omega t}. \tag{9-13}$$

Observe that each wave is defined by complex numbers V_1 and $V_2 e^{j\theta}$ which specify the respective amplitude and phase factors; further, the sum of these two sinusoids has the amplitude $V_1 + V_2 e^{j\theta}$, which is also an algebraic quantity, independent of time. Clearly, except for the implied time factor, only the amplitude and phase factors are important in algebraic operations. For this reason, the geometrical and algebraic features of complex numbers will apply. It is convenient to emphasize the amplitude and phase features by referring to the complex number representation as *phasors* or *sinors*. To see the phasor combination appropriate to Eq. (9-11), refer to Fig. 9-2. From the figure we write directly that,

$$V = \sqrt{(V_1 + V_2 \cos \theta)^2 + (V_2 \sin \theta)^2} = \sqrt{V_1^2 + V_2^2 + 2 V_1 V_2 \cos \theta},$$
$$\psi = \tan^{-1} \frac{V_2 \sin \theta}{V_1 + V_2 \cos \theta}, \tag{9-14}$$

which corresponds exactly to the results contained in Eq. (9-13).

Some discussion relative to Eq. (9-13) is important. To arrive at this equation, the amplitude and phase information of each sinusoid (and the phase information must be specified relative to the same reference point in time) was extracted. These phasors quantities were then combined according to simple algebraic rules of complex numbers to give the phasor of the resultant. To complete the result, the resultant phasor quantity must be combined with the time function that is implicit in our work. This

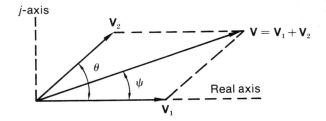

Fig. 9-2. Combination of two phasors.

means that the phasor or sinor quantities here being discussed do not have physical reality in themselves—one does not measure these complex numbers directly; it is the time function that has physical reality. Actually, not all of the phasor quantities relating to sinusoidal problems involve the time factor explicitly or implicitly, the impedance and admittance funtions are of these types. Because of this, there have been suggestions in the past that one distinguish two different phasor quantities in such problems. This is not necessary during intermediate operations that are concerned only with finding quantities which ultimately are combined to yield time information. In both cases the phasors are merely complex numbers which satisfy the rules of complex number arithmetic. Care must be exercised, of course, in the ultimate steps relating such complex to time functions. But in no case is the phasor quantity synonymous with the final time function—it is a mathematical algebraic procedure for ultimately determining amplitude and phase information.

The procedure involving the use of complex arithmetic for sinusoids can be simplified somewhat if some simple rules are established. Thus we choose *always* to write sinusoids in terms of sinusoidal time functions; thus a cosine function is written in equivalent sine form, with explicit designation of the appropriate phase angle. This allows us to drop the designation *Im*, since the imaginary part of the resulting function will always be implied. Now if we were to proceed from Eq. (9-11) we would write the quantities,

$$\mathbf{V}_1 = V_1 e^{j0} \quad \mathbf{V}_2 = V_2 e^{j\theta}, \tag{9-15}$$

and the result obtained by adding these two phasors is,

$$\mathbf{V} = \mathbf{V}_1 + \mathbf{V}_2 = V_1 e^{j0} + V_2 e^{j\theta} = V e^{j\psi},$$

which is precisely the situation illustrated in Fig. 9-2.

9-2 STEADY-STATE SYSTEM RESPONSE TO SINUSOIDAL EXCITATION FUNCTIONS

The sinusoidal excitation function is an especially important one, since it provides the basis for all of the work in a-c steady-state system response. We proceed by considering the special case of a simple series RLC circuit through which there is a current $i(t) = I_m \cos \omega t$ (the assumed steady-state response to an applied sinusoidal forcing function). We are interested in determining the corresponding voltage $v(t)$ which produces the current $i(t)$. We begin with the general equation,

$$Ri + \frac{Ldi}{dt} + \frac{1}{C}\int i\, dt = v(t),$$

where, for the current i, we use the expression,

$$i = I \cos \omega t = I\left(\frac{e^{j\omega t} + e^{-j\omega t}}{2}\right).$$

By combining these two expressions,

$$\left(R + j\omega L + \frac{1}{j\omega C}\right)\frac{I}{2}e^{j\omega t} + \left(R - j\omega L + \frac{1}{-j\omega C}\right)\frac{I}{2}e^{-j\omega t} = v(t).$$

This expression is written more conveniently as,

where

and

$$\left.\begin{array}{c} Z(j\omega)\dfrac{I}{2}e^{j\omega t} + Z(-j\omega)\dfrac{I}{2}e^{-j\omega t} = v(t) \\[6pt] Z(j\omega) = Z(s)\bigg|_{s=j\omega} = \left(R + Ls + \dfrac{1}{Cs}\right)\bigg|_{s=j\omega} \\[6pt] Z(-j\omega) = Z(s)\big|_{s=-j\omega} = \left(R + Ls + \dfrac{1}{Cs}\right)\bigg|_{s=-j\omega} \end{array}\right\} \qquad (9\text{-}16)$$

We can extend these results to a general system that is specified by the system function $T(s)$ to which is applied an excitation $v_i = V \cos \omega t$, the response being denoted $v_0(t)$. The result is,

$$v_0(t) = \frac{V}{2}[T(j\omega)e^{j\omega t} + T(-j\omega)e^{-j\omega t}]. \qquad (9\text{-}17)$$

Observe that this expression follows without recourse to the inverse Laplace transform. Of course, if we wished to find the total response function for the initially relaxed circuit, then we would take the Laplace transform of the excitation function and combine it with the system func-

Steady-State System Response to Sinusoidal Excitation Functions 351

tion, to the form,

$$v_0(t) = \mathscr{L}^{-1}\left\{\frac{V}{2}\left[T(s)\frac{1}{s-j\omega} + T(s)\frac{1}{s+j\omega}\right]\right\}. \qquad (9\text{-}18)$$

If we think in terms of a partial fraction expansion of each product term, then the particular solution (the steady-state) is obtained from those terms of the form $K_1/s-j\omega$ and $K_2/s+j\omega$ respectively. Similarly, of course, if the excitation is a sinusoidal function rather than a cosine function, the particular solution will be given by,

$$v_0(t) = \frac{V}{2j}[T(j\omega)e^{j\omega t} - T(-j\omega)e^{-j\omega t}]. \qquad (9\text{-}19)$$

We observe that the complete solution to Eq. (9-17) for an applied $V \sin \omega t$ to the initially relaxed system is given by the expression,

$$v_0(t) = \sum_{k=1}^{n}\left[(s-s_k)T(s)V\frac{\omega}{s^2+\omega^2}\right]_{s=s_k} e^{s_k t} + \frac{V}{2j}[T(j\omega)e^{j\omega t} - T(-j\omega)e^{-j\omega t}]. \qquad (9\text{-}20)$$

The foregoing can be written in different forms. From an examination of Eqs. (9-17) and (9-18) we write,

$$v_i = V \cos \omega t = \frac{V}{2}e^{j\omega t} + \frac{V}{2}e^{-j\omega t} = V'_i + V'_i{}^*. \qquad (9\text{-}21)$$

Corresponding to each excitation function is the response function,

$$V'_0 = \frac{V}{2}T(j\omega)e^{j\omega t},$$
$$V'_0{}^* = \frac{V}{2}T(-j\omega)e^{-j\omega t}. \qquad (9\text{-}22)$$

Further, we note that since,

$$v_i = V'_i + V'_i{}^* = 2Re[V'_i],$$

then

$$v_0 = V'_0 + V'_0{}^* = 2Re[V'_0], \qquad (9\text{-}23)$$

and Eq. (9-17) may be written,

$$v_0 = V Re[T(j\omega)e^{j\omega t}]. \qquad (9\text{-}24)$$

Similarly, when the excitation has the form,

$$v_i = V \sin \omega t, \qquad (9\text{-}25)$$

352 System Response to Sinusoidal Functions

the response function will be,

$$v_0 = V\,Im[T(j\omega)e^{j\omega t}]. \tag{9-26}$$

Another form follows by noting that $T(j\omega)$ is a complex function of ω and can thus be written in terms of two real functions,

$$T(j\omega) = R(\omega) + jX(\omega), \tag{9-27}$$

where $R(\omega)$ gives the real component and $jX(\omega)$ specifies the imaginary part of $T(j\omega)$. Correspondingly for systems with real coefficients,

$$T(-j\omega) = R(\omega) - jX(\omega). \tag{9-28}$$

Thus for the $V\cos\omega t$ excitation, Eq. (9-17) we can write,

$$v_0 = \frac{V}{2}[(R+jX)(\cos\omega t + j\sin\omega t) + (R-jX)(\cos\omega t - j\sin\omega t)].$$

This combines to the form,

$$v = V(R\cos\omega t - X\sin\omega t). \tag{9-29}$$

Now multiply and divide this expression by $\sqrt{R^2+X^2}$; also define,

$$\cos\psi = \frac{R}{\sqrt{R^2+X^2}} \qquad \sin\psi = \frac{X}{\sqrt{R^2+X^2}},$$
$$\psi = \tan^{-1}\frac{X}{R} = \arg T(j\omega). \tag{9-30}$$

ψ is the phase angle between the output and the input functions. Equation (9-28) can be written,

$$v_0 = V|T(j\omega)|\cos(\omega t + \psi). \tag{9-31}$$

If the excitation is $V\sin\omega t$, the component of response due to the driving function is,

$$v_0 = V|T(j\omega)|\sin(\omega t + \psi). \tag{9-32}$$

The complete solution to Eq. (9-18) for the $V\sin\omega t$ excitation function is the expression,

$$i(t) = \sum_{k=1}^{n}\left[(s-s_k)T(s)\frac{V\omega}{s^2+\omega^2}\right]_{s=s_k} e^{s_k t} + V|T(j\omega)|\sin(\omega t + \psi). \tag{9-33}$$

Example 9-2.1

Suppose that the circuit of Example 8-2.3 is driven by a source specified by $v_i = 2 \cos t$. Find the steady-state component of the response function v_2.

Solution. For the excitation function specified, the steady-state response function is,
$$v_2 = \tfrac{2}{2}[T(j1)e^{jt} + T(-j1)e^{-jt}],$$
where,
$$T(j1) = \frac{j1}{-2j-4+3j+2} = \frac{j1}{-2+j1} = \frac{1-j2}{5} = \sqrt{5}\,\underline{/-53.5°}\,.$$

Correspondingly, therefore,
$$T(-j1) = \sqrt{5}\,\underline{/53.5°}\,,$$
and so,
$$v_2 = 2\sqrt{5}\,\frac{e^{j(t-53.5°)} + e^{-j(t+53.5°)}}{2},$$
$$v_2 = 2\sqrt{5}\cos(t-53.5°).$$

9-3 POWER

The instantaneous power to a one-port system is defined by,
$$p(t) = v(t)i(t). \tag{9-34}$$
When the voltage and current are sinusoidal functions of time, say,
$$v(t) = V \sin(\omega t + \alpha),$$
$$i(t) = I \sin(\omega t + \beta), \tag{9-35}$$
then,
$$p = VI \sin(\omega t + \alpha) \sin(\omega t + \beta). \tag{9-36}$$
Consider initially the case when the two waves are in phase, with $\alpha = \beta = 0$. Under these conditions,
$$p = VI \sin^2 \omega t = \frac{VI}{2}(1 - \cos 2\omega t). \tag{9-37}$$
The time average value of the power during one period of the sinusoidal function is,
$$P = \frac{VI}{2}\left[\frac{1}{2\pi}\int_0^{2\pi}(1-\cos 2\omega t)\,d(\omega t)\right] = \frac{VI}{2}. \tag{9-38}$$

Figure 9-3 contains the significant information given above. It is observed that the power curve is a double frequency curve, with the axis of symmetry displaced by the average power $VI/2$.

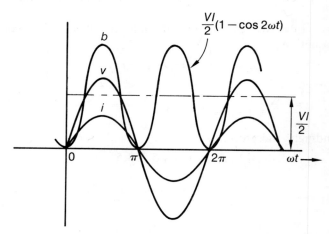

Fig. 9-3. The power curve in a sinusoidally excited system.

In the more general case the phase angles will be different from zero. To find the average power, we proceed somewhat differently from the above. We write,

$$v(t) = V \sin(\omega t + \alpha) = Im(Ve^{j\alpha} e^{j\omega t}),$$
$$i(t) = I \sin(\omega t + \beta) = Im(Ie^{j\beta} e^{j\omega t}). \tag{9-39}$$

Suppose that the phasor amplitudes of v and i are written,

$$Ve^{j\alpha} = a + jb \qquad Ie^{j\beta} = c + jd. \tag{9-40}$$

We write, using Im to denote the imaginary part of the expression,

$$p = [Im(a+jb)e^{j\omega t}][Im(c+jd)e^{j\omega t}]. \tag{9-41}$$

This expression expands to,

$$p = [Im(a+jb)(\cos \omega t + j \sin \omega t)] \cdot [Im(c+jd)(\cos \omega t + j \sin \omega t)],$$

from which,

$$p = (a \sin \omega t + b \cos \omega t)(c \sin \omega t + d \cos \omega t),$$
$$= ac \sin^2 \omega t + (bc + ad) \sin \omega t \cos \omega t + db \cos^2 \omega t. \tag{9-42}$$

The time average value of this expression is,

$$P = \tfrac{1}{2}(ac+bd) = \tfrac{1}{2}VI(\cos\alpha\cos\beta + \sin\alpha\sin\beta),$$
$$= \frac{VI}{2}\cos(\alpha-\beta). \quad (9\text{-}43)$$

The angular difference $(\alpha-\beta)$ is called the *power factor angle*, and is the difference in angle between the voltage and the current. The factor $\cos(\alpha-\beta)$ is known as the *power factor*.

We may relate P to the phasors V, V^*, I, I^*. We write,

$$VI^* = (a+jb)(c-jd) = ac+bd+j(bc-ad).$$

Also,
$$V^*I = (a-jb)(c+jd) = ac+bd-j(bc-ad). \quad (9\text{-}44)$$

Note from these expressions that the average power may be written in several different forms, namely,

$$P = \tfrac{1}{2}Re(VI^*) = \tfrac{1}{2}Re(V^*I) = \tfrac{1}{4}(VI^* + V^*I). \quad (9\text{-}45)$$

Moreover, since $V = ZI$ and $I = YV$ for the one-port, then it follows that,

$$P = \tfrac{1}{2}Re(V^*I) = \tfrac{1}{2}Re(|I^2|Z) = \tfrac{1}{2}|I^2|Re Z,$$
$$P = \tfrac{1}{2}Re(VI^*) = \tfrac{1}{2}Re(|V^2|Y) = \tfrac{1}{2}|V^2|Re Y. \quad (9\text{-}46)$$

These equations are related to the phasor diagram of Fig. 9-4.

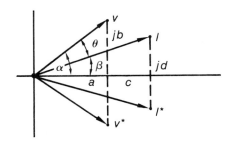

Fig. 9-4. Phasor diagram of a one-port network.

9-4 PHASOR DIAGRAMS

The discussion in Section 9-2 has shown that if a system is excited by a sinusoidal excitation function $v = V\sin\omega t$, then the response is also a sinusoidal excitation function $i = I\sin(\omega t+\theta)$. In terms of the phasor

356   System Response to Sinusoidal Functions

notation with $\sin \omega t$ as the reference base, we write,

$$v = V e^{j\omega t} \qquad i = I e^{j\omega t}. \tag{9-47}$$

Now let us consider specifically an *RLC* circuit. From Eq. (8-9) we know that the impedance function $Z(s)$ is, from Laplace transform considerations,

$$Z(s) = Ls + R + \frac{1}{sC},$$

which is, for the sinusoidal excitation, by writing $s = j\omega$,

$$Z(j\omega) = j\omega L + R + \frac{1}{j\omega C} = |Z(j\omega)| e^{j\theta}. \tag{9-48}$$

But since,

$$Z(j\omega) = Z(s)|_{s=j\omega} = \frac{V}{I}, \tag{9-49}$$

then we have that,

$$i = \frac{V}{Z(j\omega)} e^{j\omega t} = \frac{V}{|Z(j\omega)|} e^{j(\omega t - \theta)}. \tag{9-50}$$

It is interesting to show graphically the phasor relations contained in these equations. This is done in Fig. 9-5 which is called a phasor diagram.

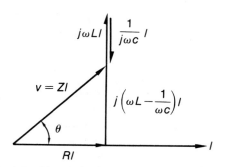

Fig. 9-5. Phasor diagram for the series *RLC* circuit.

The phasor diagram is often of considerable convenience and clarification as a graphical aid in analysis. Here it is only necessary to remember that the voltage across an inductor leads the current through it by 90 deg. since $V = sLI|_{s=j\omega} = j\omega LI = \omega LI \underline{/90°}$ (equivalently, the current through the inductor lags the voltage by 90 deg.); the voltage and current are in phase through a resistor; and the voltage lags the current by 90 deg. through a capacitor, since $V = I/sC|_{s=j\omega} = I/j\omega C = I/\omega C \underline{/-90°}$ (or equivalently, the current leads the voltage by 90 deg. through a capacitor). If the phasor diagram is drawn roughly to scale, then it may be used as a tool of analysis, since it presents precisely the same information as do the equations.

To understand the situation, let us consider the circuit of Example 8-2.3. This is here redrawn for convenience as Fig. 9-6. To draw the

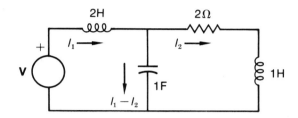

Fig. 9-6. A two loop circuit for detailed study.

phasor diagram for this network, we shall choose arbitrarily that I_2 is along the axis of reals. We shall discuss this selection below. A phasor diagram appropriate to Fig. 9-6 is given in Fig. 9-7. The system phasor diagram is built up in successive steps. Thus, with I_2 designated in direction and magnitude, this establishes the voltage across $R(2\Omega)$ and $L(1H)$,

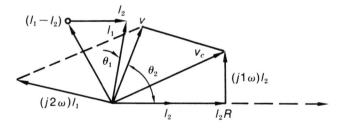

Fig. 9-7. A phasor diagram for the circuit of Fig. 9-6.

the sum of which is the voltage across the capacitor $C(1F)$. The current through the capacitor leads the voltage by 90 deg. This establishes the direction for $I_1 - I_2$, from which the magnitude and direction of I_1 results, by adding I_2 to $I_1 - I_2$. With I_1 known, the voltage across $L(2H)$ is established (it leads I_1 by 90 deg.). The applied V is the sum of the voltage across $L(2H)$ and $C(1F)$. In Fig. 9-7 I_1 leads V, and I_2 lags V, by θ_1 and θ_2 respectively.

Now a word as to the initial selection of I_2 along the axis of reals. Observe that the construction is independent of the particular direction of I_2, the subsequent construction being dictated by known magnitude and relative angle factors. Thus one can add any desired angle to the system without affecting the details of the construction, the only effect being a rotation of the entire figure about the origin. If it is assumed, therefore, that $V = V\underline{/0°}$ initially, then the line containing V is the reference line. The entire figure could be rotated to have this line horizontal, if so desired, but without otherwise affecting any of the construction or any of the subsequent phasors in it.

Example 9-4.1

Refer to the coupled circuit shown. Find the currents I_1, I_2, $i_1(t)$ and $i_2(t)$. Draw the phasor diagram.

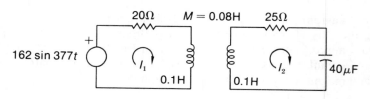

Fig. 9-4.1-1

Solution. From the specified ω and the designated circuit parameters, we write,

$$X_L = \omega L = 377 \times 0.1 = 37.7 \text{ ohm},$$

$$X_C = \frac{1}{\omega C} = \frac{10^6}{40 \times 3.77} = 66.2 \text{ ohm},$$

$$X_m = \omega M = 377 \times 0.08 = 30.2 \text{ ohm},$$

$$V = 162 \text{ peak} = 115 \text{ rms}.$$

From the figure, the equilibrium equations (KVL) are,

$$115 = (20 + j37.7)I_1 - j30.2 I_2,$$

$$0 = -j30.2 I_1 + (25 + j37.7 - j66.2)I_2.$$

These are solved for I_1 and I_2,

$$I_1 = \frac{\begin{vmatrix} 115\underline{/0} & -j30.2 \\ 0 & 25-j28.5 \end{vmatrix}}{\begin{vmatrix} 20+j37.7 & -j30.2 \\ -j30.2 & 25-j28.5 \end{vmatrix}} = \frac{115(25-j28.5)}{(20+j37.7)(25-j28.5)+30.2^2}$$

$$= \frac{115 \times 37.9 \underline{/-48.8}}{2.52 \times 10^3 \underline{/8.5}} = 1.73 \underline{/-57.3}$$

$$I_2 = \frac{\begin{vmatrix} 20+j37.7 & 115\underline{/0} \\ -j30.2 & 0 \end{vmatrix}}{\Delta} = \frac{115 \times 30.2 \underline{/90}}{2.52 \times 10^3 \underline{/8.5}} = 1.385 \underline{/81.5}$$

From these we write the explicit time functions,

$$i_1(t) = 1.73\sqrt{2} \sin(377t - 57.3°),$$
$$i_2(t) = 1.385\sqrt{2} \sin(377t + 81.5°).$$

The phasor diagram is given in the accompanying figure.

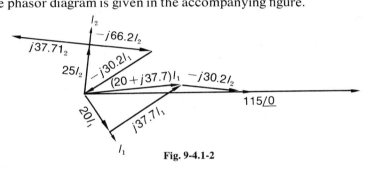

Fig. 9-4.1-2

9-5 Q-VALUE AND BANDWIDTH

Equation (8-9) gives the driving point impedance function for the series *RLC* circuit. Certain features of this function are discussed in the foregoing section. We wish to examine the function further since it possesses interesting properties when the frequency is varied, as is evident from Fig. 9-5 which shows that since $j\omega LI$ and $I/j\omega C$ vary in different ways with frequency, the system phase angle can vary from positive to negative, and can have a zero value. This variation is related to the *resonance* characteristics of the circuit.

a. The Series Circuit. Based on Eq. (9-50), we proceed by considering the driving point admittance function

$$Y(s) = \frac{1}{Z(s)} = \frac{1}{Ls + R + (1/Cs)} = \frac{1}{L} \frac{s}{[s^2 + (Rs/L) + (1/CL)]}. \quad (9\text{-}51)$$

This is written in a form that displays the p-z configuration for the case when the roots are complex,

$$Y(s) = \frac{1}{L} \frac{s}{(s-s_a)(s-s_a^*)},$$

where, (9-52)

$$s_a, s_a^* = -\frac{R}{2L} \pm j\sqrt{\frac{1}{LC} - \left(\frac{R}{2L}\right)^2}.$$

The roots are written,

$$s_a, s_a^* = -\zeta\omega_n \pm j\omega_n\sqrt{1-\zeta^2},$$

where, (9-53)

$$\omega_n = \frac{1}{\sqrt{LC}} \qquad \zeta = \frac{R}{2}\sqrt{\frac{C}{L}}.$$

ω_n is the natural undamped frequency of the system; ζ is the dimensionless damping ratio. These factors (actually the inverse of $\omega_n = 1/\tau$) appeared in Section 5-2 for the complementary function of this same series RLC circuit. We shall show below the influence of these factors on the response. In the form shown in Eq. (9-52) we are assuming that $\zeta^2 < 1$, which is the underdamped case.

The variation of the pole locations of $Y(s)$ for constant ω_n and adjustable ζ are shown in Fig. 9-8. The locus of the poles is a circle. In addition,

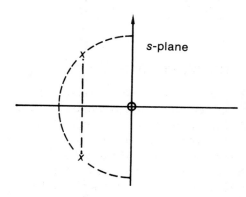

Fig. 9-8. The p-z configuration for $Y(s)$ of the RLC series circuit.

there is a zero at the origin. We wish to study the variation of this function with frequency. The details of the construction are contained in Fig. 9-9, as discussed in Section 8-5. The resulting value of the function, which is

Q-Value and Bandwidth 361

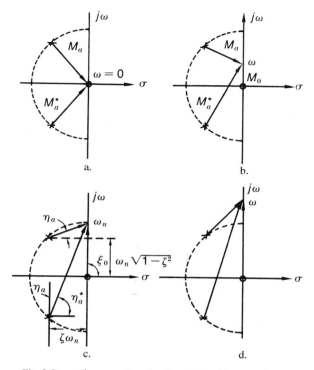

Fig. 9-9. s-plane construction for different frequencies.

specified by Eq. (8-24) is,

$$Y(j\omega) = \frac{1}{L} \frac{M_0}{M_a M_a^*} e^{j(\xi_0 - \eta_a - \eta_a^*)}. \tag{9-54}$$

The frequency response characteristic has the form illustrated in Fig. 9-10. Note from Eq. (9-54) that $Y(j\omega)$ will be a maximum near the frequency at which M_a or M_a^* is a minimum (when M_a is a minimum for positive frequency). This occurs at that frequency at which the pole M_a is the shortest distance to the $j\omega$ axis. Moreover, the smaller the distance to the $j\omega$ axis, the larger will be the value of $Y(j\omega)$ at this frequency since this factor appears in the denominator of the expression for $Y(j\omega)$. Furthermore, both M_0 and M_a^* will be slowly varying in the range when M_a is changing rapidly. The frequency at which $Y(j\omega)$ is a maximum is defined as the *resonant* frequency.

Certain features of the shape of the curves in Fig. 9-10 are better dis-

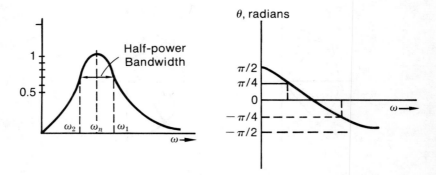

Fig. 9-10. The frequency response characteristic of the series RLC circuit.

cussed analytically. We thus examine the magnitude function,

$$Y(j\omega) = \frac{1}{\sqrt{R^2 + [(\omega L - (1/\omega C)]^2}}. \tag{9-55}$$

Observe that this function has a maximum value $= 1/R$ when the quantity,

$$\omega_n L - \frac{1}{\omega_n C} = 0,$$

or when,

$$\omega_n = \frac{1}{\sqrt{LC}}. \tag{9-56}$$

This shows that resonance occurs when $\omega = \omega_n$, the situation in Fig. 9-9c. When resonance occurs $\eta_a + \eta_a^* = 90°$, and the total phase angle is,

$$\xi_0 - (\eta_a + \eta_a^*) = 0. \tag{9-57}$$

Resonant circuits are often discussed in terms of the circuit Q, sometimes called the "quality" of the circuit. The Q of a series RLC circuit is defined by the relation,

$$Q = \frac{\omega_n L}{R} = \frac{1}{2}\frac{\omega_n}{R/2L} = \frac{1}{2}\frac{\omega_n}{\zeta\omega_n} = \frac{1}{2\zeta}. \tag{9-58}$$

We make the following observations, noting first that ω_n is the radius of the semicircle of the poles, and $\zeta\omega_n$ is the distance from the pole to the $j\omega$ axis. Note that,

1. Shorter distances $\zeta\omega_n$ from the poles s_a and s_a^* to the $j\omega$ axis results in high values of Q.

2. The value of Q varies inversely with the damping ratio ζ. Thus low damping is accompanied by high values of Q. In the limit as $R \to 0$, the Q of the circuit will be infinite.

Practically, completely lossless passive circuits are not possible. Typical values are roughly the following: circuits composed of wire-wound coils, the Q values will range from perhaps 10 to 100; coils made of hollow tubing, as used radio transmitters, a Q of 1000 may be possible; microwave resonators may have Q values as high as 50,000; crystal filters have Q values as high as 100,000. An oscillator circuit possesses an infinite Q since it provides an output without an input. These circuits are of considerable importance, but involve active devices. Some further comments will be made in Chapter 13 on the relation of stability and the location of the poles of the system. The effects of $\zeta = \frac{1}{2}Q$, the damping ratio, on the frequency response curve of the series RLC network are shown in Fig. 9-11.

It is clear from Fig. 9-11 that the amplitude of $|Y(j\omega)|$ increases with reduced damping, or with higher Q. We can use this feature to describe additional properties of the series RLC circuit. We begin with the expression for $Y(j\omega)$ in the form,

$$Y(j\omega) = \frac{1}{R + j[\omega L - (1/\omega C)]}. \tag{9-59}$$

Observe that the phase angle of this function will be ± 45 deg. when,

$$\omega_p L - \frac{1}{\omega_p C} = \mp R. \tag{9-60}$$

This equation may be written,

$$\omega_p^2 \pm \frac{R}{L}\omega_p - \frac{1}{LC} = 0,$$

from which we find that,

$$\omega_p = \mp \frac{R}{2L} \pm \sqrt{\left(\frac{R}{2L}\right)^2 + \frac{1}{LC}},$$

which may be written,

$$\omega_p = \omega_n(\mp \zeta \pm \sqrt{\zeta^2 + 1}). \tag{9-61}$$

But in the case of circuits with high Q, ζ is small, and so, approximately,

$$\omega_p = \omega_n(1 \mp \zeta). \tag{9-62}$$

Furthermore, for the values specified in Eq. (9-60), $|Y(j\omega)|$ has the value

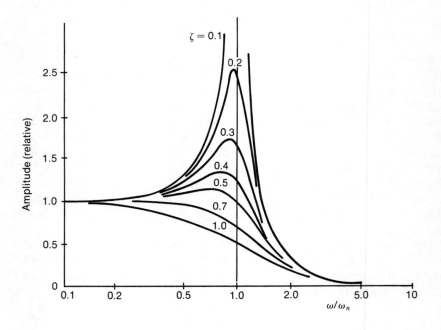

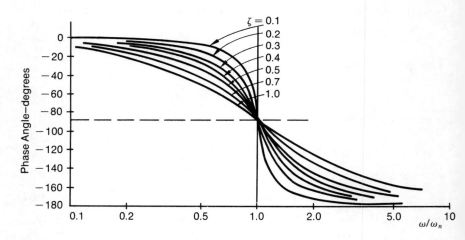

Fig. 9-11. Amplitude and phase plots of $1/[1 + 2j(\zeta\omega/\omega_n) - (\omega/\omega_n)^2]$.

0.707 of its maximum. Now, since $I = VY$, then,

$$I = \frac{V}{\sqrt{2}R} = 0.707 \frac{V}{R}. \tag{9-63}$$

From the fact that the power is $I^2R = 0.5 V^2/R$, these points are also referred to as the half-power points.

We designate the higher half-power frequency as ω_1 and the lower half-power frequency as ω_2. These are, respectively,

$$\begin{aligned}\omega_1 &= \omega_n(1+\zeta), \\ \omega_2 &= \omega_n(1-\zeta).\end{aligned} \tag{9-64}$$

The total frequency spread $\omega_1 - \omega_2$ is defined as the bandwidth of the circuit. Then,

$$B_w = \omega_1 - \omega_2 = 2\zeta\omega_n = \frac{\omega_n}{Q}. \tag{9-65}$$

From this expression we see that the bandwidth is narrow for high Q circuits, and is broad for low Q circuits. This means that to increase the bandwidth of a series circuit, the Q must be reduced, or correspondingly, the loading of the circuit must be increased. The relation of B_w to $|Y(j\omega)|$ is shown in Fig. 9-10.

Another view of the physical meaning of Q follows from Eq. (5-34), which is the transient response term of the series RLC circuit. This equation is here written in the form,

$$i_t = Be^{-\zeta\omega_n t} \sin(\omega_n\sqrt{1-\zeta^2}\,t + \theta). \tag{9-66}$$

For the case when $\zeta^2 = 1/4Q^2$ is small compared with unity, this equation becomes,

$$i_t \simeq Be^{-\zeta\omega_n t} \sin(\omega_n t + \theta). \tag{9-67}$$

This wave has the general characteristics shown in Fig. 9-12 (also Fig.

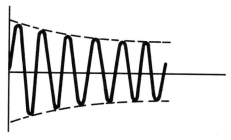

Fig. 9-12. Damped sine waves.

5-14). The amplitude of the wave falls in accordance with the exponential damping function $e^{-\zeta\omega_n t}$. The time between successive crossings of the axis is specified by,

$$T = \frac{2\pi}{\omega_n}. \tag{9-68}$$

Also, the time constant of the envelope is given by,

$$\tau = \frac{1}{\zeta\omega_n}. \tag{9-69}$$

We note that over one time constant of the envelope there are,

$$\frac{\tau}{T} = \frac{1}{\zeta\omega_n}\frac{\omega_n}{2\pi} = \frac{Q}{\pi} \text{ cycles.} \tag{9-70}$$

That is, there are Q/π cycles per time constant of the envelope. But since an exponential decays nearly to zero in 3 time constants, then it appears that the transient contains approximately Q cycles.

Still another interpretation of Q is possible. Begin with Eq. (9-58) and multiply and divide by πi^2. This is rearranged slightly to the following,

$$Q = \frac{\omega_n L}{R} = \frac{(2\pi)\frac{1}{2}Li^2}{\pi \cdot i^2 R/\omega_n} = \frac{2\pi(\frac{1}{2}Li^2)}{(i^2R/2) \cdot (2\pi/\omega_n)}. \tag{9-71}$$

Observe that the numerator is the peak stored energy (multiplied by 2π), and the denominator is the energy dissipated per cycle. Thus,

$$Q = 2\pi \frac{\text{peak stored energy}}{\text{energy dissipated per cycle}}. \tag{9-72}$$

As a final feature of the series RLC circuit resonant conditions, we note that,

$$V = V_R \qquad V_L = V_C,$$

from which it follows that,

$$V_L = \omega_n Li = \frac{\omega_n L}{R} RI = QV. \tag{9-73}$$

This shows that the voltage across the inductor and the capacitor can be many times the input voltage, and in this sense, the series RLC circuit under resonant or near resonant conditions will act as a voltage transformer with a step-up ratio equal to the circuit Q.

b. The Parallel Circuit. Many of the foregoing ideas that were developed from considerations of the series RLC circuit apply to the parallel GCL circuit. There are several aspects of these dual situations that warrant specific attention. Refer to Fig. 9-13 which shows the simple parallel GCL circuit. We write directly that,

$$Y(s) = G + Cs + \frac{1}{sL}. \qquad (9\text{-}74)$$

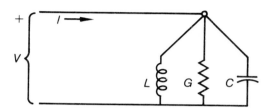

Fig. 9-13. The simple parallel GCL circuit.

We proceed with considerations of the inverse of this function, which is,

$$Z(s) = \frac{1}{Y(s)} = \frac{1}{C} \frac{s}{[s^2 + (Gs/C) + (1/CL)]}, \qquad (9\text{-}75)$$

which is the dual of Eq. (9-51). We define the quantities,

$$\omega_n = \frac{1}{\sqrt{LC}} \qquad \zeta = \frac{G}{2}\sqrt{\frac{L}{C}}. \qquad (9\text{-}76)$$

Then the discussion previously relating to $Y(j\omega)$ for the series RLC circuit applies to $Z(j\omega)$ for the parallel GCL circuit. Observe that while $Y(j\omega)$ for the *series resonant* circuit is a maximum (hence in the driving point impedance is a minimum $= R$), now we have that $Z(j\omega)$ for the *parallel antiresonant* circuit is a maximum $= 1/G$, which is often referred to as the "shunt" resistance. Antiresonant circuits are also discussed in terms of the circuit Q, which is defined by the relation,

$$Q = \frac{1}{\omega_n LG} = \frac{1}{G}\sqrt{\frac{C}{L}} = \frac{1}{2\zeta}. \qquad (9\text{-}77)$$

The bandwidth considerations are discussed in the same manner as above, and Eq. (9-65) remains valid for the antiresonant circuit.

As a final feature of the parallel GCL circuit under resonant conditions, refer to Fig. 9-14 which is merely a redrawing of Fig. 9-13. Also shown is a phasor diagram of the currents. The currents I_L and I_C at antiresonance are equal and opposite and the driving point current is a minimum $= I_G$;

368   System Response to Sinusoidal Functions

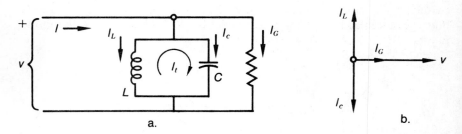

Fig. 9-14. (a) Figure 9-13 redrawn to show the circulating current in the tank. (b) The phasor relations of the currents.

thus,
$$I = I_G \qquad I_L = I_C,$$
from which it follows that,
$$I_L = \frac{V}{\omega_n L} = \frac{VG}{\omega_n LG} = QI_G. \tag{9-78}$$

This shows that the current through the inductor and capacitor can be many times the input current. In this sense, the parallel QCL circuit under resonant or near resonant conditions will act as a current transformer with a step-up ratio equal to the circuit Q.

9-6 THE $T(j\omega)$ PLANE

The system function $T(j\omega)$ assumes special importance in analysis because it is the basis for sinusoidal excitation studies. As already noted $T(j\omega)$ is a special case of $T(s)$ for $s = j\omega$ and can be written,
$$T(j\omega) = H \frac{(j\omega - s_1)(j\omega - s_2) \cdots}{(j\omega - s_a)(j\omega - s_b) \cdots}. \tag{9-79}$$

Clearly, as ω varies, $T(j\omega)$ will change its magnitude and argument. To study the general features of this variation, it is convenient to write $T(j\omega)$ in the form,
$$T(j\omega) = R(\omega) + jX(\omega), \tag{9-80}$$
where $R(\omega)$ denotes the real part, and $X(\omega)$ denotes the imaginary part of the system function, as already noted in Eq. (9-27). As ω varies, both $R(\omega)$ and $X(\omega)$ will vary, in general. It is convenient to display the variation graphically by plotting each value of $T(j\omega)$ as a point in the R, X

plane. As ω varies, the point R,X traces a curve in the R,X plane. This curve is the locus of R,X as ω varies. Because $T(j\omega)$ is appropriate to variations of a point at ω along the $j\omega$-axis in the s-plane, it is known as the image of the axis of imaginaries of the complex s-plane for the function $T(s)$, as already discussed in Section 8-5. Often it is known simply as the image-frequency diagram, the Nyquist diagram, or the $T(j\omega)$ plane diagram.

Refer to Fig. 9-15 which is a representative frequency locus diagram.

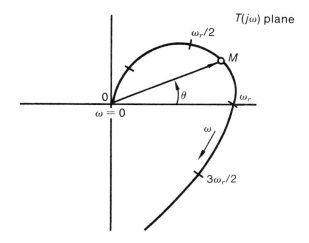

Fig. 9-15. The image of the imaginary axis for $T(s)$ — the $T(j\omega)$ locus.

The $(T(j\omega)$ locus possesses some special characteristics when $T(s)$ denotes a *driving point* impedance or admittance function. Now $R(\omega)$ is proportional to the active power dissipated in the network under sinusoidal conditions (*see* Section 9-3). For the system to remain passive, the absorbed power must be positive, which requires that the image must always lie to the right of the $X(\omega)$ axis.

9-7 MAGNITUDE-PHASE AND BODE PLOTS

Again refer to Fig. 9-15. For a sinusoidal excitation of unit amplitude, the line segment $0M$ represents the magnitude of the output, and angle θ denotes the phase angle between input and output. Thus the variation of magnitude and phase of the output as the frequency is varied is obtained from the curve in the manner illustrated. Often these data are presented as a pair of magnitude and phase versus ω diagrams, as illustrated in Fig. 9-16. We shall have frequent occasion to refer to such curves.

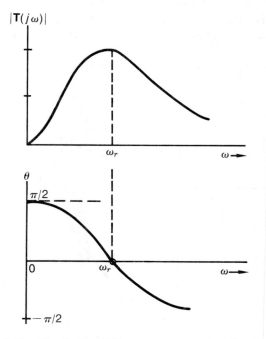

Fig. 9-16. Magnitude and Phase versus ω variations of $T(j\omega)$.

Often, instead of plotting $|T(j\omega)|$ and θ to linear scales, as in Fig. 9-16, it is convenient to plot these data on a logarithmic scale in the following manner,

$$\text{decibels (dB)} = 20 \log_{10}|T(j\omega)| \text{ versus } \log_{10}\omega,$$
$$\theta \text{ versus } \log_{10}\omega$$

as shown in Fig. 9-17. When plotted in this way, these are referred to as Bode plots. To understand the significant feature of the Bode plot, refer to the general expression for $T(s)$ which we now write,

$$T(s) = K \frac{(1+s/s_1)(1+s/s_2) \cdots}{(1+s/s_a)(1+s/s_b) \cdots},$$

where $K = H(s_1 s_2 \cdots / s_a s_b \cdots)$. We write this in decibel form, writing for

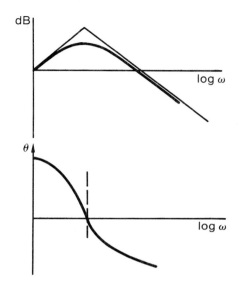

Fig. 9-17. Bode plot corresponding to Fig. 9-16.

each $s_k = 1/\tau_k$ which leads to the magnitude equation,

$$20 \log |T(j\omega)| = 20 \log K + 20 \log |1+j\omega\tau_1| + 20 \log |1+j\omega\tau_2| + \cdots$$
$$- 20 \log |1+j\omega\tau_a| - 20 \log |1-j\omega\tau_b| - \cdots \quad (9\text{-}81)$$
$$= 20 \log K + 20 \sum_{k=1}^{m} \log |1+j\omega\tau_k| - 20 \sum_{k=a}^{n} \log |1+j\omega\tau_k|.$$

This shows that a plot of $|T(j\omega)|$ versus frequency can be made by a simple point by point addition of each individual term in the summation on the right of Eq. (9-81). The corresponding angle function is,

$$\arg T(j\omega) = \sum_{k=1}^{m} \zeta_{rk} - \sum_{k=1}^{n} \eta_{rk}. \quad (9\text{-}82)$$

An approximate solution can be effected graphically by observing the asymptotic curves for the individual terms in Eq. (9-81). These curves are shown in Fig. 9-18 and are based on the fact that,

a. For $\omega\tau \ll 1$, $|1+j\omega\tau| \simeq 1$ so that $20 \log |1+j\omega\tau| \cong 0$ dB,
b. For $\omega\tau \gg 1$, $|1+j\omega\tau| \simeq \omega\tau$ $20 \log |1+j\omega\tau| \cong 20 \log \omega\tau$.

From b. we see that a factor of 10 in $\omega\tau$ means a 20 dB change per decade.

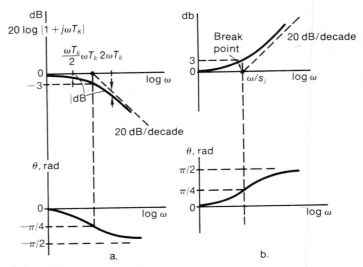

Fig. 9-18. Gain and Phase characteristics for; (a) a pole, (b) a zero, of $T(j\omega)$ at $\omega/s_k = \omega T_k$.

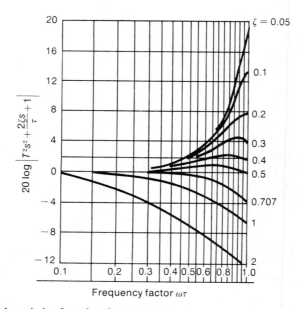

Fig. 9-19. Bode variation for pairs of conjugate poles.
(From *Automatic Feedback Control System Synthesis*, by J. G. Truxal. Copyright © 1955 McGraw-Hill Inc., used with permission of McGraw-Hill Book Company, New York.)

The intersection of these two asymptotes occurs at $\omega\tau = 1$. We note also,

c. The asymptotes for conjugate zeros or poles, which reflect themselves in terms of the form $\pm 20 \log |1 + j2\omega\zeta\tau - \omega^2\tau^2|$, will be roughly as shown in Fig. 9-19 but with a final slope of 20 dB/decade (12 dB/octave) since for $\omega\tau \gg 1$, $20 \log (\omega\tau)^2 = 40 \log \omega\tau$.

The exact variation of each term in Eq. (9-81) with relation to the asymptotic curves, and also the angle variation, are also illustrated in Fig. 9-18.

Example 9-7.1

Draw the Bode diagram for the function,

$$T(j\omega) = \frac{j\omega - 2}{j\omega}.$$

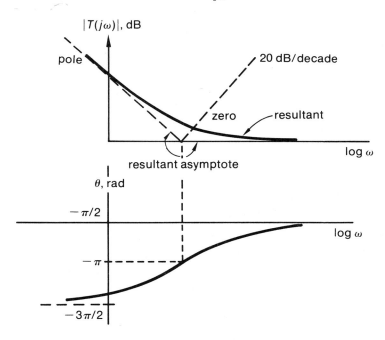

Fig. 9-7.1-1

Solution. The pole occurs at $\omega = 0$ and the zero occurs at $\omega = 2$. The required diagram has the features shown.

Example 9-7.2

The so-called integrating circuit is illustrated in the accompanying sketch. Show s-plane, $T(j\omega)$ plane, and Bode diagrams of the voltage transfer function.

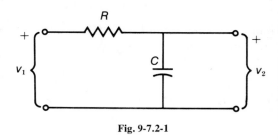

Fig. 9-7.2-1

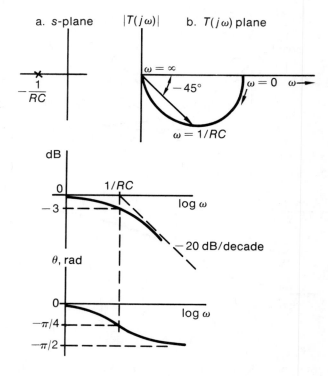

Fig. 9-7.2-2

Solution. The voltage ratio system function is,

$$\frac{V_2}{V_1} = T(s) = \frac{1/Cs}{R + 1/Cs} = \frac{1}{RCs + 1} = \frac{1/RC}{s + 1/RC}.$$

As a function of frequency, the system function is,

$$T(j\omega) = \frac{1}{j\omega RC + 1} = \frac{1/RC}{j\omega + 1/RC}.$$

We now examine three critical ranges,

a. For low frequencies $T(j\omega) \rightarrow 1$, with phase angle approximately zero.
b. For high frequencies $T(j\omega) \rightarrow 1/j\omega RC$, with angle approximately -90 deg.

It is this characteristic that accounts for the name, integrating circuit.

c. For frequency $\omega = 1/RC$, $T(j\omega) = 1/(1+j1) = 0.707 \underline{/-45}$.

The appropriate figures are given in Fig. 9-7.2-2.

Example 9-7.3

The transfer function of a particular system is,

$$T(s) = \frac{s(s+3)}{(s+7)(s^2 + 2s + 2)}.$$

Draw the overall characteristic from considerations of the asymptotic plot.

Solution. The function $T(s)$ is rewritten to make the constant term of each factor equal to unity, so that,

$$T(s) = \frac{3}{7 \times 2} \frac{s(0.33s + 1)}{(0.143s + 1)(0.5s^2 + s + 1)}.$$

The break frequencies are,

Break up at 20 dB/decade at $\omega = 3$ because of $(0.33s + 1)$.
Break down at 20 dB/decade at $\omega = 7$ because of $(0.143s + 1)$.
Break down at 40 dB/decade at $\omega = 1.414$ because of $(0.5s^2 + s + 1)$; also from $2\omega\zeta\tau = 1$, $\zeta = 0.354$.

The gain factor $\tfrac{3}{14}$ and the critical frequency at the origin require special attention. There is a 20 dB/decade rise owing to the factor s in the numerator, and the constant gain $\frac{3}{14}$. These two factors together, $j3\omega/14$ intersects 0 dB at $\omega = \frac{14}{3} = 4.67$. Thus the asymptotic plot starts at low frequencies as a line rising at 20 dB/decade toward 0 dB at $\omega = 4.67$.

The asymptotic plot has the form shown in the accompanying figure.

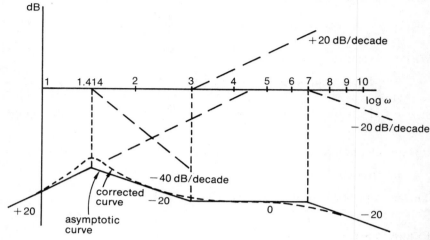

Fig. 9-7.3-1

PROBLEMS

9-1. A recurring wavetrain, as illustrated, is to be measured using voltmeters that respond as follows: full-wave average (a galvanometer type instrument), positive average (a rectifier instrument), negative average, positive peak (electronic instrument), negative peak, rms (root mean square). If the scale of each instrument is drawn to read $0.707 V_{\text{peak}}$ (the rms value of a sine wave) what will be the scale reading of each instrument?

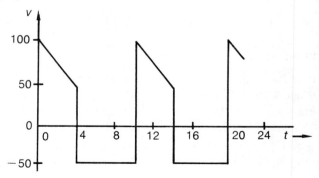

9-2. Determine the steady-state response for each of the following circuits. Draw phasor diagrams for each.

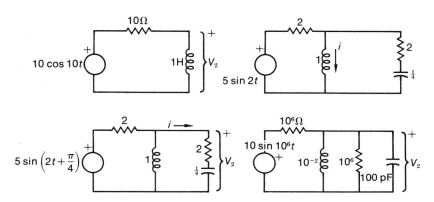

9-3. a. Determine expressions for the sinusoidal steady-state current in each capacitor in the networks given.
b. Draw phasor diagrams for each network.

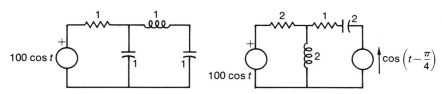

9-4. Determine v_2 and i in the network shown. Draw a phasor diagram for the circuit.

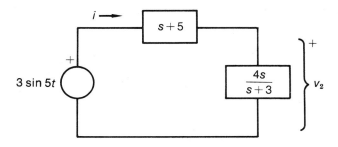

9-5. Refer to the accompanying figure.
a. Show that a state equation description is possible using the variables i_1, v_2, v_3.

378 System Response to Sinusoidal Functions

b. From a. find the corresponding phasor state vector X, and write $X(j\omega)$ as a function of $V = V\underline{/0}$.

c. Obtain expressions for the functions $I_1/V, I_2/V, V_3/V$.

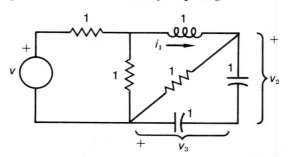

9-6. Illustrated is a 3-phase delta connected load to a balanced 3-phase Y-connected line (here the voltages are in fixed 120 deg. phase relation to each other).

a. Determine the line currents I_1, I_2, I_3 and the load currents I_a, I_b, I_c.
b. Determine the average power by each phase.
c. Determine the average power to the load.

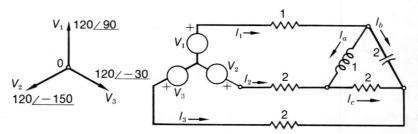

9-7. A 3-phase delta-connected supply provides power to a Y-connected load, as shown.

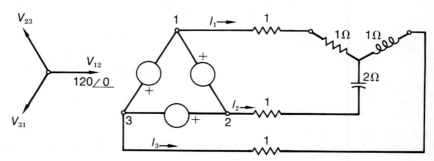

a. Determine the line currents I_1, I_2, I_3.
b. Determine the average power by each phase.
c. Determine the average power to the load.

9-8. Refer to the simple series RLC circuit, with circuit constants as follows: $R = 10$ ohm, $L = 7$ mH, $v = 2 \cos 2\pi 10^4 t$.
  a. Calculate the value of C for series resonance.
  b. Calculate the current through the circuit under these conditions.
  c. Calculate the voltage (amplitude and phase) across each element of the circuit.
  d. Determine the bandwidth of the circuit.

9-9. In the circuit shown, the parallel combination is in parallel (antiresonant) resonance.
  a. Calculate the value of C for parallel resonance.
  b. What is the voltage across the parallel resonant circuit?
  c. Determine the current through each element of the circuit. Comment on the current through the series R and that through the elements in the parallel circuit.
  d. Determine the bandwidth of the circuit.

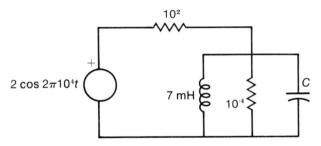

9-10. Consider a series RLC circuit with constant ω_n and adjustable ζ. Show that the locus of the roots of $Y(s)$ is a circle on the s-plane.

9-11. Show: s-plane, $T(j\omega)$ plane, magnitude and phase diagrams for the driving point impedance and admittance of the series RL circuit.

9-12. Consider the bridged-T network shown.

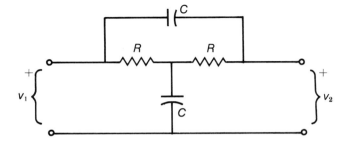

a. Find an expression for the system function V_2/V_1. Set up the problem using both the loop and the nodal analysis. In the latter case, a simple source transformation is of considerable help in simplifying the problem.

b. Plot the results as a function of frequency from $\omega = 0.1\omega_r$ to $\omega = 2\omega_r$ where $\omega_r = 1/RC$.

9-13. a. Deduce the system function $T(s) = V_2/I_1$ of the network given.

b. Plot $|T(j\omega)|$. What can be said about the transmission characteristics of this network?

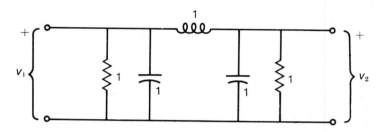

9-14. The p-z pattern of Butterworth filters of orders 3, 4, 5 are shown.

a. Deduce the analytic form for $T(s)$ for each.

b. Sketch $|T(j\omega)|$ for each, and discuss the filtering properties as the order increases.

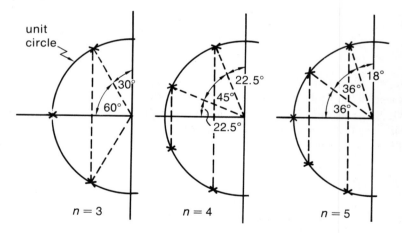

9-15. By graphical construction, deduce the frequency variation of networks that have the following specified pole-zero patterns: (1) poles $-2 \pm j5$; (2) poles $-2 \pm j5$, zero -4; (3) poles $-2 \pm j5$, zero -1. Plot the results on a polar diagram (Nyquist plot); magnitude and angle versus frequency; Bode diagram.

9-16. Asymptotic amplitude characteristics of $|Z(j\omega)|$ are shown. From these determine $Z(s)$.

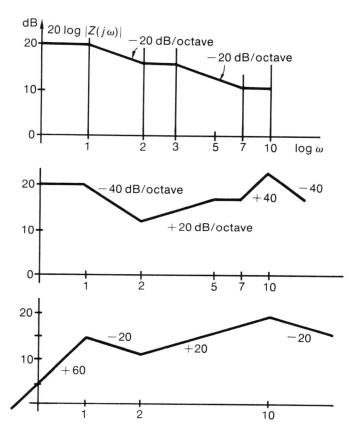

10

Special Topics in Systems Analysis

A number of special techniques and theorems have been developed as aids in the study of particular steady-state features of one-port and two-port linear, time invariant networks. This chapter will examine a number of these, since they are of considerable importance in our subsequent studies.

10-1 THÉVENIN AND NORTON THEOREMS†

The Thévenin and Norton theorems provide procedures for evaluating the voltage (the across variable) or the current (the through variable) at any point in a circuit which may be excited by a number of different sources at various points in the circuit, without setting up the network equations for the entire system. Actually, because of essential intermediate steps, the total computational effort is not often materially reduced when using these theorems. However, the theorems do provide important steps for studying general network properties.

Refer to Fig. 10-1 which represents a complex interconnected system of L loops and N node pairs. Furthermore, it is supposed that voltage sources may exist in any or all loops, and current sources may exist across any or all node pairs. Let us confine our attention to the current in a single branch, say the impedance Z which forms part of loop L, as indicated. The Thévenin theorem states that: "in so far as the perfor-

†The Thévenin theorem was developed by Helmholtz more than fifty years before it was rediscovered by Thévenin.

Thévenin and Norton Theorems

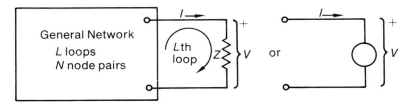

Fig. 10-1. One-port active network.

mance at the terminals of Z is concerned, the entire active network that is contained within the 'black box' can be replaced by a simple equivalent linear source consisting of a voltage driver and a series internal impedance." The proof of the theorem and the rules for finding the driver V_t and the internal impedance Z_t will be given. The essence of the Thévenin theorem is that Fig. 10-1 can be replaced by the network of Fig. 10-2.

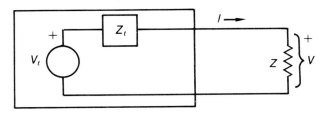

Fig. 10-2. Thévenin representation of Fig. 10-1.

A proof of this theorem follows from the general network equations, Eq. (4-45), which supposes that current sources are replaced by equivalent voltage sources which are associated with the loops. From the L loop equations, the solution for the current I in the Lth loop is given by Cramer's rule, and is,

$$I = V_1 \frac{\Delta_{L1}}{\Delta} + V_2 \frac{\Delta_{L2}}{\Delta} + \cdots + V_L \frac{\Delta_{LL}}{\Delta}. \qquad (10\text{-}1)$$

The determinants Δ_{Lj} are the appropriate cofactors of the system, and Δ is the system determinant. Since we are interested in the voltage across Z, it is convenient to write the Lth loop impedances as,

$$Z_{LL} = Z^0_{LL} + Z, \qquad (10\text{-}2)$$

where Z^0_{LL} denotes the Lth loop impedance Z_{LL} less impedance Z. This will have an effect on the form of the system determinant, which is written,

$$\Delta = Z_{1L}\Delta_{1L} + Z_{2L}\Delta_{2L} + \cdots + (Z_{LL}^0 + Z)\Delta_{LL}. \tag{10-3}$$

It is convenient to write this in the form,

$$\Delta = \Delta^0 + Z\Delta_{LL}, \tag{10-4}$$

where Δ^0 denotes the determinant corresponding to Δ, but with Z_{LL} replaced by Z_{LL}^0.

Equation (10-4) is combined with Eq. (10-1), with the result,

$$I(\Delta^0 + Z\Delta_{LL}) = V_1\Delta_{L1} + V_2\Delta_{L2} + \cdots + V_L\Delta_{LL}. \tag{10-5}$$

We write from this,

$$V = IZ = \frac{V_1\Delta_{L1} + V_2\Delta_{L2} + \cdots + V_L\Delta_{LL}}{\Delta_{LL}} - I\frac{\Delta^0}{\Delta_{LL}}. \tag{10-6}$$

Observe that this expression may now be written in the form,

$$V = V_t - Z_t I,$$

where, $\tag{10-7}$

$$V_t = \frac{V_1\Delta_{L1} + \cdots + V_L\Delta_{LL}}{\Delta_{LL}} \qquad Z = \frac{\Delta^0}{\Delta_{LL}}.$$

This equation permits a simple graphical representation, and is precisely the situation that is illustrated in Fig. 10-2.

It would be difficult to interpret the forms for V_t and Z_t from Eq. (10-6). However, they allow ready interpretation from the first of Eq. (10-7). Let us suppose that each internal source in the network is inactivated (voltage sources are reduced to zero and current sources are open-circuited, but associated impedances, if any, are to remain in the circuit). It then follows that,

$$-\frac{V}{I} = \frac{\Delta^0}{\Delta_{LL}}. \tag{10-8}$$

This shows that the impedance looking back into the output terminals under the prescribed conditions of inactivated sources is precisely Z_t. Again consider the original system, but suppose that the impedance Z is removed, thereby leaving the output terminals to be open circuited. We examine the voltage across the open-circuited terminals. From Eq. (10-7), if $I = 0$ as required, then $V = V_t$. Thus V_t is the voltage across the output terminals under open-circuited conditions. This thus specifies the Thévenin source characteristics.

The Norton theorem is the dual of the Thévenin theorem. It specifies that in so far as the performance at the terminals of Z is concerned, the active circuit of Fig. 10-1 can be replaced by a current source shunted by a passive admittance, as shown in Fig. 10-3. The result follows readily from a rearrangement of Eq. (10-7). We write,

$$I = \frac{V_t}{Z_t} - \frac{V}{Z_t}, \tag{10-9}$$

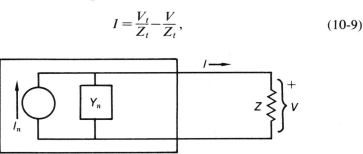

Fig. 10-3. Norton representation of Fig. 10-1.

which is represented by the equation,

$$I = I_n - Y_n V,$$

where, $\tag{10-10}$

$$I_n = \frac{V_t}{Z_t} \qquad Y_n = \frac{1}{Z_t}.$$

A physical interpretation of Eq. (10-10) is possible in much the same way that we interpreted Eq. (10-7). Suppose that the output terminals are short-circuited, with all excitation sources being maintained at their normal values. This requires that $V = 0$, and clearly, the current $I = I_n$. Thus it is seen that I_n is the current through a short-circuit across the output terminals. To interpret Y_n we note that since $Y_n = 1/Z_t$, then Y_n is the admittance of the network at the output terminals when all internal sources are inactivated, and with Z removed.

10-2 MAXIMUM POWER TRANSFER THEOREMS

Refer again to Fig. 10-1 and suppose that all sources are sinusoidal of frequency ω, so that $s = j\omega$. The question now to be considered is the following: What should be the value of Z in order that it absorb the maximum power from the connected circuit? An answer is readily deduced if the general network is replaced by the Thévenin equivalent of

Fig. 10-2. Suppose that the essential quantities have the forms,

$$Z = Ze^{j\theta} \qquad Z_t = Z_t e^{j\theta_t} \qquad V_t = V_t e^{j\beta}. \qquad (10\text{-}11)$$

The average power by the equivalent Thévenin source is given by Eq. (9-75) and is,

$$P = \tfrac{1}{2} Re\,(VI^*) = \tfrac{1}{2}|I^2|R \qquad (V \text{ is peak value})$$

$$= \frac{1}{2}\left(\frac{V_t}{Z_s}\right)^2 Z \cos\theta,$$

where,

$$Z_s = Z_t + Z = |Z_s|e^{j\theta_s}. \qquad (10\text{-}12)$$

From the fact that Z_s is the sum of two complex numbers, we write for the magnitude,

$$Z_s^2 = Z_t^2 + Z^2 + 2Z_t Z \cos(\theta - \theta_t). \qquad (10\text{-}13)$$

Because of this, Eq. (10-12) becomes,

$$P = \frac{\tfrac{1}{2}V_t^2 Z \cos\theta}{Z_t^2 + Z^2 + 2Z_t Z \cos(\theta - \theta_t)}. \qquad (10\text{-}14)$$

We observe that this equation for P is a function of the two variables Z and θ, and we shall use this function to provide an answer to our present query.

To find the conditions under which the power is a maximum, we examine the expression for dP. This is,

$$dP = \frac{\partial P}{\partial Z}\,dZ + \frac{\partial P}{\partial \theta}\,d\theta. \qquad (10\text{-}15)$$

To deduce the conditions for P to be a maximum, we set the derivative to zero (it is readily proved that this is a maximum and not a minimum.) We write,

$$\frac{\partial P}{\partial Z} = 0 \qquad \frac{\partial P}{\partial \theta} = 0. \qquad (10\text{-}16)$$

Several conditions are possible.

1. If both conditions are satisfied simultaneously, the power is a maximum.

2. If one derivative is zero but the other is not, then the power is a maximum under the particular constraint. For example, if $\partial P/\partial Z$ only is zero, then P is a maximum for fixed value of θ. Likewise for fixed value of Z, the power is a maximum when $\partial P/\partial \theta$ is zero. We examine the two possibilities.

Maximum Power Transfer Theorems

We first assume that θ is maintained constant, and so we examine the quantity $\partial P/\partial Z$. We now have that,

$$\frac{\partial P}{\partial Z} = \frac{1}{2} \frac{Z^2 + Z_t^2 + 2Z_t Z \cos(\theta - \theta_t) - Z[2Z + 2Z_t \cos(\theta - \theta_t)]}{[Z^2 + Z_t^2 + 2ZZ_t \cos(\theta - \theta_t)]^2} V_t^2 \cos\theta. \tag{10-17}$$

When this expression is set equal to zero, we find the requirement to be,

$$Z = Z_t. \tag{10-18}$$

This condition means that if the angle of the load is a constant, the power is a maximum when the magnitude of the load equals the magnitude of the internal impedance of the source. An important example of this would be the use of a pure resistance load on a circuit with a reactive internal impedance.

Now suppose that Z is held constant, but that the phase angle may be controllable. Now we find that,

$$\frac{\partial P}{\partial \theta} = \frac{1}{2} \frac{[Z^2 + Z_t^2 + 2Z_t Z \cos(\theta - \theta_t)](-Z \sin\theta) + Z \cos\theta[2Z_t Z \sin(\theta - \theta_t)]}{[Z^2 + Z_t^2 + 2ZZ_t \cos(\theta - \theta_t)]^2} V_t^2. \tag{10-19}$$

When this expression is set equal to zero, for maximum constrained power, the results require that,

$$\sin\theta = -\frac{2ZZ_t}{Z^2 + Z_t^2} \sin\theta_t. \tag{10-20}$$

A most important case occurs when both conditions for maximum power are met simultaneously. Under these circumstances,

$$Z = Z_t,$$

and

$$\sin\theta = -\sin\theta_t, \tag{10-21}$$

or equivalently,

$$\theta = -\theta_t.$$

Under these conditions Z is the complex conjugate of Z_t, which thus specifies the existence of series resonance. The circuit is then purely resistive, with the external resistance matching the internal resistance. Now one half of the power is dissipated in Z_t, the remaining half being supplied to the load. The maximum power under conditions is,

$$P_{\text{max,max}} = \frac{1}{2}\left(\frac{V}{2R}\right)^2 R = \frac{1}{8}\frac{V_{\text{peak}}^2}{R} = \frac{1}{4}\frac{V_{\text{rms}}^2}{R}. \tag{10-22}$$

10-3 SOURCE TRANSFORMATION

The idea of source transformations was introduced in Sections 1-10, 3-7 and 3-8. The discussion relating to the Thévenin and Norton theorems helps in an understanding of equivalent sources. The value of source transformations rests in the ability to transform a voltage driver into an equivalent current driver, or vice versa, a transformation that can be of considerable convenience in a specific case.

The essential idea of the source transformation rests in the fact that relative to a pair of terminals in a network, a voltage driver plus a series impedance can be replaced by a current driver and shunting admittance. The situation is illustrated in Fig. 10-4. For a specified V and I, Fig. 10-4a leads to the relation,

$$V_s - Z_s I = V. \tag{10-23}$$

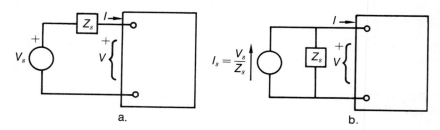

Fig. 10-4. Equivalent sources in a network.

If we now define the quantity,

$$V_s = Z_s I_s, \tag{10-24}$$

then we may write,

$$I_s = \frac{V}{Z_s} + I = \frac{V_s}{Z_s}, \tag{10-25}$$

which corresponds to Fig. 10-4b. The general similarity of Figs. 10-4 with Figs. 10-2 and 10-3 is evident. However, here one is not representing a complete active circuit by a simplified representation, but is concerned with individual sources and their transformation.

10-4 TWO-PORT PASSIVE NETWORKS; y-SYSTEM EQUATIONS

A problem of some interest is the response of a network at one pair of terminals when an excitation is applied to another pair of terminals else-

Two-Port Passive Networks; y-System Equations 389

where in the network. The network is assumed to be passive. Here, as in the discussion of the Thévenin theorem, there is the implication that the total network, except for the input and output ports can be represented by a relatively simple equivalent network. The situation is illustrated in Fig. 10-5. We may write directly, on the basis of previous discussion, equations on a KVL basis, and then solve for the currents. The results are,

$$I_1 = \frac{\Delta_{11}}{\Delta} V_1 + \frac{\Delta_{12}}{\Delta} V_2,$$
$$I_2 = \frac{\Delta_{21}}{\Delta} V_1 + \frac{\Delta_{22}}{\Delta} V_2, \qquad (10\text{-}26)$$

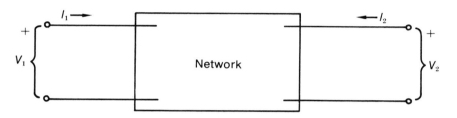

Fig. 10-5. A general network with two isolated ports.

where Δ is the network determinant, and $\Delta_{12} = \Delta_{21}$ are the first co-factors. We note specifically that the *reciprocity* condition $\Delta_{12} = \Delta_{21}$ is valid, in general, only for passive networks. We shall later discuss the case for active networks, and in this case the reciprocity conditions will no longer apply. The meaning of the term "active" network and its relation to the presence of dependent sources will be considered in Section 10-10.
We here introduce the quantities,

$$y_{11} = \frac{\Delta_{11}}{\Delta} \qquad y_{12} = y_{21} = \frac{\Delta_{12}}{\Delta} = \frac{\Delta_{21}}{\Delta} \qquad y_{22} = \frac{\Delta_{22}}{\Delta}, \qquad (10\text{-}27)$$

in which case, Eqs. (10-26) become,

$$I_1 = y_{11}V_1 + y_{12}V_2,$$
$$I_2 = y_{21}V_1 + y_{22}V_2. \qquad (10\text{-}28)$$

These are often written in matrix form,

$$\begin{bmatrix} I_1 \\ I_2 \end{bmatrix} = \begin{bmatrix} y_{11} & y_{12} \\ y_{21} & y_{22} \end{bmatrix} \cdot \begin{bmatrix} V_1 \\ V_2 \end{bmatrix}. \qquad (10\text{-}29)$$

These equations are called the *y*-system equations of the equivalent two-port network. The coefficients y_{jk} have direct interpretation from Eq. (10-28) and are,

$$y_{11} = \left.\frac{I_1}{V_1}\right|_{V_2=0} \quad \text{Driving point admittance at port 1 when port 2 is short-circuited.}$$

$$y_{22} = \left.\frac{I_2}{V_2}\right|_{V_1=0} \quad \text{Driving point admittance at port 2 when port 1 is short-circuited.}$$

$$y_{12} = \left.\frac{I_1}{V_2}\right|_{V_1=0} \quad \text{Transfer admittance when input is short-circuited.}$$

$$y_{21} = \left.\frac{I_2}{V_1}\right|_{V_2=0} \quad \text{Transfer admittance when output is short-circuited.}$$

(10-30)

Because of the form of these coefficients, the *y*-system description is often called the short-circuit description of the network.

Example 10-4.1

Determine the *y*-system parameters of the accompanying network.

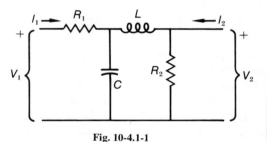

Fig. 10-4.1-1

Solution. We proceed with the calculations according to the definitions given in Eq. (10-30). From the reduced network as shown in Fig. 10-4.1-2,

$$y_{11} = \frac{1}{R_1 + \dfrac{L/C}{Ls + 1/Cs}}$$

$$= \frac{LCs^2 + 1}{R_1 LCs^2 + Ls + R_1}$$

$$= \frac{LCs^2 + 1}{D}.$$

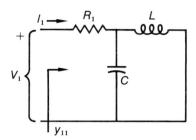

Fig. 10-4.1-2

For y_{22}, use Fig. 10-4.1-3,

$$y_{22} = \frac{1}{R_2 \| \left[Ls + \dfrac{R_1/Cs}{R_1 + 1/Cs} \right]} = \frac{1}{R_2 \| \left[\dfrac{D}{R_1Cs+1} \right]} = \frac{1}{\dfrac{R_2[D/(R_1Cs+1)]}{R_2 + D/(R_1Cs+1)}},$$

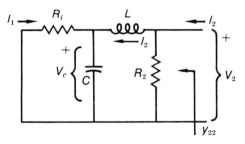

Fig. 10-4.1-3

where $\|$ denotes the parallel combination.

$$y_{22} = \frac{R_1LCs^2 + (L + R_1R_2C)s + (R_1 + R_2)}{R_2 D}$$

For y_{12}, use Fig. 10-4.1-3,

$$I_L = \frac{V_2}{Ls + R_1/(R_1Cs+1)} = \frac{V_2(R_1Cs+1)}{D},$$

$$V_c = I_L \frac{R_1}{R_1Cs+1} = V_2 \frac{R_1}{D},$$

$$I_1 = -\frac{V_c}{R_1} = -\frac{V_2}{D}.$$

Then,
$$y_{12} = \frac{I_1}{V_2} = -\frac{1}{D}.$$

10-5 z-SYSTEM EQUATIONS

The equation set given in Eq. (10-28) can be inverted to yield equations for the V-s in terms of the I-s. The result of solving these equations is the set,
$$\begin{aligned} V_1 &= z_{11}I_1 + z_{12}I_2, \\ V_2 &= z_{21}I_1 + z_{22}I_2, \end{aligned} \tag{10-31}$$

which, in matrix form, is,
$$\begin{bmatrix} V_1 \\ V_2 \end{bmatrix} = \begin{bmatrix} z_{11} & z_{12} \\ z_{21} & z_{22} \end{bmatrix} \begin{bmatrix} I_1 \\ I_2 \end{bmatrix}. \tag{10-32}$$

That is, beginning with Eq. (10-29), which is written in the form,
$$I = YV,$$

and by premultiplying by Y^{-1}
$$Y^{-1}I = V = ZI,$$

so that,
$$Z = Y^{-1},$$

where, $\tag{10-33}$
$$\begin{bmatrix} z_{11} & z_{12} \\ z_{21} & z_{22} \end{bmatrix} = \begin{bmatrix} y_{11} & y_{12} \\ y_{21} & y_{22} \end{bmatrix}^{-1}.$$

It follows from this that the z-coefficients, in terms of the y-coefficients, are,
$$z_{11} = \frac{y_{22}}{\Delta y} \quad z_{12} = z_{21} = -\frac{y_{12}}{\Delta y} \quad z_{22} = \frac{y_{11}}{\Delta y},$$

with $\tag{10-34}$
$$\Delta y = \begin{vmatrix} y_{11} & y_{12} \\ y_{21} & y_{22} \end{vmatrix} = y_{11}y_{22} - y_{12}y_{21}.$$

The z-s may be interpreted directly from Eq. (10-31) in a manner that parallels Eq. (10-30). We write,
$$z_{11} = \frac{V_1}{I_1}\bigg|_{I_2=0} = \text{Driving point impedance at port 1 when port 2 is open-circuited.}$$

$$z_{22} = \left.\frac{V_2}{I_2}\right|_{I_1=0} = \text{Driving point impedance at port 2 when port 1 is open-circuited.}$$

$$z_{12} = \left.\frac{V_1}{I_2}\right|_{I_1=0} = \text{Transfer impedance when input is open-circuited (sometimes called the reverse transfer impedance).} \quad (10\text{-}35)$$

$$z_{21} = \left.\frac{V_2}{I_1}\right|_{I_2=0} = \text{Transfer impedance when output is open-circuited (sometimes called the forward transfer impedance).}$$

Because of the definition of these quantities, they are known as the open-circuit impedance parameters.

Example 10-5.1

Deduce the z-system parameters for the network of Example 10-4.1.

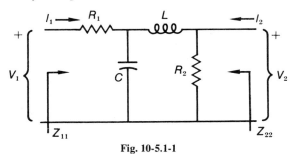

Fig. 10-5.1-1

Solution. Use will be made of the basic definitions in Eq. (10-35). Refer to Fig. 10-5.1-1. We write,

$$z_{11} = R_1 + \frac{1}{Cs} \| (R_2 + Ls) = R_1 + \frac{(R_2 + Ls)/Cs}{R_2 + Ls + 1/Cs}$$
$$= \frac{R_1 C L s^2 + (R_1 R_2 C + L)s + (R_1 + R_2)}{D_1},$$

where $D_1 = LCs^2 + R_2 Cs + 1$,

$$z_{22} = R_2 \| \left(Ls + \frac{1}{Cs}\right) = \frac{R_2(LCs^2 + 1)}{LCs^2 + R_2 Cs + 1} = \frac{R_2(LCs^2 + 1)}{D_1}.$$

To deduce z_{12}, we note from Fig. 10-5.1-2,

$$V_1 = \frac{V_2}{Ls + 1/Cs} 1/Cs = \frac{V_2}{LCs^2 + 1},$$

$$I_2 = \frac{V_2}{R_2} + \frac{V_2}{Ls+1/Cs} = V_2\left(\frac{1}{R_2} + \frac{Cs}{LCs^2+1}\right) = V_2\frac{D_1}{R_2\cdot(LCs^2+1)},$$

$$z_{12} = \frac{V_1}{I_2} = \frac{1}{D_1}.$$

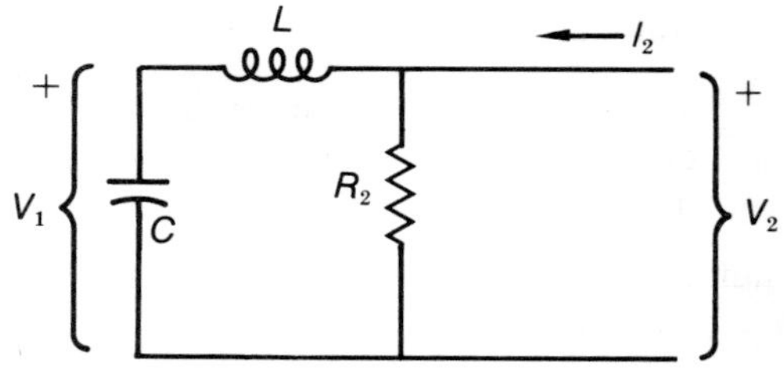

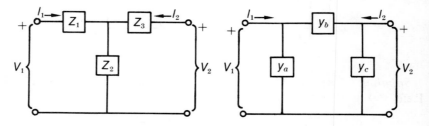

Fig. 10-5.1-2

10-6 T AND Π EQUIVALENT NETWORKS

The foregoing two sections have shown that a two-port network may be expressed in terms of the y-system or the z-system sets of parameters. We shall use these results to show that the T and Π networks are natural representations for the two-port. Let us refer to Fig. 10-6. The values of the y- and z-coefficients, defined respectively by Eqs. (10-30) and (10-35),

Fig. 10-6. T and Π two-port networks.

are tabulated for both networks. These are contained in Table 10-1. Clearly, either the T or the Π network may be used to represent the two-port network. However, it is seen that the Π network provides a natural representation for the y-system, whereas the T-network is a natural representation for the z-system set.

Observe that y-parameters and the z-parameters relative to the specified two-port have been defined in terms of the original network parameters, and as such, they are physically realizable quantities. For the equivalent Π- and T-networks; even though the y- and z-coefficients are

TABLE 10-1. System parameters for T and Π networks.

Parameter	T section	Π section
y_{11}	$\dfrac{y_1(y_2+y_3)}{y_1+y_2+y_3}$	y_a+y_b
y_{22}	$\dfrac{y_3(y_1+y_2)}{y_1+y_2+y_3}$	y_b+y_c
$y_{21}=y_{12}$	$\dfrac{y_1 y_3}{y_1+y_2+y_3}$	y_b
z_{11}	z_1+z_2	$\dfrac{z_a(z_b+z_c)}{z_a+z_b+z_c}$
z_{22}	z_2+z_3	$\dfrac{z_c(z_a+z_b)}{z_a+z_b+z_c}$
$z_{12}=z_{21}$	z_2	$\dfrac{z_a z_c}{z_a+z_b+z_c}$

physically realizable, this does not require that the elements of the Π- and T-networks must necessarily be physically realizable, in order to represent the original network. To be specific, we note that y_{11} must be physically realizable, but the relation $y_{11} = y_a+y_b$ does not carry with it the requirement that y_b must necessarily be realizable by real passive elements (e.g., y_b can, in the general case, involve negative resistance, inductance or capacitance). For y_a, y_b, y_c to be separately realizable by passive elements requires that special conditions be met. It is important to realize that physical realizability is not essential for computational purposes. Note, however, that negative impedances are possible using active circuits.

10-7 HYBRID PARAMETERS

As already discussed in Section 1-19, the choice of I_1 and V_2 as independent variables leads to a network description that is used extensively in transistor circuit analysis. In this case the network description is written,

$$V_1 = h_{11}I_1 + h_{12}V_2,$$
$$I_2 = h_{21}I_1 + h_{22}V_2. \tag{10-36}$$

These equations are called hybrid equations, and the h-coefficients are called *hybrid parameters* because both open-circuit and short-circuit conditions are specified in defining the h-parameters from the general

396 Special Topics in Systems Analysis

network. The interpretation of the h-s appearing in these equations proceeds directly from Eq. (10-36). A feature of the h-parameters is that they lend themselves to direct measurement. We see that,

$$h_{11} = h_i = \frac{V_1}{I_1}\bigg|_{V_2=0} = \text{Driving point impedance at port 1 when port 2 is short-circuited.}$$

$$h_{12} = h_r = \frac{V_1}{V_2}\bigg|_{I_1=0} = \text{Reverse voltage ratio for open-circuited port 1.}$$

$$h_{21} = h_f = \frac{I_2}{I_1}\bigg|_{V_2=0} = \text{Forward current ratio for short-circuited port 2.}$$

$$h_{22} = h_o = \frac{I_2}{V_2}\bigg|_{I_1=0} = \text{Driving point admittance of port 2 when port 1 is open-circuited.}$$

(10-37)

The notation that is common in transistor circuit applications is also included in these equations.

Example 10-7.1

Deduce the hybrid parameters for the network of Example 10-4.1.

Solution. Refer to the definitions given in Eq. (10-37). We see that $h_{11} = (y_{11})^{-1}$ by comparison of Eq. (10-37) with Eq. (10-30). Thus,

$$h_{11} = \frac{D}{LCs^2 + 1},$$

$h_{22} = 1/z_{22}$ by comparison of Eq. (10-37) with Eq. (10-35),

$$h_{22} = \frac{R_2(LCs^2 + 1)}{D_1}.$$

By reference to Fig. 10-4.1-1 with $I_1 = 0$ when $V_1 = V_c$,

$$V_1 = \frac{V_2}{Ls + 1/Cs}\frac{1}{Cs} = \frac{V_2}{LCs^2 + 1},$$

from which,

$$h_{12} = \frac{1}{LCs^2 + 1}.$$

Clearly, of course, since the h-parameters arise in an alternate description of the two-port, they are related to the z- and y-parameters. The explicit relations among these coefficients are readily deduced. For example, by solving Eqs. (10-36) and (10-31) we find the results to be,

$$h_{11} = h_i = z_{11} - \frac{z_{12}z_{21}}{z_{22}} \qquad h_{12} = h_r = \frac{z_{12}}{z_{22}},$$
$$h_{21} = h_f = -\frac{z_{21}}{z_{22}} \qquad h_{22} = h_o = \frac{1}{z_{22}}. \qquad (10\text{-}38)$$

The inverse relations are the following,

$$z_{11} = h_i - \frac{h_r h_f}{h_o} \qquad z_{12} = \frac{h_r}{h_o},$$
$$z_{21} = -\frac{h_f}{h_o} \qquad z_{22} = \frac{1}{h_o}. \qquad (10\text{-}39)$$

These parameters will receive further attention in Section 10-10.

A second set of hybrid parameters involving the complementary quantities I_1 and V_2 as dependent variables can be written by inverting Eq. (10-36). This set has no particularly special characteristics, and moreover, is of little interest in our subsequent studies. For this reason, we shall not give these any attention.

10-8 CASCADE PARAMETERS, abcd COEFFICIENTS

Two important cascade or chain type representations of the two-port network exist. The first of these is referred to as the *abcd* representation. In this description, the network equations are written,

$$V_1 = aV_2 + bI_2,$$
$$I_1 = cV_2 + dI_2, \qquad (10\text{-}40)$$

which, in matrix form, is,

$$\begin{bmatrix} V_1 \\ I_1 \end{bmatrix} = \begin{bmatrix} a & b \\ c & d \end{bmatrix} \cdot \begin{bmatrix} V_2 \\ I_2 \end{bmatrix}. \qquad (10\text{-}41)$$

We can proceed, as in the previous network descriptions, to define the network coefficients. These are,

$$a = \frac{V_1}{V_2}\bigg|_{I_2=0} = \text{Reverse voltage ratio when port 2 is open-circuited.}$$

$$b = \frac{V_1}{I_2}\bigg|_{V_2=0} = \text{Transfer impedance when port 2 is short-circuited.}$$

$$c = \frac{I_1}{V_2}\bigg|_{I_2=0} = \text{Transfer admittance when port 2 is open-circuited.} \qquad (10\text{-}42)$$

$$d = \frac{I_1}{I_2}\bigg|_{V_2=0} = \text{Reverse current ratio for port 2 under short-circuited conditions.}$$

Special Topics in Systems Analysis

These network parameters can, of course, be written in terms of other sets of parameters that describe the same network. The following relations can be shown to exist,

$$a = \frac{z_{11}}{z_{21}} = \frac{y_{22}}{y_{21}} \qquad b = \frac{\Delta_z}{z_{21}} = \frac{1}{y_{21}},$$

$$c = \frac{1}{z_{21}} = \frac{\Delta_y}{y_{21}} \qquad d = \frac{z_{22}}{z_{21}} = \frac{y_{11}}{y_{21}}, \qquad (10\text{-}43)$$

$$\Delta_z = z_{11}z_{22} - z_{12}z_{21} \qquad \Delta_y = y_{11}y_{22} - y_{12}y_{21}.$$

Now let us consider the case when two two-port networks are connected in cascade, as illustrated in Fig. 10-7. Suppose that each network is specified by its $abcd$ coefficients. The second network is denoted by primed quantities. Corresponding to the two networks are the equations,

$$\begin{bmatrix} V_1 \\ I_1 \end{bmatrix} = \begin{bmatrix} a & b \\ c & d \end{bmatrix} \begin{bmatrix} V_2 \\ I_2 \end{bmatrix} \qquad \begin{bmatrix} V_1' \\ I_1' \end{bmatrix} = \begin{bmatrix} a' & b' \\ c' & d' \end{bmatrix} \begin{bmatrix} V_2' \\ I_2' \end{bmatrix}. \qquad (10\text{-}44)$$

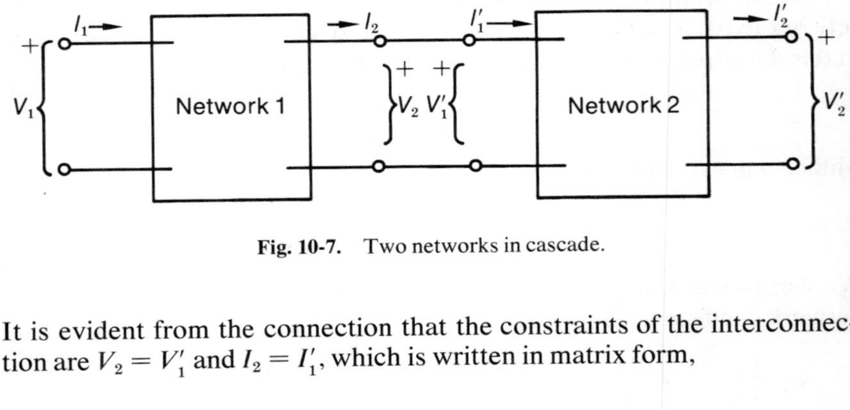

Fig. 10-7. Two networks in cascade.

It is evident from the connection that the constraints of the interconnection are $V_2 = V_1'$ and $I_2 = I_1'$, which is written in matrix form,

$$\begin{bmatrix} V_2 \\ I_2 \end{bmatrix} = \begin{bmatrix} V_1' \\ I_1' \end{bmatrix}. \qquad (10\text{-}45)$$

It is evident therefore that we may write for the networks in cascade,

$$\begin{bmatrix} V_1 \\ I_2 \end{bmatrix} = \begin{bmatrix} a & b \\ c & d \end{bmatrix} \begin{bmatrix} a' & b' \\ c' & d' \end{bmatrix} \begin{bmatrix} V_2' \\ I_2' \end{bmatrix}. \qquad (10\text{-}46)$$

The *abcd* formulation is of importance in the study of filters from an image parameter point of view, a subject that is not to be studied here.

An alternative approach to the analysis of networks in cascade is in terms of the voltage or current gain. This is a particularly useful form and is used extensively in active network analysis, whether it employs transistors, tubes, or other active devices. In this description, the transistor or tube is represented in the circuit by its z-, y- or h-parameter form, but the total circuit performance is discussed in terms of the signal properties, either voltage, current, or some related quantities. Refer again to Fig. 10-7, and suppose that each block represents an amplifier stage. In this circuit V_1 denotes the input voltage applied to the input terminals of the first stage. If each stage is considered as being independent of the following stage but with the output of one stage being the input to the next stage, then it follows that,

$$A_1 = [T_1(s)] = \frac{V_2}{V_1} = \frac{\text{output voltage of first stage}}{\text{input voltage to first stage}},$$

and

$$A_2 = [T_2(s)] = \frac{V_2'}{V_1'} = \frac{\text{output voltage of second stage}}{\text{input voltage to second stage}},$$
$$= \frac{\text{output voltage of second stage}}{\text{output voltage of first stage}}.$$

The resultant overall gain is,

$$A = [T(s)] = \frac{\text{output voltage of second stage}}{\text{input voltage to first stage}}. \quad (10\text{-}47)$$

It follows from these expressions that,

$$A = A_1 A_2. \quad (10\text{-}48)$$

The overall frequency characteristic of the cascaded network is the product of the separate network frequency characteristics, and is, therefore,

$$T(s) = T_1(s) T_2(s), \quad (10\text{-}49)$$

which is just the product of the appropriate system functions of each stage. Attention is called to the fact that this property was implicit in the SFG and in the subsequent analog computer programming.

10-9 INPUT, OUTPUT, AND TRANSFER IMPEDANCES

Often there is interest in the driving point or input impedance to a terminated two-port. This is the impedance at the input of the terminated two-port, without regard for the properties of the driving stage. To find this, refer to Fig. 10-8. Observe that this figure shows the two-port to be terminated in the impedance,

$$Z_2 = -\frac{V_2}{I_2}. \tag{10-50}$$

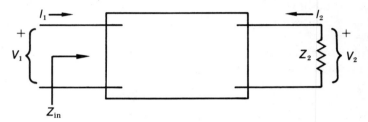

Fig. 10-8. To determine the input impedance of a network.

We employ Eq. (10-31) to describe the network, and we write,

$$\begin{aligned} V_1 &= z_{11}I_1 + z_{12}I_2, \\ 0 &= z_{21}I_1 + (z_{22} + Z_2)I_2. \end{aligned} \tag{10-51}$$

These equations are solved for I_1, which is,

$$I_1 = \frac{\Delta_1}{\Delta} = \frac{\begin{vmatrix} V_1 & z_{12} \\ 0 & z_{22} + Z_2 \end{vmatrix}}{\begin{vmatrix} z_{11} & z_{12} \\ z_{21} & z_{22} + Z_2 \end{vmatrix}}.$$

This expands to,

$$I_1 = \frac{V_1(z_{22} + Z_2)}{z_{11}(z_{22} + Z_2) - z_{12}z_{21}}. \tag{10-52}$$

From this we can write,

$$Z_{\text{in}} = \frac{V_1}{I_1} = z_{11} - \frac{z_{12}z_{21}}{z_{22} + Z_2}. \tag{10-53}$$

This expression shows that the input impedance under load equals the open-circuit input impedance modified by the coupled effect of the output circuit and the load.

The output impedance is deduced in a somewhat similar way, but now we refer to Fig. 10-9. Observe that the input properties of the input source are included, but any source voltage is reduced to zero. An excitation source is now applied to the output terminals. The result is readily found to be,

$$Z_{out} = z_{22} - \frac{z_{12}z_{21}}{z_{11}+Z_1}. \qquad (10\text{-}54)$$

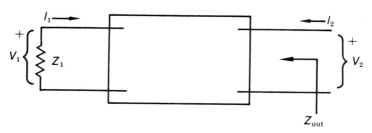

Fig. 10-9. To determine the output impedance of a network.

The interpretation of this expression parallels that of Eq. (10-53) and shows that the output impedance of the loaded network is the open-circuited output impedance modified by the coupled effect of the input circuit. Note that the load impedance Z_2 is not involved in this expression.

The transfer admittance between input and output is given by the ratio $Y_T = I_2/V_1$. However, from Eq. (10-51) we can find readily that,

$$I_2 = \frac{\Delta_2}{\Delta} = \frac{-V_1 z_{21}}{z_{11}(z_{22}+Z_2) - z_{12}z_{21}}, \qquad (10\text{-}55)$$

so that,

$$Y_T = \frac{1}{Z_T} = \frac{1}{z_{12} - [z_{11}(z_{22}+Z_2)/z_{21}]}. \qquad (10\text{-}56)$$

This expression shows the influence of the load in modifying the open circuit transfer admittance.

10-10 ACTIVE NETWORKS

We have found, as shown in Fig. 10-6, that it is possible to represent a passive network as a T or Π equivalent passive circuit. The feature of passive networks is that the network equations are reciprocal, that is, $z_{12} = z_{21}$. If we are prepared to relax the condition of reciprocity, then it is possible to establish a number of other important networks, which prove to be essential in the representation of active networks, i.e., those that include dependent sources.

We begin our discussion by considering Eqs. (10-31), the z-system description of a network.

$$V_1 = z_{11}I_1 + z_{12}I_2,$$
$$V_2 = z_{21}I_1 + z_{22}I_2. \quad (10\text{-}57)$$

We know from the discussion in Section 10-6 that for the reciprocal networks, this can be represented by a T or Π network made up of passive elements. These equations can also be given the network representation shown in Fig. 10-10. The two sources that appear in this network configuration are seen to depend on currents that appear elsewhere in the

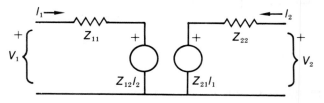

Fig. 10-10. A network realization of Eqs. (10-57).

network. These are dependent sources, in this case current-dependent voltage sources. Actually, dependent sources may be of four general types: current dependent voltage sources, current dependent current sources, voltage dependent voltage sources, and voltage dependent current sources. Observe that Fig. 10-10 is a valid network representation of Eqs. (10-57) for both the reciprocal and the nonreciprocal cases.

We now rewrite Eqs. (10-57) in the following form,

$$V_1 = (z_{11} - z_{12})I_1 + z_{12}(I_1 + I_2),$$
$$V_2 = z_{12}I_1 + z_{22}I_2 + (z_{21} - z_{12})I_1. \quad (10\text{-}58)$$

These equations have the network configuration shown in Fig. 10-11. This is a T-network with one dependent source, which would reduce, for the reciprocal network, to the simple T structure shown in Fig. 10-6. It is convenient to define the quantities,

$$\begin{array}{ll} z_a = z_{11} - z_{12} & z_b = z_{12}, \\ z_c = z_{22} - z_{12} & z_m = z_{21} - z_{12}, \\ a = \dfrac{z_m}{z_c} = \dfrac{z_{21} - z_{12}}{z_{22} - z_{12}}. & \end{array} \quad (10\text{-}59)$$

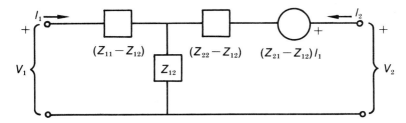

Fig. 10-11. A network equivalent to Fig. 10-10.

By Eq. (10-59), Eq. (10-58) can be written in two forms,

a. $V_1 = (z_a + z_b)I_1 + z_b I_2,$
$V_2 = (z_b + z_m)I_1 + (z_b + z_c)I_2,$

and (10-60)

b. $V_1 = (z_a + z_b)I + z_b I_2,$
$V_2 = (az_c + z_b)I_1 + (z_b + z_c)I_2.$

These equations have the equivalent network representations in Fig. 10-12. Observe that Fig. 10-12b is obtained from Fig. 10-12a by a simple source transformation, with a current controlled voltage source going into a current controlled current source.

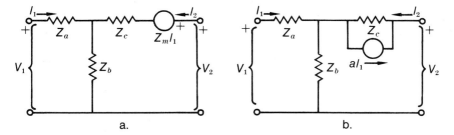

Fig. 10-12. Two networks equivalent to Fig. 10-11.

If we were to begin with Eq. (10-28), then we can draw potentially equivalent networks in the forms shown in Fig. 10-13. The equations that describe these two circuits are,

Figure 10-13a

a. $I_1 = y_{11}V_1 + y_{12}V_2,$
$I_2 = y_{21}V_1 + y_{22}V_2.$ (10-61)

Figure 10-13b

b. $I_1 = (y_a + y_b)V_1 - y_b V_2,$
$I_2 = (-y_b + b)V_1 + (y_b + y_c)V_2,$

where,

$y_a = y_{11} + y_{12}$ $y_b = -y_{12},$
$y_c = y_{22} + y_{12}$ $b = y_{21} - y_{12}.$

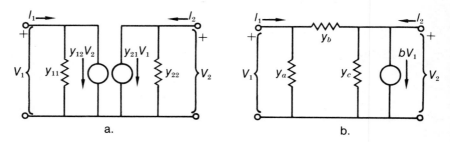

Fig. 10-13. Equivalent circuits using y-parameters.

It is observed that dependent voltage sources appear in these figures. Of course, additional source-transformed equivalents can be used to introduce new forms.

The hybrid parameter representation proves to be of particular importance in the modeling of transistors. Equations (10-36) which give the description are given graphical representation in Fig. 10-14. As drawn,

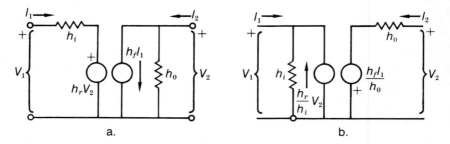

Fig. 10-14. Equivalent circuits using h-parameters.

these hybrid forms include both a dependent voltage source and a dependent current source. A number of different equivalent circuits are possible simply by source-transforming procedures, as in going from Fig. 10-12a to Fig. 10-12b. For example, the *hybrid* Π circuit which is used extensively in high frequency modeling of transistors, is easily developed in this way (*see* Problem 10-12).

Example 10-10.1

Deduce the Π and the hybrid-Π equivalent networks of the accompanying circuit.

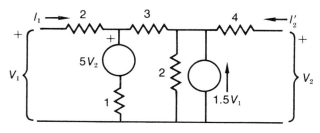

Fig. 10-10.1-1

Solution. Refer to Fig. 10-10.1-2 which is a source-transformed version of Fig. 10-10.1-1. From this figure, we write the set of controlling equations,

$$V_1 - 5V_2 = 3I_1 - I_3,$$
$$-3V_1 + 5V_2 = -I_1 + 6I_3 - 2I_2,$$
$$3V_1 - V_2 = -2I_3 + 6I_2.$$

Fig. 10-10.1-2

From these equations, we write expressions for the currents,

$$I_1 = \frac{\begin{vmatrix} (V_1 - 5V_2) & -1 & 0 \\ (-3V_1 + 5V_2) & 6 & -2 \\ (3V_1 - V_2) & -2 & 6 \end{vmatrix}}{\begin{vmatrix} 3 & -1 & 0 \\ -1 & 6 & -2 \\ 0 & -2 & 6 \end{vmatrix}} = \frac{20}{90} V_1 - \frac{132}{90} V_2.$$

The corresponding expression for I_2 is,

$$I_2 = \frac{\begin{vmatrix} 3 & -1 & (V_1 - 5V_2) \\ -1 & 6 & (-3V_1 + 5V_2) \\ 0 & -2 & (3V_1 - V_2) \end{vmatrix}}{90} = \frac{35}{90}V_1 + \frac{3}{90}V_2.$$

The two-port equations for this network are,

$$I_1 = \frac{20}{90}V_1 - \frac{132}{90}V_2,$$

$$I_2' = -I_2 = -\frac{35}{90}V_1 - \frac{3}{90}V_2.$$

It follows from these equations that,

$$y_{11} = \frac{20}{90} \qquad y_{12} = -\frac{132}{90},$$

$$y_{21} = -\frac{35}{90} \qquad y_{22} = -\frac{3}{90}.$$

For the Π equivalent,

$$y_a = y_{11} + y_{12} = -\frac{112}{90} \qquad b = y_{21} - y_{12} = \frac{97}{90},$$

$$y_b = -y_{12} = \frac{132}{90} \qquad y_c = y_{22} + y_{12} = -\frac{135}{90}.$$

The Π equivalent circuit is given in Fig. 10-10.1-3.

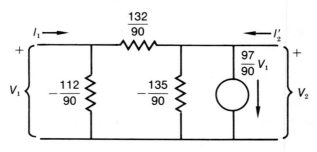

Fig. 10-10.1-3

To find the hybrid equivalent circuit, begin with the y-system equations above. Rearrange these equations with V_1 and I_2 as the dependent variables.

First we write,
$$V_1 = \frac{90}{20}I_1 + \frac{132}{20}V_2,$$
and so,
$$I'_2 = -\frac{35}{90}\left[\frac{90}{20}I_1 + \frac{132}{20}V_2\right] - \frac{3}{90}V_2,$$
$$= -\frac{35}{20}I_1 - \frac{7.8}{3}V_2.$$

The appropriate figure is given in Fig. 10-10.1-4.

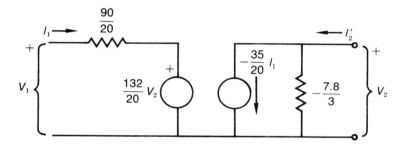

Fig. 10-10.1-4

10-11 TELLEGEN'S THEOREM (21)

It is a fundamental property of an interconnected system consisting of passive elements and independent sources that the instantaneous power and also the average power satisfies the principle of conservation of energy. This means that any given instant the total power to the system — that being provided by sources, that being stored in inductors and capacitors, and that being dissipated in resistors, must balance. This total energy balance can be written, "the algebraic sum of power flow into all elements of a circuit must be zero." What can be said about the situation if controlled sources exist in the system? To examine this question, we consider the following example.

Example 10-11.1

Examine the input impedance to the circuit shown, and discuss the results in relation to conservation of energy.

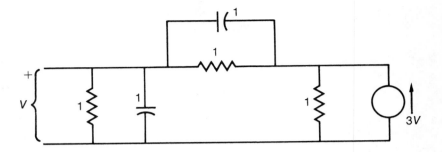

Solution. The circuit is redrawn for analysis. From this, we write the controlling equations,

$$V = \frac{I-I_1}{(s+1)} \qquad -2V = I_1 \frac{(s+2)}{(s+1)}.$$

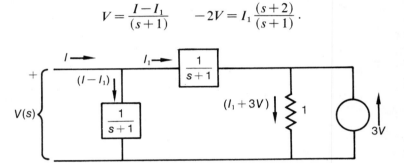

These equations are solved together to yield $Z(s)$, the input impedance,

$$Z(s) = \frac{s+2}{s^2+s}.$$

When written as a function of frequency, with $s = j\omega$, we find that,

$$Z(j\omega) = \frac{-1}{\omega^2+1} - j\omega \frac{(2+\omega^2)}{\omega^4+\omega^2}.$$

This shows that the real part of the impedance, $\text{Re } Z(j\omega)$ is negative for all frequencies. This means that power is actually being provided to the source by the network. Since R and C are passive elements, then it follows that the power must be coming from the controlled source.

A conclusion to be drawn from this example is that when considering power flow in a network, all elements, active or passive, linear or nonlinear, time varying or time invariant, must be considered. The situation is fully covered by Tellegen's theorem which can be written, "Consider a network of one-port elements (if controlled sources exist, each source is considered as a one-port element also). Let v_k be the voltages across the elements, and let i_k be the currents through them. Assuming that the v_k-s satisfy all constraints imposed by the KVL, and that the i_k-s satisfy the constraints imposed by the KCL, then,

$$\sum_{k=1}^{n} v_k i_k = 0.$$

A feature of this theorem is that it is valid for any lumped network, linear or nonlinear, passive or active, time varying or time invariant.

PROBLEMS

10-1. Draw Thévenin and Norton networks for each of the following:

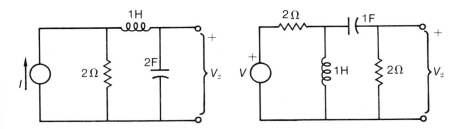

10-2. Employ the Thévenin theorem in finding the current i in the network illustrated.

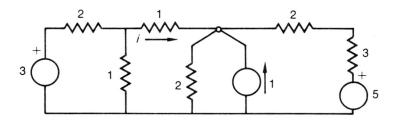

10-3. The voltage and current sources in the given network have the same frequency, say both are of the form Ae^{st}. Find an expression for the current I, using both the Thévenin and the Norton theorems.

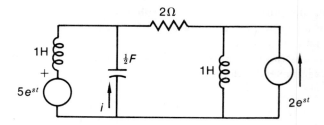

10-4. Refer to the accompanying figure.
   a. Employ the superposition theorem to find the loop currents.
   b. Apply source transformations to both sources and find the current through the 2 ohm resistor.

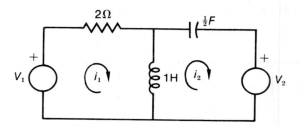

10-5. Using simple source transformations and superposition, find the current i in the network shown.

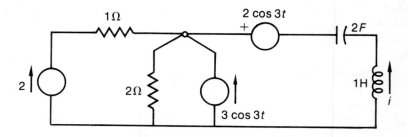

10-6. Measurements are made on a passive two-port. The results are:
Input admittance when port 2 is open-circuited $= 1/(s+1)$.
Input admittance when port 2 is short-circuited $= (s+1)/(s^2+s+1)$.
Input admittance to port 2 when port 1 is open-circuited $= s/(s+1)$.
   a. Find an expression for V_1/I_2 when port 2 is short-circuited.
   b. The input impedance when port 2 is terminated in an impedance $s+2$.

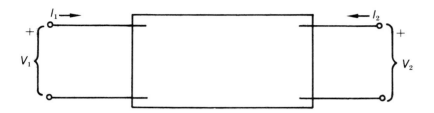

10-7. Write expressions for the T and Π equivalents of the following networks. Figure a is a ladder filter, b and c are notch filters.

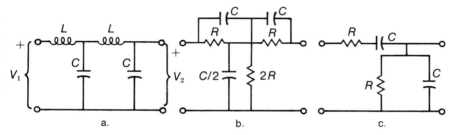

a.   b.   c.

10-8. Determine y_{11}, y_{22} and z_{22} for the network shown.

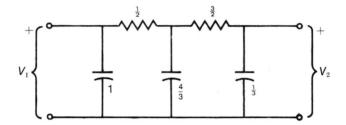

10-9. Determine the z- and h-parameters of the active two-port shown in the accompanying figure.

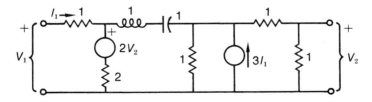

10-10. In its most general form, the voltage-current relations of a three-terminal device is given by the algebraic model,

$$\begin{bmatrix} a_{11} & a_{12} & -b_{11} & -b_{21} \\ a_{21} & a_{22} & -b_{12} & -b_{22} \end{bmatrix} \begin{bmatrix} v_1 \\ v_2 \\ i_1 \\ i_2 \end{bmatrix} = 0.$$

By rearranging columns of the matrix and then by applying appropriate operations on the coefficient matrix, obtain the following: short-circuit, open-circuit, and cascade parameter form.
HINT: the form for the hybrid parameters is,

$$\begin{bmatrix} i_1 \\ v_2 \\ v_1 \\ i_2 \end{bmatrix}.$$

10-11. A state model is to be found for the network shown. Here it is to be assumed that the system is composed of two cascaded networks, as indicated.

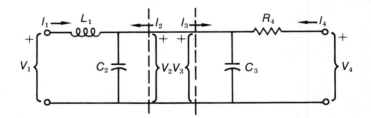

Write the component relations,

$$\begin{bmatrix} i_1 \\ v_4 \end{bmatrix} \quad \text{and} \quad \begin{bmatrix} i_2 \\ v_3 \end{bmatrix},$$

and then combine to find an expression for,

$$\begin{bmatrix} i_1 \\ v_3 \end{bmatrix}.$$

10-12. Begin with Fig. 10-14a, and write for h_i the sum $h_i = Z_a + Z_b$. By a source transformation of the portion in the broken line box and using the transformation from Fig. 10-13a to Fig. 10-13b show that the accompanying figure a transforms to b. Specify the network parameters in terms of the h-s.

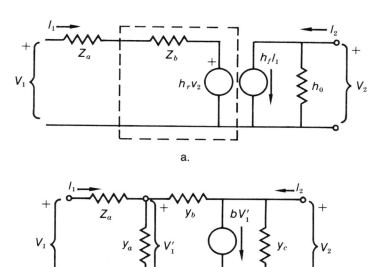

a.

b.

11

General Excitation Functions

The studies in Chapter 9 are concerned with the system response to sinusoidal excitation functions. This chapter will be an extension of these studies to periodic nonsinusoidal functions. It will be shown that the periodic nonsinusoidal function can be represented by an infinite sum of sinusoidal functions, and so a doubly infinite set of exponential functions. For the case of nonperiodic functions, we shall find that this can be represented as a continuous spectrum of sinusoids. Despite this added complexity we shall find that the procedure allows the solution of problems that might otherwise not be possible.

11-1 PERIODIC EXCITATION FUNCTION—FOURIER SERIES

We wish to show that periodic nonsinusoidal excitation functions can be expressed in terms of sinusoidal excitation functions. This requires that the periodic nonsinusoidal function be expressed in a Fourier series expansion, each term of which is a sinusoid. Because of this, the response of a system to a periodic nonsinusoidal excitation is the sum of the responses to each component in the Fourier series expansion.

The Fourier series relates to periodic functions. A function $f(x)$ is periodic with period 2π if,

$$f(x) = f(x + 2n\pi). \tag{11-1}$$

The Fourier series permits the representation of a periodic nonsinusoidal function in terms of an infinite sum of harmonically related sinusoids,

$$f(x) = a_0 + a_1 \cos x + a_2 \cos 2x + \cdots + b_1 \sin x + b_2 \sin 2x + \cdots, \tag{11-2}$$

where the coefficients $a_0, a_1, \ldots, b_1, b_2, \ldots$ are constants. The problem is that of determining the a_k and b_k coefficients for a given $f(x)$. This is possible only if the function $f(x)$ is Fourier series transformable. For a function to be Fourier series transformable, it must satisfy the Dirichlet conditions, namely:

1. $f(x)$ must be defined at every point in the interval of definition $0 < x < 2\pi$.
2. $f(x)$ must be finite, single-valued, and piecewise continuous, i.e., a finite number of discontinuities are permissible, but any two consecutive discontinuities must be separated by a finite interval.
3. $f(x)$ must have a finite number of maximum and minimum values in the interval of definition. At a point of discontinuity $g, f(g)$ is defined as having the value,

$$f(g) = \tfrac{1}{2}[f(g+\delta) + f(g-\delta)]_{\delta \to 0}.$$

We now proceed with the determination of the constants in Eq. (11-2). In this and in subsequent evaluations a number of standard form definite integrals are involved. The important ones are contained in Table 11-1.†

TABLE 11-1. Standard form integrals.

1.	$\int_0^{2\pi} \cos nx \, dx = 0$	$n \neq 0$
2.	$\int_0^{2\pi} \sin nx \, dx = 0$	$n \neq 0$
3.	$\int_0^{2\pi} \cos nx \cos mx \, dx = 0$	$n \neq m$
4.	$\int_0^{2\pi} \sin nx \sin mx \, dx = 0$	$n \neq m$
5.	$\int_0^{2\pi} \cos nx \sin mx \, dx = 0$	for all n, m
6.	$\int_0^{2\pi} \cos^2 nx \, dx = \pi$	$n \neq 0$
7.	$\int_0^{2\pi} \sin^2 nx \, dx = \pi$	$n \neq 0$

†A set of functions $\varphi_n(x)$ is called orthogonal in the interval $a < x < b$ with respect to a weighting function $w(x)$ if,

$$\int_a^b w(x)\varphi_n(x)\varphi_m(x) \, dx = \begin{cases} 0 & \text{if } (n \neq m) \\ \text{const.} & \text{if } (n = m) \end{cases}.$$

For the Fourier series $w(x) = 1$, and the functions $\varphi_n(x)$ are of the form $\sin nx$ and $\cos nx$.

First we write the series in the more compact form,

$$f(x) = a_0 + \sum_{n=1}^{\infty} a_n \cos nx + \sum_{n=1}^{\infty} b_n \sin nx. \qquad (11\text{-}3)$$

Now multiply each term in the series by dx and integrate over the interval 0 and 2π. This gives,

$$\int_0^{2\pi} f(x)\, dx = a_0 \int_0^{2\pi} dx + \sum_{n=1}^{\infty} a_n \int_0^{2\pi} \cos nx\, dx + \sum_{n=1}^{\infty} b_n \int_0^{2\pi} \sin nx\, dx. \qquad (11\text{-}4)$$

Using the appropriate entries in Table 11-1, we find that all terms in Eq. (11-4) vanish except the first. Thus we have that,

$$\int_0^{2\pi} f(x)\, dx = a_0 \int_0^{2\pi} dx = a_0 2\pi,$$

From which it follows that,

$$a_0 = \frac{1}{2\pi} \int_0^{2\pi} f(x)\, dx. \qquad (11\text{-}5)$$

Observe therefore that a_0 is the full-cycle average value of $f(x)$.

Again begin with the series, Eq. (11-3), and multiply each term by $\cos mx\, dx$ and integrate over the interval 0 and 2π. This gives,

$$\int_0^{2\pi} f(x) \cos mx\, dx = a_0 \int_0^{2\pi} \cos mx\, dx + \sum_{n=1}^{\infty} a_n \int_0^{2\pi} \cos nx \cos mx\, dx$$

$$+ \sum_{n=1}^{\infty} b_n \int_0^{2\pi} \sin nx \cos mx\, dx.$$

All terms in this series vanish with the exception of the one for $m = n$ in the second term on the right, and this yields,

$$\int_0^{2\pi} f(x) \cos nx\, dx = a_n \int_0^{2\pi} \cos^2 nx\, dx.$$

By carrying out the indicated integration on the right and rearranging the expression, we have that,

$$a_n = \frac{1}{\pi} \int_0^{2\pi} f(x) \cos nx\, dx. \qquad (11\text{-}6)$$

Finally, suppose that each term in Eq. (11-3) is multiplied by $\sin mx\, dx$ and the resulting integrals are integrated over the interval 0 and 2π. This gives,

$$\int_0^{2\pi} f(x) \sin mx \, dx = a_0 \int_0^{2\pi} \sin mx \, dx + \sum_{n=1}^{\infty} a_n \int_0^{2\pi} \cos nx \sin mx \, dx$$

$$+ \sum_{n=1}^{\infty} b_n \int_0^{2\pi} \sin nx \sin mx \, dx.$$

Here all terms vanish with one exception, that for $m = n$ in the last term, with the result that,

$$\int_0^{2\pi} f(x) \sin nx \, dx = b_n \int_0^{2\pi} \sin^2 nx \, dx.$$

By carrying out the integration on the right, and then rearranging terms, the result is,

$$b_n = \frac{1}{\pi} \int_0^{2\pi} f(x) \sin nx \, dx. \tag{11-7}$$

In summary, therefore, any function that is Fourier series transformable (i.e., that satisfies the Dirichlet conditions in the interval of definition) possesses a converging Fourier series in that interval of the general form,

$$f(x) = a_0 + \sum_{n=1}^{\infty} a_n \cos nx + \sum_{n=1}^{\infty} b_n \sin nx, \tag{11-8}$$

where,

$$a_0 = \frac{1}{2\pi} \int_0^{2\pi} f(x) \, dx,$$

$$a_n = \frac{1}{\pi} \int_0^{2\pi} f(x) \cos nx \, dx,$$

$$b_n = \frac{1}{\pi} \int_0^{2\pi} f(x) \sin nx \, dx.$$

The integrations involved can be carried out analytically; they can also be carried out numerically using a digital computer. If $f(x)$ possesses symmetries over the interval, certain of the coefficients may vanish. In the absence of symmetries, determining all of the significant coefficients may be quite time consuming. Clearly, of course, unless the series converges rapidly, the detail becomes unmanageable, and the practical advantage of the Fourier series representation of the function is lost.

Example 11-1.1

Deduce the Fourier series representation of the recurring square wave illustrated in the accompanying sketch.

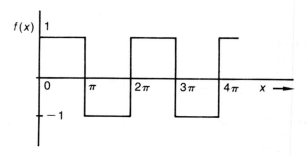

Fig. 11-1.1-1

Solution. The given waveshape is expressed analytically over the total period,

$$f(x) = 1 \quad (0 < x < \pi),$$
$$f(x) = -1 \quad (\pi < x < 2\pi).$$

We now determine the coefficients specified in Eq. (11-8). These are,

$$a_0 = \frac{1}{2\pi}\left\{\int_0^\pi dx - \int_\pi^{2\pi} dx\right\} = 0,$$

$$a_n = \frac{1}{\pi}\left\{\int_0^\pi \cos nx\, dx - \int_\pi^{2\pi} \cos nx\, dx\right\} = 0,$$

$$b_n = \frac{1}{\pi}\left\{\int_0^\pi \sin nx\, dx - \int_\pi^{2\pi} \sin nx\, dx\right\},$$

$$b_n = \frac{1}{\pi}\left\{-\frac{\cos nx}{n}\bigg|_0^\pi + \frac{\cos nx}{n}\bigg|_\pi^{2\pi}\right\} = \frac{4}{\pi}\frac{1}{n} \quad n \text{ odd}$$
$$= 0 \quad n \text{ even.}$$

The Fourier series is,

$$f(x) = \frac{4}{\pi}\{\sin x + \tfrac{1}{3}\sin 3x + \tfrac{1}{5}\sin 5x + \cdots\}.$$

11-2 EFFECT OF SYMMETRY—CHOICE OF ORIGIN

If certain symmetries are recognized in the function, then a number of the constants in the Fourier series automatically vanish. Two of the more obvious and more important symmetries are illustrated in Fig. 11-1. For the symmetries illustrated, we note,

$$\begin{array}{lll} \text{zero axis symmetry} & f(x) = f(-x) & \text{even function,} \\ \text{zero point symmetry} & f(x) = -f(-x) & \text{odd function.} \end{array} \quad (11\text{-}9)$$

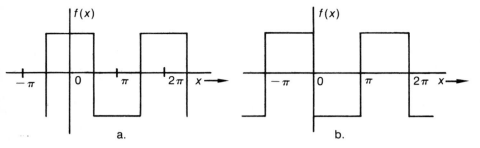

Fig. 11-1. Waves with (a) zero axis (even) symmetry, (b) zero point (odd) symmetry.

The effects of these symmetries are tabulated in Table 11-2. It is observed that Fig. 11-1a has zero point symmetry, whence only sine terms remain in the Fourier series expansion (*see* Example 11-1.1). However, this same wave with its axis shifted, as in Fig. 11-1b contains only cosine

TABLE 11-2. Symmetries and their effects on the Fourier coefficients.

Conditions	Effect on Fourier coefficients
Even function, $f(x) = f(-x)$	$b_n = 0$ for all n
Odd function, $f(x) = -f(-x)$	$a_n = 0$ for all n, including $n = 0$
Equal areas above and below the axis over one cycle	$a_0 = 0$

terms in the expansion. This means that the form of the Fourier series depends on the choice of axis about which the series is developed.

To examine the effects due to the choice of axis, refer to Fig. 11-2, which shows a simple sinusoid, with various reference points. The representations of this simple waveform relative to various axis are given

420 General Excitation Functions

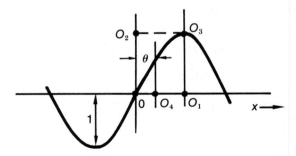

Fig. 11-2. A sinusoid and various reference points.

in Table 11.3. Based on the entries in this table, it follows that if the Fourier series is known relative to one axis, the series may be written relative to any other point by making the appropriate variable transformation and shift in axis, if any.

TABLE 11-3. Effect of origin on description of waveform.

Axis	Describing function	Effect relative to 0
0	$\sin x$	
0_1	$\cos x$	$x \to x + \pi/2$
0_2	$-1 + \sin x$	$x \to x$
0_3	$-1 + \cos x$	$x \to x + \pi/2$
0_4	$\sin(x + \theta)$	$x \to x + \theta$

Example 11-2.1

Consider the square wave given in Example 11-1.1, which is redrawn for convenience. Specify the Fourier series relative to axes through points 0_1 and 0_2.

Solution. We begin with the Fourier series relative to the origin 0, which is the result developed in Example 11-1.1.

$$f(x) = \frac{4}{\pi}[\sin x + \tfrac{1}{3}\sin 3x + \tfrac{1}{5}\sin 5x + \cdots].$$

Relative to an origin through 0_1, we write,

$$x \to x + \pi/2,$$

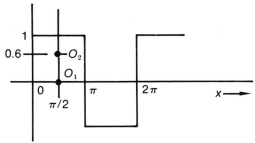

Fig. 11-2.1-1

so that,

$$f(x) = \frac{4}{\pi}\left[\sin\left(x+\frac{\pi}{2}\right) + \frac{1}{3}\sin 3\left(x+\frac{\pi}{2}\right) + \cdots\right],$$
$$= \frac{4}{\pi}[\cos x - \tfrac{1}{3}\cos 3x + \tfrac{1}{5}\cos 5x - \cdots].$$

Relative to an origin through O_2, we write,

$$f(x) = -0.6 + \frac{4}{\pi}[\cos x - \tfrac{1}{3}\cos 3x + \tfrac{1}{5}\cos 5x - \cdots].$$

A direct physical interpretation of the Fourier series expansion is possible by reference to Fig. 11-3. This shows that the single periodic nonsinusoidal excitation function has been replaced by the infinite set of functions which are sinusoids. The value of this step is that we can employ the techniques already discussed for the response of the network to each sinusoid and so, in principle, at least, obtain a response function for the system, which is the sum of the responses of the network to each component term in the Fourier series expansion.

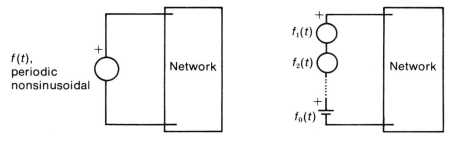

Fig. 11-3. The Fourier series representation of a periodic driving force applied to a network.

As a practical matter, this approach may be of limited computational value unless the number of significant terms in the expansion of $f(t)$ is relatively small. Of course, if the details of the calculation are programmed for computer evaluation, then a larger number of terms in the expansion can be tolerated. Often, however, other methods may be preferred.

While our discussion has required that the function $f(x)$ must be periodic in order that it be Fourier series transformable, with care, one can, in fact, represent a function that has a finite duration in its variable. That is, one can find a Fourier series representation of $f(x)$ which exists over a finite interval by assuming that $f(x)$ is periodic, with only a fixed range of validity, with the results being undefined outside of the specified range. The situation is illustrated in Fig. 11-4. Both Figs. 11-4b and 11-4c satisfy the Dirichlet conditions and so are Fourier series transformable.

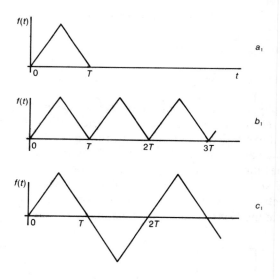

Fig. 11-4. (a) a function $f(x)$ of finite definition, (b) the same function used to define a recurrent even function, (c) the same function used to define a recurrent odd function.

Over the interval 0 to $2\pi/T$ the Fourier series for either of these latter functions represent the original $f(x)$. Outside of this range of original definition, the resulting functions do not represent the specified function. Such expansions are known as half-range expansions.

11-3 COMPLEX FOURIER SERIES

We shall find it very convenient to resolve the function $f(x)$ which satisfies the Dirichlet conditions into a series of exponential terms rather than in the form given in Eq. (11-8). This can be accomplished by writing the sine and cosine terms in terms of exponential functions. Thus from Eq. (11-8) we write,

$$f(x) = a_0 + \sum_{n=1}^{\infty} \left(a_n \frac{e^{jnx} + e^{-jnx}}{2} + b_n \frac{e^{jnx} - e^{-jnx}}{2j} \right),$$

which becomes, upon rearrangement of the terms,

$$f(x) = a_0 + \sum_{n=1}^{\infty} \left(\frac{a_n - jb_n}{2} e^{jnx} + \frac{a_n + jb_n}{2} e^{-jnx} \right). \quad (11\text{-}10)$$

We now define the complex coefficient c_n, which is,

$$c_n = \frac{a_n - jb_n}{2}. \quad (11\text{-}11)$$

By Eq. (11-8) this coefficient is,

$$c_n = \frac{1}{2\pi} \int_0^{2\pi} f(x) \cos nx \, dx - \frac{j}{2} \int_0^{2\pi} f(x) \sin nx \, dx. \quad (11\text{-}12)$$

This can be combined to,

$$c_n = \frac{1}{2\pi} \int_0^{2\pi} f(x) (\cos nx - j \sin nx) \, dx,$$

or finally,

$$c_n = \frac{1}{2\pi} \int_0^{2\pi} f(x) e^{-jnx} \, dx. \quad (11\text{-}13)$$

In the same way, we define the coefficient,

$$c_{-n} = \frac{a_n + jb_n}{2}. \quad (11\text{-}14)$$

This becomes, following precisely the same procedure as for c_n,

$$c_{-n} = \frac{1}{2\pi} \int_0^{2\pi} f(x) e^{jnx} \, dx. \quad (11\text{-}15)$$

Equations (11-15), (11-14) and (11-13) are combined with Eq. (11-10). This gives,

$$f(x) = c_0 + \sum_{n=1}^{\infty} (c_n e^{jnx} + c_{-n} e^{-jnx}). \quad (11\text{-}16)$$

General Excitation Functions

Observe, however, that this may be written in the form,

$$f(x) = c_n e^{jnx}|_{n=0} + \sum_{n=1}^{\infty} c_n e^{jnx} + \sum_{n=-1}^{-\infty} c_n e^{jnx}. \tag{11-17}$$

On the basis of this development, the Fourier series is written as the pair of equations,

$$f(x) = \sum_{-\infty}^{\infty} c_n e^{jnx},$$

$$c_n = \frac{1}{2\pi} \int_0^{2\pi} f(x) e^{-jnx} \, dx. \tag{11-18}$$

where n takes on all integral values from $-\infty$ to ∞ including the value zero. This set of equations is sometimes called a *Fourier series transform pair*. Clearly, given $f(x)$ then c_n can be found, and vice versa, given c_n, then $f(x)$ can be found. This idea of a transform pair was discussed in connection with the Laplace transform.

The steady-state component of the response of a linear network which is described by the system function $T(s)$ to an excitation function which is Fourier series transformed is simply the sum of the responses of the network to each term in the Fourier series expansion. The response is written,

$$v(t) = \sum_{n=-\infty}^{\infty} c_n T(jn\omega) e^{jn\omega t}. \tag{11-19}$$

Here, as discussed in connection with Fig. 11-3, if the Fourier series expansion requires more than a relatively few terms for a reasonably good representation of the applied excitation function, the details of the numerical work may make the solution impractical, unless possibly the problem has been programmed for digital computer solution. Some discussion of the truncated Fourier series will be given in Section 11-5.

A very interesting representation exists for a function $f(x)$ which has been expanded according to Eq. (11-18). This follows from the fact that each amplitude factor c_n is a complex number, in general, and so has both amplitude and phase for each value of n. If one plots the amplitude versus order n and phase versus order n, the results will be of the form shown in Fig. 11-5, which is known as the line spectrum of the function $f(x)$. Observe that this figure consists of a set of discrete amplitude and phase angle values for each value of n. For negative n, since, for a real signal $f(x)$,

$$c_n = c_{-n}^*, \tag{11-20}$$

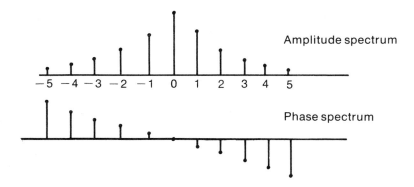

Fig. 11-5. The line spectrum for the function $f(x)$.

then the amplitude spectrum is an even function and the phase spectrum is an odd function of n, as illustrated.

Example 11-3.1

Find the Fourier series expansion of the pulse train shown in the accompanying figure, and plot the amplitude spectrum of the results.

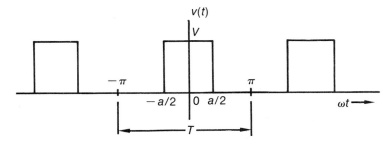

Fig. 11-3.1-1

Solution. It is convenient to change the limits in the Fourier series expansion from 0 to 2π to the range $-\pi$ to $+\pi$. This may be done since one complete cycle is specified in both cases. For the function illustrated, c_n is given by the expression,

$$c_n = \frac{1}{2\pi} \int_{-a/2}^{a/2} V e^{-jn\omega_0 t} d(\omega_0 t),$$

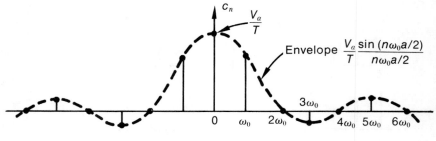

Fig. 11-3.1-2

where ω_0 denotes the discrete value of ω and t is the continuous variable. This integrates to,

$$c_n = -\frac{V}{2\pi}\frac{1}{jn}e^{-jn\omega_0 t}\Big|_{-a/2}^{a/2} = \frac{V}{\pi n}\left(\frac{e^{jn\omega_0 a/2} - e^{-jn\omega_0 a/2}}{2j}\right),$$

so that,

$$c_n = V\frac{\omega_0 a}{2\pi}\frac{\sin(n\omega_0 a/2)}{n\omega_0 a/2}.$$

However, the period $T = 2\pi/\omega_0$, and the expression for c_n is then,

$$c_n = V\frac{a}{T}\frac{\sin(n\omega_0 a/2)}{n\omega_0 a/2}. \qquad (11\text{-}21)$$

The envelope of c_n is a continuous function, which, by replacing $n\omega_0 a/2$ by x is seen to be a $\sin x/x$ type variation. The function $\sin x/x$ plays an important role in frequency analysis, and is known as the *sampling* function. The spectrum has the form illustrated. In this example c_n is real, and the single graph shows both the amplitude and the phase spectrums. In this figure, negative values of c_n correspond to values of phase of $-\pi$ rad. Note also that for $n = 0$, $c_0 = Va/T$, which is the average value of the pulse train. Note specifically that the spectrum c_n exists only at the discrete values $\omega = n\omega_0$, the envelope of the discrete values being a function of the continuous variable ω.

Example 11-3.2

Use the results of Example 11-3.1 to deduce the Fourier series of an impulse train.

Solution. An impulse train consists of a set of impulses (with area under the curve equal to unity) of zero time duration. This condition can

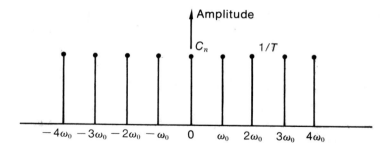

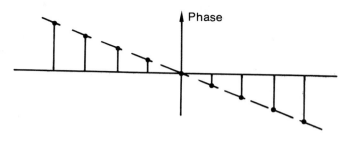

Fig. 11-3.2-1

be met by using the pulse train with amplitude equal to $1/a$ and then taking the limit as $a \to 0$. The result is then,

$$c_n = \lim_{a \to 0} \frac{1}{T} \frac{\sin(n\omega_0 a/2)}{n\omega_0 a/2} = \frac{1}{T}.$$

This shows that the spectrum is constant for all frequencies. Of course, this same result follows directly from the expression,

$$c_n = \frac{1}{2} \int_{-\pi}^{\pi} u_0(t) e^{-jn\omega_0 t} d(\omega_0 t) = \frac{1}{T}, \quad (11\text{-}22)$$

where $u_0(t)$ is the representation for the impulse function.

If the train of impulses are so phased that one impulse occurs d sec. after the time origin ($d \leq T/2$) then,

$$\begin{aligned} c_n &= \frac{1}{2\pi} \int_{-\pi}^{\pi} u_0(t-d) e^{-jn\omega_0 t} d(\omega_0 t) \\ &= \frac{1}{T} e^{-jn\omega_0 d}. \end{aligned} \quad (11\text{-}23)$$

428 General Excitation Functions

Here the amplitude spectrum is the same as before, but the phase spectrum is linearly decreasing, with an envelope slope $-d$. The results are shown in the accompanying figure.

11-4 PROPERTIES OF FOURIER SERIES

There are two properties of Fourier series that we wish to discuss at this time. One of these is Parseval's theorem, and the second can be called the least-squares approximation.

a. Parseval's theorem. This theorem relates to the mean square value of a function that has been expanded into a Fourier series. Suppose that $x(t)$ represents a voltage or current in an electric circuit. The average power to a resistor is, as discussed in Section 9-5, proportional to the mean square value of $x(t)$, which is written mathematically,

$$\langle x^2(t) \rangle = \frac{1}{T} \int_{-T/2}^{T/2} x^2(t)\, dt, \tag{11-24}$$

which assumes, of course, that the integral is finite. In conjunction with the evaluation of this integral, we consider two periodic functions having the same period. Thus we consider,

$$\frac{1}{T} \int_{-T/2}^{T/2} x_1(t) x_2(t)\, dt = \frac{1}{T} \int_{-T/2}^{T/2} \left[\sum_{n=-\infty}^{\infty} c_n^{(1)} e^{jn\omega_0 t} \right] x_2(t)\, dt,$$

where $c_n^{(1)}$ is the complex coefficient appropriate to the Fourier series expansion of $x_1(t)$. This equation is rearranged as follows,

$$= \sum_{n=-\infty}^{\infty} c_n^{(1)} \left[\frac{1}{T} \int_{-T/2}^{T/2} x_2(t) e^{jn\omega_0 t}\, dt \right],$$

which is written,

$$= \sum_{n=-\infty}^{\infty} c_n^{(1)} \left[\frac{1}{T} \int_{-T/2}^{T/2} x_2(t) e^{-j(-n)\omega_0 t}\, dt \right].$$

This allows us to write,

$$= \sum_{n=-\infty}^{\infty} c_n^{(1)} c_n^{(2)}, \tag{11-25}$$

where $c_n^{(2)}$ is the complex coefficient appropriate to the Fourier series expansion of $x_2(t)$. If we set $x_1(t) = x_2(t) = x(t)$ we then have that,

$$\frac{1}{T} \int_{-T/2}^{T/2} x^2(t)\, dt = \sum_{n=-\infty}^{\infty} c_n c_{-n}. \tag{11-26}$$

For real $x(t)$ by Eqs. (11-11) and (11-13),

$$c_{-n} = c_n^*,$$

so that we have finally,

$$\langle x^2(t) \rangle = \frac{1}{T} \int_{-T/2}^{T/2} x^2(t)\, dt = \sum_{n=-\infty}^{\infty} c_n c_n^* = \sum_{n=-\infty}^{\infty} |c_n|^2, \quad (11\text{-}27)$$

for a real signal. This result states that "the mean square value of a periodic real signal is the sum of the squares of the amplitudes of the harmonics of which it is composed."

Example 11-4.1

Show the equivalent form of Parseval's theorem in terms of the a and b coefficients.

Solution. From the fact that,

$$c_0 = a_0 \qquad c_n = \tfrac{1}{2}(a_n - jb_n) \qquad c_{-n} = \tfrac{1}{2}(a_n + jb_n),$$

then,

$$c_0^2 = a_0^2 \qquad |c_n|^2 = \tfrac{1}{4}(a^2 + b^2) = |c_{-n}|^2.$$

But we write,

$$\sum_{n=-\infty}^{\infty} |c_n|^2 = c_0^2 + \sum_{n=1}^{\infty} |c_n|^2 + \sum_{n=1}^{\infty} |c_{-n}|^2,$$

which may be written,

$$= c_0^2 + 2 \sum_{n=1}^{\infty} |c_n|^2,$$

and finally,

$$\langle x^2(t) \rangle = a_0^2 + \tfrac{1}{2} \sum_{n=1}^{\infty} (a_n^2 + b_n^2),$$

which relates the mean square of the function with the square of the amplitudes of the coefficients in the Fourier series expansion of the function.

b. The *least-squares approximation* property of the Fourier series relates to the difference between the specified function $x(t)$ and its Fourier series approximation. Suppose that a real function $x(t)$ is approximated by a *finite* or *truncated* series of exponentials, not necessarily of the Fourier type,

$$x_N(t) = \sum_{n=-N}^{N} d_n e^{jn\omega_0 t}. \quad (11\text{-}28)$$

We wish to select the coefficients d_n in a manner to make the mean square error resulting from approximating $x(t)$ by $x_N(t)$ be a minimum. We designate the error $\epsilon(t)$ as,

$$\epsilon(t) = x(t) - x_N(t),$$

so that,

$$\epsilon(t) = \sum_{n=-\infty}^{\infty} c_n e^{jn\omega_0 t} - \sum_{n=-N}^{N} d_n e^{jn\omega_0 t}.$$

We observe that $\epsilon(t)$ is a periodic function, and we can write in the form of a Fourier series.

$$\epsilon(t) = \sum_{n=-\infty}^{\infty} g_n e^{jn\omega_0 t},$$

where,

$$g_n = \begin{cases} c_n - d_n & (-N \leq n \leq N) \\ c_n & (|n| > N) \end{cases}.$$

(11-29)

We wish to examine the mean square value of $\langle \epsilon^2(t) \rangle$. By Eq. (11-29) we can write,

$$\langle \epsilon^2 \rangle = \sum_{n=-N}^{N} |c_n - d_n|^2 + \sum_{|n|>N} |c_n|^2. \tag{11-30}$$

Each term in this expression is positive. Clearly, to minimize $\langle \epsilon^2 \rangle$, we should set,

$$d_n = c_n, \tag{11-31}$$

since this will make the first summation vanish. The result is, therefore,

$$\langle \epsilon_{\min}^2 \rangle = \sum_{|n|>N} |c_n|^2. \tag{11-32}$$

This result shows that the mean square error is minimized by making the coefficient d_n in the finite exponential series, Eq. (11-28), identical with the Fourier coefficient c_n. Thus, if the Fourier series expansion of a function $x(t)$ is truncated at any given value of N, it approximates $x(t)$ with a smaller mean square error than any other exponential series with the same number N of terms. Furthermore, since the error is the sum of positive terms, the error is a monotonically decreasing function of the number of harmonics used in the approximation.

Example 11-4.1

Apply Parseval's theorem to find $\langle v^2(t) \rangle$ of the pulse train of Example 11-3.1. Use this result to find the mean square error resulting from approximating $v(t)$ by the first N harmonics.

Solution. A direct application of Eq. (11-27) using Eq. (11-21) gives,

$$\langle v^2(t) \rangle = \sum_{n=-\infty}^{\infty} \left(\frac{Va}{T}\right)^2 \left[\frac{\sin(n\omega_0 a/2)}{n\omega_0 a/2}\right]^2.$$

Furthermore, by direct calculation,

$$\langle v^2(t) \rangle = \frac{1}{T} \int_{-a/2}^{a/2} V^2 \, dt = V^2 \frac{a}{T}.$$

Thus from Eq. (11-32),

$$\langle \epsilon_{\min}^2 \rangle = \sum_{|n|>N} |c_n|^2 = \sum_{n=-\infty}^{\infty} |c_n|^2 - \sum_{n=-N}^{N} |c_n|^2,$$

$$= \langle v^2(t) \rangle - \sum_{n=-N}^{N} |c_n|^2.$$

But this can be conveniently written,

$$\langle \epsilon_{\min}^2 \rangle = \frac{Va}{T} \left\{ 1 - \sum_{n=-N}^{N} \frac{a}{T} \left[\frac{\sin(n\omega_0 a/2)}{n\omega_0 a/2}\right]^2 \right\}.$$

From this we write the fractional error,

$$\frac{\langle \epsilon_{\min}^2 \rangle}{\langle v^2(t) \rangle} = 1 - \sum_{n=-N}^{N} \frac{a}{T} \left[\frac{\sin(n\omega_0 a/2)}{n\omega_0 a/2}\right]^2,$$

$$= 1 - \sum_{n=-N}^{N} \frac{a}{T} \frac{\sin^2(n\pi a/T)}{(n\pi a/T)^2}.$$

An approximate value of this expression is possible. If a/T is sufficiently small so that $n\pi a/T \ll 1$ for $n \leq N$, then each term in the series is approximately equal to a/T, whence,

$$\frac{\langle \epsilon_{\min}^2 \rangle}{\langle v^2(t) \rangle} \simeq 1 - \frac{a}{T}(2N+1).$$

11-5 NUMERICAL DETERMINATION OF FOURIER COEFFICIENTS

If the function $f(x)$ is known analytically and is of simple form, the coefficients a_k and b_k, or correspondingly c_k, are easily determined. Other times, the form of $f(x)$ may be such that analytical integration is difficult or even impossible. An important practical situation arises when $f(x)$ is given in graphical form or when $f(x)$ is given in tabular form at discrete

432 General Excitation Functions

(and usually equally spaced) points spanning the range of integration. In these cases, numerical methods must be employed to find the coefficients. When $f(x)$ is specified as sampled data points, it may be written as the sequence $\{f(x_j)\}$ where,

$$\{f(x_j); \ j = 0, 1, \ldots, N; \ x_j \text{ equally spaced at intervals of size } h\},$$
(11-33)

where $x_j = jh$, and h is the uniform spacing interval, with j an index that ranges from 0 to N for the $N+1$ data points.

In numerical integration methods, it is customary to use sampled data points to form Riemann sums to approximate the integral. For the unweighted Riemann sum, we write that,

$$\begin{aligned} a_k &= \frac{1}{\pi} \int_0^{2\pi} f(x) \cos kx \, dx \simeq \frac{h}{\pi} \sum_{j=0}^{N} f(x_j) \cos kx_j, \\ &= \frac{h}{2} \sum_{j=0}^{N} f(x_j) \cos kjh. \end{aligned}$$
(11-34)

We note that this expression becomes quite poor for higher order coefficients. This arises because any uniform sampling of an oscillatory function will become less meaningful as the frequency of oscillation increases. Quantitatively, the integration will be relatively poor once $kh > \pi/4$, and becomes meaningless when $kh > 2\pi$. This conclusion is placed in visual evidence in Fig. 11-6. The resultant of curve $(2) \times f(x)$ will be zero, yet is clear that the coefficient for b_k is not zero for this index k. More will be said about this in Section 11-7c when the sampling theorem is discussed.

The method of Filon for integrals involving periodic functions that generate integration formulas by the method of undetermined coefficients avoids the problem of sampled data for high frequency trigonometric terms. In this method the original function is approximated in a piecewise manner by a low order polynomial, say a quadratic. This approximation polynomial is so chosen that it passes through the sampled data points over some subinterval. The situation for the function of Fig. 11-6 might then be as shown in Fig. 11-7. With a knowledge of the approximating polynomials, the important Fourier coefficient integrals over appropriate intervals may be evaluated analytically. The results, by successive piecewise polynomial approximation, are then summed over the entire interval

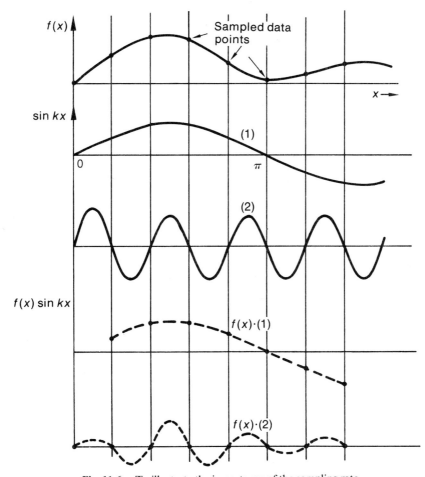

Fig. 11-6. To illustrate the importance of the sampling rate.

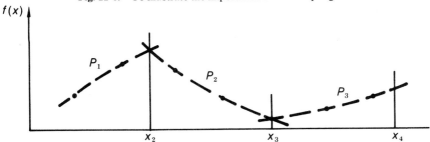

Fig. 11-7. Piecewise quadratic approximation to $f(x)$.

General Excitation Functions

to give the Fourier coefficients,

$$a_k = \frac{1}{\pi} \sum_{i=1}^{M} \int_{x_i}^{x_{i+1}} f_i(x) \cos kx \, dx,$$

$$b_k = \frac{1}{\pi} \sum_{i=1}^{M} \int_{x_i}^{x_{i+1}} f_i(x) \sin kx \, dx. \tag{11-35}$$

Without going through the details of the development, the results can be summarized as follows. Let,

b,a be the upper and lower limits of integration,
N be the number of data points, assumed to be odd,

$$h = \frac{b-a}{N}, \tag{11-36}$$
$$\theta = kh.$$

$f(a), f(a+h), f(a+2h), \ldots, f(b)$, be the sampled values of the function,

$C_{2n} = \frac{1}{2} f(a) \cos ka + f(a+2h) \cos k(a+2h) + \cdots + \frac{1}{2} f(b) \cos kb,$

$C_{2n-1} = f(a+h) \cos k(a+h) + f(a+3h) \cos k(a+3h) + \cdots$
$\quad + f(b-h) \cos k(b-h),$

$S_{2n} = \frac{1}{2} f(a) \sin ka + f(a+h) \sin k(a+2h) + \cdots + \frac{1}{2} f(b) \sin kb,$

$S_{2n-1} = f(a+h) \sin k(a+h) + f(a+3h) \sin k(a+3h) + \cdots$
$\quad + f(b-h) \sin k(b-h).$

Then,

$$\int_a^b f(x) \cos kx \, dx = h\{\alpha[f(b) \sin kb - f(a) \cos ka] + \beta C_{2n} + \gamma C_{2n-1}\},$$
$$\tag{11-37}$$
$$\int_a^b f(x) \sin kx \, dx = h\{-\alpha[f(b) \cos kb - f(a) \cos ka] + \beta S_{2n} + \gamma S_{2n-1}\},$$

where,

$$\alpha = \frac{\theta^2 + \theta \sin\theta \cos\theta - 2\sin^2\theta}{\theta^3} \simeq \frac{2\theta^3}{45} - \frac{2\theta^5}{315} + \cdots,$$

$$\beta = \frac{2[\theta(1-\cos^2\theta) - 2\sin\theta\cos\theta]}{\theta^3} \simeq \frac{2}{3} + \frac{2\theta^2}{12} + \cdots,$$

$$\gamma = \frac{4(\sin\theta - \theta\cos\theta)}{\theta^3} \simeq \frac{4}{3} - \frac{2\theta^2}{15} + \cdots.$$

By adapting the range of integration to the particular case and multiplying by $1/\pi$, this method can be used to compute the desired Fourier coefficients with the accuracy depending primarily on the mesh size of the original sampled data and the degree to which this can be approximated over adjoining intervals by quadratic functions. In any case, the accuracy will not deteriorate rapidly as the index increases.

This approach to the numerical determination of Fourier coefficients has been programmed for digital computation, with very satisfactory results.

11-6 THE FOURIER TRANSFORM AND CONTINUOUS FREQUENCY SPECTRUMS

When a function is defined over the infinite time range and is not periodic, the Fourier series representation of Section 11-3 is no longer possible, except in the manner discussed in relation to Fig. 11-4. If the function, which we now write as $f(t)$ satisfies the Dirichlet conditions, and in addition, satisfies the requirement that the integral,

$$\int_{-\infty}^{\infty} |f(t)|\,dt \quad \text{is finite},$$

then it is possible to resolve $f(t)$ into exponentials of the form $e^{j\omega t}$, where the continuous variable ω takes the place of the discrete variable $n\omega_0$.

As a step in the development of the analytic description of $f(t)$, let us refer to Example 11-3.1. We wish to raise two questions; (1) how does c_n change for constant width of the pulse, but as the period T increases, (2) what happens if T/a becomes infinite, i.e., if the period becomes infinite, thereby resulting in a single, nonrecurring pulse? The answer to question (1) is suggested by finding c_n for a pulse train for which the pulse period is increased so that instead of the value in Example 11-3.1, say, the ratio a/T is $\frac{1}{5}$ of that already examined. If the calculation is carried out, the result is found to be,

$$c_n = \frac{Va}{5T} \frac{\sin(n\omega_0 a/2)}{n\omega_0 a/2}. \tag{11-38}$$

A plot of this function is given in Fig. 11-8. A comparison of this figure with that which illustrates Eq. (11-21) shows that because the period is increased by 5, the amplitude is decreased by 5. Furthermore, increasing the period by 5 reduces the fundamental frequency by 5. Thus the resulting spectrum now has components occurring at multiples of the new

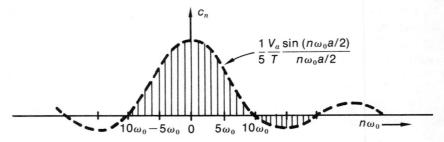

Fig. 11-8. Spectrum of pulse train of Example 11-3.1 with $a/T = \frac{1}{5}$ of original value.

fundamental frequency $\omega = \omega_0/5$. Note, however, that the envelope of the spectrum retains the same shape as before. That is, with increased T and constant a, the amplitude of the envelope decreases, and the number of spectral lines increases, but the shape of the envelope function remains unchanged.

To answer question (2), we would expect that as $T \to \infty$ that the spectrum would become continuous, but that the amplitudes will approach zero. That is, as the period becomes infinite, the spacing between frequency components becomes infinitesimal, and $n\omega_0$ approaches the continuous variable. If we were to begin with the Fourier series given in Eq. (11-18) and then adopt the limiting process, we should deduce a description of the results. We carry this suggested procedure out by writing Eq. (11-18) in the form

$$f(t) = \sum_{n=-\infty}^{\infty} \left[\frac{1}{2\pi} \int_{-\pi}^{\pi} f(t) e^{-jn\omega_0 t} d(\omega_0 t) \right] e^{jn\omega_0 t}. \tag{11-39}$$

Now focus on the arbitrary interval $-\omega_0/\pi$ and ω_0/π and write

$$f(t) = \sum_{n=-\infty}^{\infty} \left[\frac{\omega_0}{2\pi} \int_{-\pi/\omega_0}^{\pi/\omega_0} f(t) e^{-jn\omega_0 t} dt \right] e^{jn\omega_0 t}.$$

In the limit as $\omega_0 \to 0$, $n \to \infty$, assuming that the limit exists,

$$\omega_0 \to d\omega,$$
$$n\omega_0 \to \omega,$$

and the summation becomes an integral, with the result that

$$f(t) = \int_{-\infty}^{\infty} \frac{1}{2\pi} \left[\int_{-\infty}^{\infty} f(t) e^{-j\omega t} dt \right] e^{j\omega t} d\omega. \tag{11-40}$$

If the bracketed quantity is written $F(j\omega)$, then we have the pair of equations,

$$F(j\omega) = \int_{-\infty}^{\infty} f(t)e^{-j\omega t}\, dt,$$
$$f(t) = \frac{1}{2\pi} \int_{-\infty}^{\infty} F(j\omega)e^{j\omega t}\, d\omega. \tag{11-41}$$

The first of these is called the *Fourier transform* of $f(t)$, and often is written symbolically,

$$F(j\omega) = \mathscr{F}[f(t)]. \tag{11-42}$$

The second expression, which is called the *inverse Fourier transform* of $F(j\omega)$ may be written symbolically,

$$f(t) = \mathscr{F}^{-1}[F(j\omega)]. \tag{11-43}$$

Attention is called to the fact that the foregoing is not a mathematically rigorous derivation of the Fourier integral theorem (22). A rigorous approach is possible from considerations of the two-sided Laplace transform, somewhat the reverse of the discussion in Section 11-7b.

Example 11-6.1

Find the Fourier transform of a pulse of unit amplitude and width a.
Solution. We write, from Eq. (11-40),

$$F(j\omega) = \int_{-a/2}^{a/2} e^{-j\omega t}\, dt = a\, \frac{\sin(\omega a/2)}{\omega a/2}.$$

Observe that this spectral density has the same shape as the envelope of Fig. 11-8.

11-7 PROPERTIES OF FOURIER TRANSFORMS

We wish to discuss some of the more important properties of Fourier transforms. A number of these will be considered.

a. Parseval's Theorem. An analogous relation exists between the mean squared value of a nonperiodic signal to its Fourier transform and that of the mean squared value of a periodic signal and its Fourier series. Let us denote $f^*(t)$ as the conjugate of a signal $f(t)$ in terms of its Fourier transform. Then we may write,

$$f(t)f^*(t) = f(t)\frac{1}{2\pi}\int_{-\infty}^{\infty} F^*(j\omega)e^{-j\omega t}\,d\omega.$$

We multiply by dt and integrate over the infinite time range to find,

$$\int_{-\infty}^{\infty} f(t)f^*(t)\,dt = \frac{1}{2\pi}\int_{-\infty}^{\infty} f(t)\int_{-\infty}^{\infty} F^*(j\omega)e^{-j\omega t}\,d\omega\,dt.$$

Inverting the order of integration (which assumes uniform convergence of the inner integral), then,

$$\int_{-\infty}^{\infty} f(t)f^*(t)\,dt = \frac{1}{2\pi}\int_{-\infty}^{\infty} F^*(j\omega)\,d\omega \int_{-\infty}^{\infty} f(t)e^{-j\omega t}\,dt.$$

But the second integral is the Fourier transform $F(j\omega)$ of $f(t)$, so that,

$$\int_{-\infty}^{\infty} f(t)f^*(t)\,dt = \frac{1}{2\pi}\int_{-\infty}^{\infty} F^*(j\omega)F(j\omega)\,d\omega.$$

This can be written as,

$$\int_{-\infty}^{\infty} |f(t)|^2\,dt = \frac{1}{2\pi}\int_{-\infty}^{\infty} |F(j\omega)|^2\,d\omega. \qquad (11\text{-}44)$$

If $f(t)$ represents voltage or current, then $f(t)^2$ is proportional to power and the time integral of this quantity is proportional to the energy dissipated in a resistor. On the basis of Eq. (11-44) the total energy is related to the square of the Fourier transform of $f(t)$. Thus $|F(j\omega)|^2$ may be regarded as the *energy density spectrum* of $f(t)$. On this basis, the energy contained in a band of frequencies $\omega_1 \leq \omega \leq \omega_2$ is given by,

$$\frac{1}{2\pi}\int_{\omega_1}^{\omega_2} |F(j\omega)|^2\,d\omega. \qquad (11\text{-}45)$$

b. Inversion of Fourier Transform. Now the problem to be considered is that of finding $f(t)$, given $F(j\omega)$. Often $F(j\omega)$ may be the result of algebraic operations, as would be the case if we had first resolved, say a voltage $v(t)$ into its Fourier decomposition $V(j\omega)$, which then was applied to a network having a system function $T(j\omega)$, to give a response,

$$F(j\omega) = T(j\omega)V(j\omega). \qquad (11\text{-}46)$$

Actually, this assumes that $T(j\omega)$ is the proper form for the system function of a network that is Fourier transformed. This is a result that can be justified on the basis of our previous work with the Laplace transform.

To find $f(t)$ requires the evaluation of the integral,

$$f(t) = \frac{1}{2\pi} \int_{-\infty}^{\infty} F(j\omega) e^{j\omega t} d\omega, \qquad (11\text{-}47)$$

which requires integration along the ω-axis. Often the integral does not converge, since there may be poles on the $j\omega$-axis. In such cases, other methods must be resorted to. One can ensure convergence by including a factor $e^{-\sigma t}$, with σ later being reduced to zero. In particular, therefore, we consider the Fourier integral of $fe^{-\sigma t}$, which is,

$$\int_{-\infty}^{\infty} fe^{-\sigma t} e^{-j\omega t} dt = \int_{-\infty}^{\infty} f(t) e^{-(\sigma + j\omega)t} dt = F(\sigma + j\omega). \qquad (11\text{-}48)$$

Now the need is to obtain $f(t)$ from $F(\sigma + j\omega)$. From the second part of Eq. (11-40),

$$f(t) e^{-\sigma t} = \mathscr{F}^{-1}[F(\sigma + j\omega)] = \frac{1}{2\pi} \int_{-\infty}^{\infty} F(\sigma + j\omega) e^{j\omega t} d\omega. \qquad (11\text{-}49)$$

Both sides of this equation are multiplied by $e^{\sigma t}$, which gives,

$$f(t) = \frac{1}{2\pi} \int_{-\infty}^{\infty} F(\sigma + j\omega) e^{(\sigma + j\omega)t} d\omega. \qquad (11\text{-}50)$$

As a change of variable, we introduce the complex quantity $s = \sigma + j\omega$. For the appropriate limits of integration, we see that when $\omega = \pm \infty$, $s = \sigma \pm j\infty$, and Eq. (11-50) becomes,

$$f(t) = \frac{1}{2\pi j} \int_{\sigma - j\infty}^{\sigma + j\infty} F(s) e^{st} ds. \qquad (11\text{-}51)$$

Note that the integration is performed in the s-plane along a line $s = \sigma$, where, as already noted, σ is a constant damping factor which will ensure the convergence of the integral,

$$\int_{-\infty}^{\infty} |f(t) e^{-\sigma t}| dt = \int_{-\infty}^{\infty} |f(t)| e^{-\sigma t} dt < \infty, \qquad (11\text{-}52)$$

which is the necessary condition for $f(t) e^{-\sigma t}$ to be Fourier integral transformable. We direct attention to Eq. (11-51), which is exactly Eq. (7-22), the inverse Laplace transform.

c. The Sampling Theorem. The amount of frequency space required by the frequency spectrum of a signal $f(t)$ is of importance in some studies. The extent of the frequency spectrum is given by the Fourier series for periodic signals and by the Fourier integral for nonperiodic signals. We shall consider an ideal band-limited signal whose Fourier transform is zero outside of a finite pass band, such as that illustrated in Fig. 11-9. The value of $f(t)$ corresponding to this frequency spread is given by,

$$f(t) = \frac{1}{2\pi} \int_{-W}^{W} F(j\omega) e^{j\omega t}\, d\omega. \qquad (11\text{-}53)$$

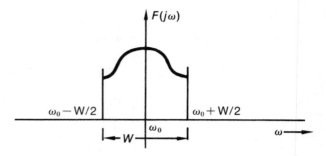

Fig. 11-9. Spectrum of band-limited signal.

Even though $F(j\omega)$ is not a periodic function in ω, nevertheless as discussed in Section 11-2, we can expand $F(j\omega)$ into a Fourier series that is valid over the range $|\omega| < W$. The desired form is given by Eq. (11-18),

$$F(j\omega) = \sum_{n=-\infty}^{\infty} c_n e^{jnT\omega},$$

$$c_n = \frac{T}{2\pi} \int_{-W}^{W} F(j\omega) e^{-jnT\omega}\, d\omega, \qquad (11\text{-}54)$$

where $T = \pi/W$. By comparing this expression for the coefficients of the Fourier series with Eq. (11-53) for the Fourier integral, we see that we can write,

$$c_n = Tf(-nT),$$

so that the first of Eq. (11-54) becomes,

$$F(j\omega) = T \sum_{n=-\infty}^{\infty} f(-nT) e^{jnT\omega} = T \sum_{n=-\infty}^{\infty} f(nT) e^{-jnT\omega}. \qquad (11\text{-}55)$$

This expression is combined with Eq. (11-53), and the order of integration and summation are interchanged. This gives,

$$f(t) = \frac{T}{2\pi} \sum_{n=-\infty}^{\infty} f(nT) \int_{-W}^{W} e^{j(t-nT)\omega}\, d\omega,$$

which integrates to,

$$f(t) = \sum_{n=-\infty}^{\infty} f(nT) \frac{\sin W(t-nT)}{W(t-nT)}. \qquad (11\text{-}56)$$

This result is often called the *sampling theorem in the time domain*. It states: "if a signal $f(t)$ has a spectrum which vanishes above W rad/sec, it is completely determined by its ordinates at the equally spaced sampling intervals $T = \pi/W$ sec."† More precisely, the sampling theorem states, "if the highest frequency contained in the signal is W Hz, then the sampling interval is $\frac{1}{2}W$ sec." This says that if $T = 10^{-4}$ sec signals up to 5 KHz can be reprocessed. To construct the signal from these ordinates, each ordinate $f(nT)$ is multiplied by the function,

$$\phi(t-nT) = \frac{\sin W(t-nT)}{W(t-nT)}, \qquad (11\text{-}57)$$

which is known as the *interpolating function*, and the product is summed over all n, as required by Eq. (11-56). Observe that the interpolating function is a $\sin x/x$ function, and so has the form of the envelope function of Fig. 11-8. The function $\phi(t-nT)$ is unity for $t = nT$ and is zero for all other values of $t = kT \neq nT$, although the function $f(t)$ will be finite, in general, for any finite time.

An important practical consequence of the sampling theorem is that a sampled signal $f(nT)$ can be transmitted over a channel of finite bandwidth (chosen according to the sampling theorem) without any loss of information content. Conversely, a continuous signal that cannot be approximated by $f(nT)$ cannot be transmitted over a channel of finite bandwidth without the loss of some information.

11-8 FREQUENCY RESPONSE CHARACTERISTICS

We wish to show that the sinusoidal steady-state response characteristic of a system can be evaluated in terms of the Fourier transform of the transient component of the solution of the state equations. These considerations are adjuncts to those discussed in Section 6-9. We begin with the state equations,

$$\begin{aligned}\dot{x} &= Ax + Bu, \\ y &= Px + Qu,\end{aligned} \qquad (11\text{-}58)$$

with the excitation vector,

$$u(t) = Ue^{j\omega t} + U^* e^{-j\omega t}. \qquad (11\text{-}59)$$

†Periodic nonuniform sampling, which is quite possible, requires special attention.

General Excitation Functions

We shall assume that the system is stable and that the response is given by,

$$y(t) = Y e^{j\omega t} + Y^* e^{-j\omega t}. \tag{11-60}$$

Thus we must have that (*see* Section 7-7),

$$Y = [P(j\omega I - A)^{-1}B + Q]U = PX + QU, \tag{11-61}$$

where the vector,

$$X = X(\omega) = (j\omega I - A)^{-1} BU,$$

denotes the frequency response of the steady-state vector. The vector QU is a constant. Here the constant matrix P relates the frequency response characteristics of the response variables to the frequency response characteristics of the state variables.

We can evaluate the frequency response of the state vector through the use of the Fourier transform of the solution of the homogeneous system,

$$\dot{x} = Ax, \tag{11-62}$$

with initial conditions $x(0) = BU$. That this is possible follows from the fact that,

$$x(t) = e^{At}BU. \tag{11-63}$$

If $x(t) = 0$ for $t < 0$, and if all eigenvalues of A have negative real parts (for the assumed stable system), the Fourier transform of $x(t)$ exists and is,

$$X(j\omega) = \int_0^\infty e^{-j\omega t} x(t)\, dt = \int_0^\infty e^{-j\omega t} e^{At} BU\, dt,$$
$$= \int_0^\infty e^{-(j\omega I - A)t} BU\, dt = (j\omega I - A)^{-1} BU. \tag{11-64}$$

It is observed, by comparing this result with Eq. (11-61), that the frequency response of the state vector can be evaluated by multiplying the solution to Eq. (11-62) by $e^{-j\omega t}$ and integrating the result over a long time interval. This integration can be carried out numerically, if necessary.

As one procedure for carrying out the integration, we can represent each function in the state vector by a staircase approximation, in the manner shown in Fig. 11-10. If we denote a single pulse by,

$$\xi_n(t) = \begin{cases} 0 & t < nh;\, t > (n+1)h \\ x_i(nh) & nh \leq t \leq (n+1)h \end{cases} \tag{11-65}$$

then,

$$x_i(t) \simeq \sum_{n=0}^\infty \xi_n(t). \tag{11-66}$$

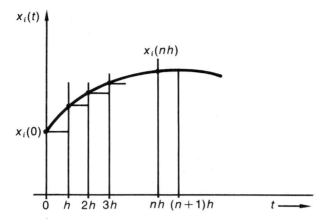

Fig. 11-10. Approximation of state functions by staircase approximation.

In this case, the Fourier transform of the state vector is,

$$X(j\omega) = \int_{-\infty}^{\infty} x(t) e^{-j\omega t}\, dt \simeq \sum_{n=0}^{\infty} x(nh) \int_{nh}^{(n+1)h} e^{-j\omega t}\, dt,$$
(11-67)

$$= \sum_{n=0}^{\infty} x(nh) \left\{ \frac{[\sin(n+1)h\omega - \sin nh\omega] + j[\cos(n+1)h\omega - \cos nh\omega]}{\omega} \right\},$$

where $x(nh)$ represents the numerical value of the state vector at $t = nh$. But it is assumed that the system is stable, and so $x_i(nh) \to 0$ as $n \to \infty$. The infinite series in Eq. (11-67) can be approximated by a truncated series. This permits us to proceed numerically to find the frequency response characteristics of the output variable, which involves combining Eq. (11-67) with Eq. (11-61). The result is,

$$Y(j\omega) = PX(j\omega) + QU(j\omega).$$
(11-68)

To improve the accuracy, one can employ a piecewise linear approximation for $x(t)$ instead of the staircase approximation shown in Fig. 11-10.

11-9 THE DISCRETE FOURIER TRANSFORM

Fourier transform techniques are used extensively in signal filtering studies. However, in important practical applications, one is faced with determining the spectral components corresponding to a signal defined by a large number of experimentally determined data samples. The processing procedures usually employ a digital computer in carrying out the integrations involved in the Fourier transforms. These integrations can be accomplished along two general lines, and these in some ways parallel the ideas used in carrying out the solution of differential equations. One method calls for carrying out the integrations by numerical means. The second method, and one that is used very extensively, calls for replacing the continuous Fourier transform by an equivalent Discrete Fourier transform (DFT). The procedure is parallel, in some measure, to the ideas discussed in Section 6-8 where it was shown that the state equations under sampled data conditions attained the form of discrete time state equations. However, the use of a DFT to replace the Fourier transform is not, in itself, of great advantage, since the direct solution of the DFT of N samples requires N^2 operations (multiplication and addition), and with say 10^4 samples (a small number in some cases) 10^8 operations are required. It was the development of the Fast Fourier Transform (FFT), a computational technique that reduces the number of operations in the evaluation of the DFT to $N \log_2 N$, that makes the DFT attractive in digital filtering studies.

The discrete Fourier transform is particularly suitable for describing phenomenons related to discrete time series. It has mathematical properties that are analogous to those of the Fourier integral transform. In particular, it defines a spectrum of a time series; also, the multiplication of the transform of two time series corresponds to convolving the time series.

If digital analysis techniques are to be used for analyzing a continuous waveform, then of course, the continuous data must be sampled so as to produce a time series of discrete samples. When the samples are equally spaced, they are known as Nyquist samples. From the discussion in Section 11-7, such a time series completely represents the continuous waveform provided that the waveform is frequency band-limited, with the sampling rate at least twice the highest frequency present in the waveform.

The DFT of a time series can be developed from the Fourier integral transform of the continuous waveform from which samples have been taken to form the time series. We shall follow these lines. We thus begin with the Fourier integral transform, which is given by Eq. (11-40). In the sampled data case given N equally spaced data points $f(t_k) = f(kT)$ these equations can be written,

$$F(j\omega_n) = \Delta t \sum_{k=0}^{N} f(t_k) e^{-j\omega_n t_k} \qquad (n = 0, 1, 2, \ldots, N/2),$$

$$f(t_k) = \frac{\Delta\omega}{2\pi} \sum_{n=-N/2}^{N/2} F(j\omega_n) e^{j\omega_n t_k} \qquad (k = 0, 1, \ldots, N-1). \tag{11-69}$$

If we now write,

$$\begin{aligned} t_k &= k\Delta t; & \Delta t &= h; & h &= T/N \\ \omega_n &= n\Delta\omega = n\omega_0; & \omega_0 &= 1/T \end{aligned} \tag{11-70}$$

then Eq. (11-69) becomes the DFT of the points $f(k)$,

$$F(n) = h \sum_{k=0}^{N-1} f(k) e^{-j2\pi nk/N} \qquad (n = 0, 1, \ldots, N-1),$$

with the inverse DFT $\qquad\qquad\qquad\qquad\qquad\qquad\qquad\qquad\qquad\qquad$ (11-71)

$$f(k) = \frac{\omega_0}{2\pi} \sum_{n=0}^{N-1} F(n) e^{j2\pi nk/N} \qquad (k = 0, 1, \ldots, N-1),$$

where the argument n takes on the values $0, 1, \ldots, N-1$ rather than $0, \pm 1, \ldots, \pm N/2$. This substitution does not alter the expression, but it does simplify the computational procedure. The term $n = N/2$ corresponds to the Nyquist folding frequency. Clearly, the DFT of a sequence of samples of $f(kT)\ 0 \leq k \leq N-1$ is another sequence of samples $F(n)$.

As noted above, a feature of the DFT is that the product of two DFT-s is equivalent to convolution in the time domain. This means that if $H(n\Omega)$ is the DFT of $h(kT)$ and $U(n\Omega)$ is the DFT of $u(kT)$, then the product of $H(n\Omega)U(n\Omega)$ is the DFT of the convolution of $h(kT)$ and $u(kT)$, that is,

$$H(n\Omega) U(n\Omega) = \text{DFT} \sum_{k=0}^{N-1} h(kT) u(m-k) T,$$

where, $\qquad\qquad\qquad\qquad\qquad\qquad\qquad\qquad\qquad\qquad\qquad\qquad$ (11-72)

$$F(n\Omega) = \sum_{k=0}^{N-1} f(kT) e^{-j\Omega Tnk} \qquad (k = 0, 1, \ldots, N-1).$$

Because of the presence of the exponential factor $e^{-j\Omega Tnk}$ which denotes that the points in the expansion are equally spaced on a unit circle, Eq. (11-72) denotes circular convolution in the time domain, as against Eq. (6-26) which denotes linear convolution. Consequently, performing such a convolution involves shifting along a circle rather than just linearly. Such a periodic convolution is somewhat more involved than a linear one, but the general ideas are the same in both.

For purposes of computation, it is convenient to define the quantity,

$$W = e^{-j2\pi/N}, \tag{11-73}$$

which denotes a unit distance at the angle $-2\pi/N$. Equation (11-71) is written,

$$F(n) = h \sum_{k=0}^{N-1} f(k) W^{nk},$$

$$f(k) = \frac{\omega_0}{2\pi} \sum_{n=0}^{N-1} F(n) W^{-nk}. \tag{11-74}$$

Here W^{nk} denotes N equally spaced points on the unit circle. These expressions may be written in matrix form. The matrix form for $F(n)$ is written,

$$[F(n)] = [W^{nk}][f_0(k)], \tag{11-75}$$

where $[F(n)]$ and $[f_0(k)]$ are $N \times 1$ column matrixes, and $[W^{nk}]$ is an $N \times N$ matrix.

Let us consider the special case for $N = 4$ sample points. Equation (11-75) would be written,

$$\begin{bmatrix} F(0) \\ F(1) \\ F(2) \\ F(3) \end{bmatrix} = \begin{bmatrix} W^0 & W^0 & W^0 & W^0 \\ W^0 & W^1 & W^2 & W^3 \\ W^0 & W^2 & W^4 & W^6 \\ W^0 & W^3 & W^6 & W^9 \end{bmatrix} \begin{bmatrix} f_0(0) \\ f_0(1) \\ f_0(2) \\ f_0(3) \end{bmatrix}. \tag{11-76}$$

This constitutes a set of simultaneous equations, and this will require 4^2 complex multiplications and additions in its solution. In the general case for an $N \times N$ matrix; N^2 complex multiplications and additions are required. This can make carrying out the details of problem solutions very time consuming, even when a substantial digital computer is used. Fortunately as already noted, an algorithm exists for the computation of the Fourier coefficients which requires much less computational effort than in the solution formulated above. This FFT method, developed by Cooley and Tukey (23), has produced major changes in computational techniques used in digital spectral analysis, filter simulation, and related fields.

11-10 THE FAST FOURIER TRANSFORM (23,24)

An important feature of the FFT algorithm is that it makes possible the computation of the coefficients of the DFT with only $N \log_2 N$ operations. For large N, this represents a very substantial saving in computational time.

To establish the ideas behind the FFT, suppose that the number of sample points of $f(t)$ are chosen according to the relation $N = 2^i$, where i is an integer (in Eq. (11-76), with $N = 4$, we have $i = 2$). It is convenient to write W^{nk} in the form,

$$W^{nk} = W^{nk \bmod N}, \qquad (11\text{-}77)$$

where $nk \bmod N$ denotes the remainder upon division of nk by N. For example, if $N = 4$, $n = 2$, $k = 3$, then $nk = 6$, and $nk \bmod N = 2$, so that $W^6 = W^2$. Based on these considerations Eq. (11-76) is written,

$$[F(n)] = \begin{bmatrix} F(0) \\ F(1) \\ F(2) \\ F(3) \end{bmatrix} = \begin{bmatrix} 1 & 1 & 1 & 1 \\ 1 & W^1 & W^2 & W^3 \\ 1 & W^2 & W^0 & W^2 \\ 1 & W^3 & W^2 & W^1 \end{bmatrix} \cdot \begin{bmatrix} f_0(0) \\ f_0(1) \\ f_0(2) \\ f_0(3) \end{bmatrix}. \qquad (11\text{-}78)$$

The next step, and this is a critical one, is to write Eq. (11-78) in the factored form,

$$\overline{[F(n)]} = \begin{bmatrix} F(0) \\ F(2) \\ F(1) \\ F(3) \end{bmatrix} = \begin{bmatrix} 1 & W^0 & 0 & 0 \\ 1 & W^2 & 0 & 0 \\ 0 & 0 & 1 & W^1 \\ 0 & 0 & 1 & W^3 \end{bmatrix} \cdot \begin{bmatrix} 1 & 0 & W^0 & 0 \\ 0 & 1 & 0 & W^2 \\ 1 & 0 & W^2 & 0 \\ 0 & 1 & 0 & W^2 \end{bmatrix} \cdot \begin{bmatrix} f_0(0) \\ f_0(1) \\ f_0(2) \\ f_0(3) \end{bmatrix}.$$

$$(11\text{-}79)$$

Observe that even though W^0 is equivalent to unity, both W^0 and 1 are used in the factored form. Observe also that rows 1 and 2 have been interchanged in Eq. (11-79). We shall discuss this matter below. Clearly, of course, the method of factorization constitutes an essential step in the FFT, and the scheme for determining the factorization constitutes the essence of the FFT.

The basis for the operation in going from Eq. (11-78) to Eq. (11-79) is best examined graphically. First we consider the two matrixes on the right, which we write,

$$\begin{bmatrix} f_1(0) \\ f_1(1) \\ f_1(2) \\ f_1(3) \end{bmatrix} = \begin{bmatrix} 1 & 0 & W^0 & 0 \\ 0 & 1 & 0 & W^2 \\ 1 & 0 & W^2 & 0 \\ 0 & 1 & 0 & W^2 \end{bmatrix} \cdot \begin{bmatrix} f_0(0) \\ f_0(1) \\ f_0(2) \\ f_0(3) \end{bmatrix}. \qquad (11\text{-}80)$$

448   General Excitation Functions

The steps in the expansion of this determinant are shown graphically (Fig. 11-11). As drawn, this figure shows that one step in the procedure reduces the 4-point DFT evaluation into the evaluation of two 2-point DFT-s. The next step in the evaluation is the matrix,

$$\begin{bmatrix} F(0) \\ F(2) \\ F(1) \\ F(3) \end{bmatrix} = \begin{bmatrix} 1 & W^0 & 0 & 0 \\ 1 & W^2 & 0 & 0 \\ 0 & 0 & 1 & W^1 \\ 0 & 0 & 1 & W^3 \end{bmatrix} \cdot \begin{bmatrix} f_1(0) \\ f_1(1) \\ f_1(2) \\ f_1(3) \end{bmatrix}. \qquad (11\text{-}81)$$

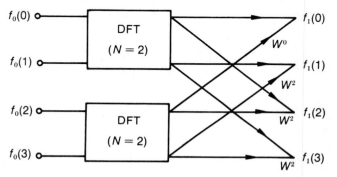

Fig. 11-11. A step in the FFT algorithm.

A graphical portrayal of this expansion is given in Fig. 11-12. The total process, and this shows that the 4-point DFT is completely reduced to complex multiplication and additions by repeated "decimation", is shown

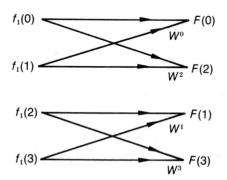

Fig. 11-12. A second step in the FFT algorithm.

in Fig. 11-13. This procedure clearly shows that the evaluation of a DFT, say with N points, is reduced into the evaluation of more and more DFT-s with fewer and fewer points.

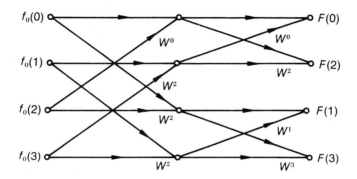

Fig. 11-13. The FFT algorithm for evaluating Eq. (11-78).

It is of interest to determine the number of additions and multiplications involved in the expansion of Eq. (11-79). To evaluate this, we first consider the two matrixes on the right, and write,

$$\begin{bmatrix} f_1(0) \\ f_1(1) \\ f_1(2) \\ f_1(3) \end{bmatrix} = \begin{bmatrix} 1 & 0 & W^0 & 0 \\ 0 & 1 & 0 & W^2 \\ 1 & 0 & W^2 & 0 \\ 0 & 1 & 0 & W^2 \end{bmatrix} \cdot \begin{bmatrix} f_0(0) \\ f_0(1) \\ f_0(2) \\ f_0(3) \end{bmatrix}. \qquad (11\text{-}82)$$

We see that $f_1(0)$ is determined by one complex multiplication and one addition, since,

$$f_1(0) = f_0(0) + W^0 f_0(2). \qquad (11\text{-}83)$$

For finding $f_1(1)$, we see that this also requires one complex multiplication and addition. $f_1(2)$ is written,

$$\begin{aligned} f_1(2) &= f_0(0) + W^2 f_0(2), \\ &= f_0(0) - W^0 f_0(2), \end{aligned} \qquad (11\text{-}84)$$

since $W^0 = -W^2$. Here, since $W^0 f_0(2)$ has already been performed, we require one complex addition. By the same reasoning $f_1(3)$ requires one complex addition. Hence the evaluation of $[f_1(k)]$ requires a total of four additions and two multiplications.

In the next step, we determine the number of computations involved

450   General Excitation Functions

in the expansion,

$$\begin{bmatrix} F(0) \\ F(2) \\ F(1) \\ F(3) \end{bmatrix} = \begin{bmatrix} 1 & W^0 & 0 & 0 \\ 1 & W^2 & 0 & 0 \\ 0 & 0 & 1 & W^1 \\ 0 & 0 & 1 & W^3 \end{bmatrix} \cdot \begin{bmatrix} f_1(0) \\ f_1(1) \\ f_1(2) \\ f_1(3) \end{bmatrix}. \qquad (11\text{-}85)$$

We find that $F(0)$ is determined by one multiplication and one addition, since,

$$F(0) = f_1(0) + W^0 f_1(1).$$

Because $W^0 = -W^2$, the determination of element $F(2)$ requires one addition. By similar reasoning, $F(1)$ requires one complex multiplication and one addition, and $F(3)$ requires only one addition. In total, therefore, vector $[\overline{F(n)}]$ has been computed by a total of $Ni/2 = 4$ complex multiplications, and $Ni = 8$ complex additions. This shows a total of 12 complex multiplications and additions, as compared with the $N^2 = 16$ such operations for finding $F(n)$ in Eq. (11-75). For large N, the savings are huge.

In its more general form, the Cooley-Tukey FFT algorithm can be considered as a method for factoring an $N \times N$ matrix into $iN \times N$ matrixes ($N = 2^i$) such that each of the factored matrixes has the special property of minimizing the number of complex multiplications. In this optimum matrix factoring process one does encounter the rearrangement from $[F(n)]$ to $[\overline{F(n)}]$. But a scheme to rearrange $[\overline{F(n)}]$ to $[F(n)]$ is not difficult. This calls for replacing the argument n in $[F(n)]$ by its binary equivalent. Thus,

$$\begin{bmatrix} F(0) \\ F(1) \\ F(2) \\ F(3) \end{bmatrix} \text{ becomes } \begin{bmatrix} F(00) \\ F(01) \\ F(10) \\ F(11) \end{bmatrix}. \qquad (11\text{-}86)$$

If the binary arguments of Eq. (11-86) are flipped or bit-reversed; that is, 01 becomes 10, 10 becomes 01, etc. then,

$$[F(n)] = \begin{bmatrix} F(0) \\ F(1) \\ F(2) \\ F(3) \end{bmatrix} \text{ flips to } \begin{bmatrix} F(00) \\ F(10) \\ F(01) \\ F(11) \end{bmatrix} = [\overline{F(n)}] \qquad (11\text{-}87)$$

This is a general result.

The critical feature of the FFT is the method for determining the

PROBLEMS

11-1. Express each of the following periodic waves in Fourier series form.

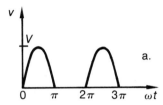
a.

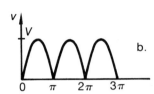
b.

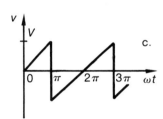
c.

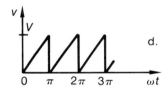
d.

11-2. A sawtooth voltage is applied to the RL circuit shown. Find the resulting current, (steady-state component only).
  a. By the methods of Chapter 7.
  b. By means of a Fourier series expansion of the driving source.

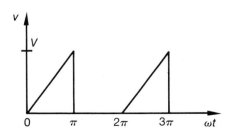
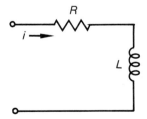

11-3. Determine and sketch the amplitude and phase spectrums of the periodic signals illustrated.

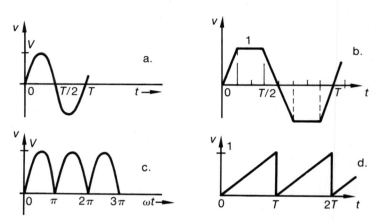

11-4. Find the mean-square error when the signals in Problem 11-3 are approximated by the first three harmonic terms in the Fourier series expansion.

11-5. Determine the Fourier transforms of the functions illustrated.

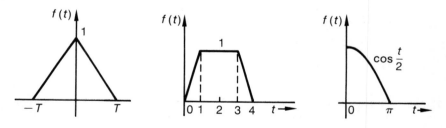

11-6. Determine and sketch the amplitude and frequency spectrum functions for the following waveforms.

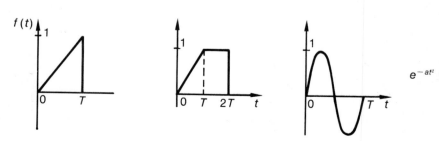

11-7. Deduce the form of the Fourier transform when $f(t)$ is a real time function and also:
  a. An even function.
  b. An odd function.

11-8. Show that the Fourier transform of the unit step function is.

$$\mathscr{F}[u_{-1}(t)] = \frac{1}{j\omega} + \pi u_0(t).$$

Note that one cannot use the direct application of the Fourier integral since the result is not defined at the upper limit. The use of the convergence factor e^{-at} gives only part of the answer.

11-9. Given the frequency functions $F(j\omega)$, find the appropriate functions $f(t)$.

$$\frac{j\omega - 1}{(j\omega + 1)(j\omega + 2)} \qquad \frac{j\omega}{(j\omega + 2)(j\omega + 1)^2} \qquad \frac{1}{(j\omega + \alpha)(j\omega + \beta)}.$$

11-10. Find the inverse Fourier transform of the functions given.

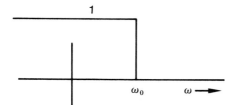

$$e^{-\omega^2} \quad \left[\text{use the fact that } \mathscr{F}[e^{-at^2}] = \frac{e^{-\omega^2/4a}}{2/\pi a}\right]$$

11-11. Find the response of the RC circuit to a single pulse, using the Fourier integral in the analysis.

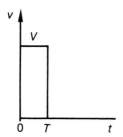
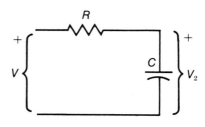

11-12. A single pulse of the type shown in Problem 11-11 is applied to a network with the system function $T(j\omega) = |T(j\omega)|e^{j\theta(\omega)}$. Find the response function.

11-13. Show that the Fourier transform of the convolution of two time functions is the product of their transforms; that is, given $\mathscr{F}[f(t)] = F(j\omega)$ and $\mathscr{F}[g(t)] =$

$G(j\omega)$, show that,

$$\mathscr{F}[f * g] = \mathscr{F}\left[\int_{-\infty}^{\infty} f(\tau)g(t-\tau)\,d\tau\right] = F(j\omega)G(j\omega).$$

11-14. A signal $f(t)$ has a Fourier transform which vanishes for $|\omega| > \Omega/2$, $\Omega = 2\pi/T$. Show that $f(t)$ can be determined from a knowledge of,

$$f(4nT), f'(4nT) \qquad f(4nT+), f'(4nT+),$$

for all integral values of n, with $0 < \tau < 4T$.

Part IV   Selected Topics

12

The Z-Transform and Discrete Time Systems

When a linear system operates by regularly sampling some signal which is used as basic input data, or when the system is such that data is supplied to it in discrete or pulse form at regular intervals (e.g., pulsed radar data) the analysis is accomplished more conveniently using Z-transform techniques than by using Laplace transform techniques. That is, the Z-transform, which can be directly related to the discrete Fourier transform, serves as a tool for the study of discrete time systems in much the same manner that the Laplace transform serves as a tool for the study of continuous time systems. Moreover, the general philosophy of Z-transform analysis is similar to that for Laplace transform analysis and involves replacing the discrete time signals and the system function by appropriate Z-transformed forms of these quantities. It is found that the Z-transform of the discrete time response of a system is given as the product of the discrete time system function and the Z-transform of the input signal, as in Laplace methods. The time response of the system is deduced from this transformed form by one of several methods; one is an inverse Z-transform procedure that provides a time function in closed form; the other provides the value of the response at the sampling times.

This chapter will consider the general features of Z-transform methods, and will show the application both to discrete time systems and to continuous time systems with sampled inputs.

12-1 TIME SAMPLING AND THE Z-TRANSFORM

One feature of sampling that requires attention is the nature of the "sampler," the device that is to convert a continuous time function into a

458   The Z-Transform and Discrete Time Systems

train of identically shaped pulses whose amplitudes are proportional to the input at the sampling instant. Different types of samplers are possible, and some of the results depend on the resulting form of the sampled data. Not all sampling methods are equally important. Among the important samplers are: impulse samplers, staircase (piecewise constant; sample and hold) samplers, piecewise linear samplers. The general features of such sampling are shown in Fig. 12-1. As illustrated, the output from the

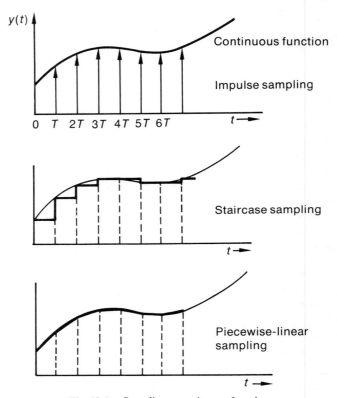

Fig. 12-1.   Sampling a continuous function.

impulse sampler is a train of impulses which is weighted by the sample values of the continuous time signal. The staircase sampler yields pulses similarly weighted by the sample values of the continuous signal.

Let us consider a function $y(t)$ to be impulse sampled every T seconds. The sequence of these sampled values is the successive values of $y(t)$, namely, $y(0), y(T), y(2T), \ldots$, and may be written,

$$\{y(kT)\} = \{y(0), y(T), y(2T), \ldots, y(nT), \ldots\}. \qquad (12\text{-}1)$$

We now recall that $u_0(t-nT)$ denotes a unit impulse at time $t = nT$, and furthermore, that the quantity $y(nT)u_0(t-nT)$ denotes a quantity that is zero everywhere in time except at $t = (nT)$ when it has the weighted value $y(nT)$. Clearly, therefore, we can write the sequence of impulse sampled values corresponding to the sequence given in Eq. (12-1), and this is denoted $y^*(t)$, as,

$$y^*(t) = [y(0)u_0(t) + y(1)u_0(t-T) + y(2)u_0(t-2T) + \cdots]. \qquad (12\text{-}2)$$

In compact form this is,

$$y^*(t) = y(t) \sum_{n=0}^{\infty} u_0(t-nT) = \sum_{n=0}^{\infty} y(nT)u_0(t-nT). \qquad (12\text{-}3)$$

We can discuss this sampling process in somewhat different terms, involving the use of the previously introduced time delay operator $1/z$. We recall that the symbol $1/z$ was used to denote a delay of one sampling period†, with $1/z^n = (1/z)^n = z^{-n}$ denoting a delay of n sampling periods. Consequently, the sequence in Eq. (12-1), when represented in terms involving the delay factor $1/z$, is written $Y(z)$, and has the form,

$$Y(z) = y(0) + \frac{y(T)}{z} + \frac{y(2T)}{z^2} + \frac{y(3T)}{z^3} + \cdots. \qquad (12\text{-}4)$$

In accordance with this discussion $y(0) = y(0)/z^0$ specifies the value of $y(t)$ at zero delay; $y(T)/z$ is the value of $y(t)$ at the end of the first sampling period, $y(nT)/z^n$ is the value of $y(t)$ at the end of the nth sampling period, etc. $Y(z)$ is called the Z-transform of $y(t)$, and in the form of Eq. (12-4) is fundamental to our subsequent work.

Example 12-1.1

Write the Z-transform of the function,

$$y(t) = Ae^{-at} \qquad (t \geq 0),$$

which is sampled every T seconds.

Solution. The sampled values are written,

$$y(kT) = A, Ae^{-aT}, Ae^{-2aT}, \ldots.$$

†This corresponds to the function e^{-sT}, the time delay function, in Laplace transform domain, as discussed in connection with Eq. (7-14).

This can be written in Z-transformed form,

$$Y(z) = A\left[1 + \left(\frac{e^{-aT}}{z}\right) + \left(\frac{e^{-aT}}{z}\right)^2 + \left(\frac{e^{-aT}}{z}\right)^3 + \cdots\right].$$

It is of interest that this expression can be written in closed form. This follows because of the known expansion,

$$\frac{1}{1-x} = 1 + x + x^2 + x^3 + \cdots.$$

Thus we have that,

$$Y(z) = \frac{A}{1 - (e^{-aT}/z)} = \frac{Az}{z - e^{-aT}}.$$

Example 12.1-2

Find the Z-transform of the function,

$$y(t) = 2e^{-t} + e^{-2t},$$

for $T = 0.1$ secs.

Solution. We write, on the basis of the results of Example 12-1.1,

$$Y(z) = \frac{2z}{z - e^{-0.1}} + \frac{z}{z - e^{-0.2}},$$

which is,

$$Y(z) = \frac{2z}{z - 0.905} + \frac{z}{z - 0.819}.$$

This is $Y(z)$ in partial fraction form. By expansion, $Y(z)$ can be written,

$$Y(z) = \frac{z\,2(z - 0.819) + (z - 0.905)}{(z - 0.905)(z - 0.819)} = \frac{3z^2 - 2.543z}{z^2 - 1.724z + 0.741}.$$

By long division, we find that,

$$\begin{array}{r}
3 + \dfrac{2.629}{z} + \dfrac{2.309}{z^2} + \cdots \\[2pt]
z^2 - 1.724z + 0.741 \overline{\big)\, 3z^2 - 2.543z } \\
3z^2 - 5.172z + 2.223 \\
\hline
2.629z - 2.223 \\
2.629z - 4.532 + \dfrac{1.948}{z} \\
\hline
2.309 - \dfrac{1.948}{z}
\end{array}$$

This process leads to,

$$Y(z) = 3 + \frac{2.629}{z} + \frac{2.309}{z^2} + \cdots .$$

12-2 THE Z-TRANSFORM (25)

As is evident from the discussion in the foregoing section, the Z-transform of a discrete signal $f(n)$ is defined as a power series in z^{-1} whose coefficients are the amplitudes of the discrete-time signal. Two Z-transforms exist,

one-sided transform $\quad Z[f(n)] = F_I(z) = \sum_{n=0}^{\infty} f(n) z^{-n},$

two-sided transform $\quad Z[f(n)] = F_{II}(z) = \sum_{n=-\infty}^{\infty} f(n) z^{-n}.$ (12-5)

The first form is valid if $f(n) = 0$ for $n < 0$; hence the one-sided transform of $f(n)$ carries information only for nonnegative n. It is suitable in most cases except for problems involving stationary random processes, since $f(n) = 0$ for $n < 0$ is not necessarily valid in such problems. Clearly, the Z-transform is the "generating signal" for the discrete time signal to which it corresponds, precisely paralleling the relation between the Fourier series and continuous signals represented thereby. Observe, from purely mathematical considerations, that,

$$F_I(z) = f(0) + \frac{f(1)}{z} + \frac{f(2)}{z^2} + \cdots , \quad (12\text{-}6)$$

is the principal part of a Laurent series expansion about the point $z = 0$, plus a constant $f(0)$. Hence convergence occurs outside of a unit circle in the z-plane centered at the origin.

Inversion of the Z-transform involves, from a practical viewpoint, the use of a table of Z-transforms. A short table of such transform pairs plus the technique of partial fraction expansion, is usually adequate for most needs. In a more formal way, the general inversion integral is,

$$f(n) = \frac{1}{2\pi j} \oint_C F(z) z^{n-1} \, dz, \quad (12\text{-}7)$$

where the contour C is a circle of radius r that encloses all the singularities of $F(z) z^{n-1}$. To carry out such a contour integral requires a knowledge of complex function theory. A short table of Z-transform pairs, also including the corresponding Laplace transform expressions, is given in Table

462 The Z-Transform and Discrete Time Systems

12-1. Proofs of these entries can be accomplished by proceeding from Eq. (12-6).

It is of interest to compare Eq. (12-5) with Eq. (11-71). This shows, if we take $z^{-n} = W^{-nk} = e^{j\pi nk/N}$, that the discrete Fourier transform of the finite sequence $f(nT)$ at the N points $\{0, 1, 2, \ldots, (N-1)\}$ is the Z-transform of $f(n)$ evaluated at N equally spaced points on the unit circle in the z-plane.

TABLE 12-1. Short Table of Z-Transforms.

Time Function	Z-transform	$\mathscr{L}$-transform
$u_0(t)$	$I_0 = 1$	1
$u_{-1}(t)$	$I_1 = \dfrac{z}{z-1}$	$\dfrac{1}{s}$
$u_{-2}(t) = nT$	$I_2 = \dfrac{Tz}{(z-1)^2}$	$\dfrac{1}{s^2}$
$u_{-3}(t) = \dfrac{(nT)^2}{2!}$	$I_3 = \dfrac{T^2}{2}\dfrac{z(z+1)}{(z-1)^3}$	$\dfrac{1}{s^3}$
$u_{-4}(t) = \dfrac{(nT)^3}{3!}$	$I_4 = \dfrac{T^3}{6}\dfrac{z(z^2+4z+1)}{(z-1)^4}$	$\dfrac{1}{s^4}$
e^{-anT}	$\dfrac{z}{z - e^{-aT}}$	$\dfrac{1}{s+a}$
$e^{-anT} \cos \beta nT$	$\dfrac{z(z - e^{-aT} \cos \beta T)}{z^2 - 2ze^{-aT} \cos \beta T + e^{-2aT}}$	$\dfrac{s+a}{(s+a)^2 + \beta^2}$
$e^{-anT} \sin \beta nT$	$\dfrac{ze^{-aT} \sin \beta T}{z^2 - 2ze^{-aT} \cos \beta T + e^{-2aT}}$	$\dfrac{\beta}{(s+a)^2 + \beta^2}$
$e^{-[(1/T) \log \beta] nT}$	$\dfrac{z}{z - \beta}$	$\dfrac{1}{s + (1/T) \ln \beta}$

It is noted that higher order singularity functions $u_n(t)$, $n > 0$, are not defined in the z-domain.

12-3 PROPERTIES OF THE Z-TRANSFORM

We wish to develop some of the important properties of the Z-transform.

a. Differences. In this connection, we consider the Z-transform of several functional forms. From Eq. (12-4) we write,

$$Z[f(n)] = F(z) = \sum_{n=0}^{\infty} f(n)z^{-n} = f(0) + \frac{f(1)}{z} + \frac{f(2)}{z^2} + \cdots,$$

$$Z[f(n+1)] = \sum_{n=0}^{\infty} f(n+1)z^{-n} = f(1) + \frac{f(2)}{z} + \frac{f(3)}{z^2} + \cdots, \quad (12\text{-}8)$$

$$= z[F(z) - f(0)],$$

$$Z[f(n+m)] = z^m \left[F(z) - \sum_{i=0}^{m-1} f(i)z^{-i} \right] \quad m \geq 0,$$

$$Z[f(n-m)] = z^{-m}F(z).$$

The Z-transform of the first forward difference is defined as,

$$Z[\Delta f(n)] = Z[f(n+1)] - Z[f(n)], \quad (12\text{-}9)$$

using the appropriate forms from Eq. (12-8) this becomes,

$$Z[\Delta f(n)] = z[F(z) - f(0)] - F(z),$$
$$= (z-1)F(z) - zf(0). \quad (12\text{-}10)$$

The Z-transform of the second forward difference follows in a similar way. Now we have,

$$Z[\Delta^2 f(n)] = Z\{\Delta[\Delta f(n)]\} = (z-1)Z[\Delta f(n)] - z\Delta f(0),$$
$$= (z-1)^2 F(z) - z(z-1)f(0) - z\Delta f(0). \quad (12\text{-}11)$$

b. Convolution. The discrete convolution property plays a key part in discrete time analysis, just as the analogous theorem in Laplace transform analysis plays a key part in continuous time analysis. To study this property, we begin with the convolution summation (*see* Section 6-4)

$$y(n) = \sum_{k=0}^{n} h(n-k)u(k). \quad (12\text{-}12)$$

The Z-transform is taken of both sides of this expression. This gives,

$$Z[y(n)] = Y(z) = \sum_{n=0}^{\infty} z^{-n} \sum_{k=0}^{n} h(n-k)u(k),$$

since h(negative) is zero, we can replace n by ∞ in the summation in k so that,

$$Z[y(n) = Y(z) = \sum_{n=0}^{\infty} z^{-n} \sum_{k=0}^{\infty} h(n-k)u(k).$$

Invert the order of summation,

$$Y(z) = \sum_{k=0}^{\infty} u(k) \sum_{n=0}^{\infty} h(n-k)z^{-n} = \sum_{k=0}^{\infty} u(k)z^{-k} \sum_{l=-k}^{\infty} h(l)z^{-l},$$

$$= \sum_{k=0}^{\infty} u(k)z^{-k} \sum_{l=0}^{\infty} h(l)z^{-l} = U(z)H(z), \qquad (12\text{-}13)$$

since $h(l) = 0$ for $l < 0$. Thus the Z-transform of the convolution summation is the product of the Z-transforms of the two-functions.

12-4 DISCRETE TIME SYSTEM FUNCTION

The discrete time system function for a fixed discrete time system is defined as the Z-transform of the unit impulse response $h(n)$ of the system; thus,

$$T(z) = Z\{h(n)\} = H(z). \qquad (12\text{-}14)$$

By Eq. (12-13), therefore,

$$Y(z) = T(z)U(z). \qquad (12\text{-}15)$$

This result means that a system $T(z)$ to which the input sequence, $\{\ldots z^{-2}, z^{-1}, 1, z, z^2, \ldots\}$ is applied has the response,

$$y(n) = \sum_{k=-\infty}^{n} h(n-k)z^k. \qquad (12\text{-}16)$$

By a change of index, $m = n - k$,

$$y(n) = \sum_{m=0}^{\infty} h(m)z^{n-m} = z^n T(z), \qquad (12\text{-}17)$$

from which we may write,

$$T(z) = \frac{\text{response to } z^n}{z^n}. \qquad (12\text{-}18)$$

Example 12-4.1

A filter possesses the magnitude-frequency function (zero phase shift everywhere) shown. Find the weighting function of this filter.

Solution. The weighting function is, by definition, the response of the network to a unit pulse $I_0(z)$ at $n = 0$. We thus write, by Eq. (12-7),

$$f(nT) = \frac{1}{2\pi j} \oint_C F(z) z^{n-1} \, dz.$$

Write $z = e^{j\omega T}$, and inserting the appropriate limits,

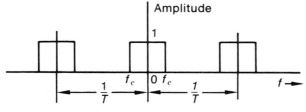

Fig. 12-4.1-1

$$f(nT) = \frac{1}{\pi} \int_{-\omega_c}^{\omega_c} e^{jn\omega T} d\omega = \frac{\sin n\omega_c T}{n\pi}.$$

This result can be compared with Eq. (11-57). Additional forms for the discrete time system function $T(z)$ are developed in Section 12-6.

12-5 Z-REPRESENTATION OF DIFFERENTIATION

Reference is made to Section 2-10 and to the difference equation forms for the derivatives. When we wrote,

$$\frac{dv}{dt} = \frac{v_{n+1} - v_n}{T},$$

$$\frac{d^2v}{dt^2} = \frac{v_{n+2} - 2v_{n+1} + v_n}{T^2}, \quad (12\text{-}19)$$

$$\frac{d^3v}{dt^3} = \frac{v_{n+3} - 3v_{n+2} + 3v_{n+1} - v_n}{T^3} \text{ etc.},$$

it was implicitly assumed that derivatives of all orders are piecewise linear forward in time. Writing the Z-transform of derivatives by using these expressions, or equivalently, using the same assumption by taking the Z-transform of v as $V(z)$ and that $zV(z)$ specifies the next value in the sequence of $V(z)$ (one sample period forward in time), then,

$$Z\left[\frac{dv}{dt}\right] = \frac{zV(z) - V(z)}{T} = \frac{(z-1)}{T} V(z). \quad (12\text{-}20)$$

Similarly, we can write,

$$Z\left[\frac{d^2y}{dt^2}\right] = Z\left[\frac{d}{dt}\left(\frac{dv}{dt}\right)\right] = \frac{z-1}{T}\left(\frac{z-1}{T}\right)V(z) = \left(\frac{z-1}{T}\right)^2 V(z), \quad (12\text{-}21)$$

or from the difference equation form in Eq. (12-19),

$$Z\left[\frac{d^2y}{dt^2}\right] = \frac{z^2 - 2z + 1}{T^2} V(z) = \left(\frac{z-1}{T}\right)^2 V(z).$$

466   The Z-Transform and Discrete Time Systems

For the third derivative,

$$Z\left[\frac{d^3v}{dt^3}\right] = \frac{z^3 - 3z^2 + 3z - 1}{T^3} V(z) = \left(\frac{z-1}{T}\right)^3 V(z). \quad (12\text{-}22)$$

If, instead of assuming that $v(t)$ and all derivatives are linear forward in time, we assume that $v(t)$ is piecewise linear, then we must approach the determination of derivatives somewhat differently. If $v(t)$ is piecewise-linear, it is represented as a sum of ramp functions of appropriately chosen amplitude and starting time, as illustrated in Fig. 12-1. But since the operation of time differentiation is linear and time invariant, superposition can be employed to simplify the evaluation of the Z-transform of derivatives of $v(t)$. This allows us to view the differentiator as a device to which a ramp function is applied with a step function output. This process is illustrated in Fig. 12-2. Thus from Table 12-1 we write,

$$Z[\text{input}] = Z[u_{-2}(t)] = I_2 = \frac{Tz}{(z-1)^2},$$

$$Z[\text{output}] = Z[u_{-1}(t)] = I_1 = \frac{z}{z-1}.$$

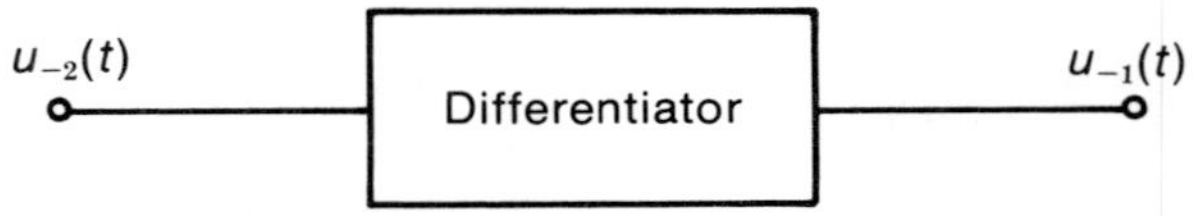

```
u_{-2}(t)    ┌──────────────────┐    u_{-1}(t)
   o─────────┤  Differentiator  ├─────────o
             └──────────────────┘
```

Fig. 12-2. Ramp function input to a differentiator.

The corresponding $T(z)$ of the differentiator is then,

$$T(z) = \frac{I_1}{I_2} = \frac{z}{z-1} \Big/ \frac{Tz}{(z-1)^2} = \frac{z-1}{T},$$

from which we conclude that,

$$Z\left[\frac{dv}{dt}\right] = \frac{z-1}{T} V(z).$$

This same procedure may be used for finding higher order derivatives. For the Z-transform of the second derivative function, we use the fact that a ramp function applied to the input of the differentiator yields an impulse in the output. Thus we have, using the appropriate entries in Table 12-1,

$$Z\left[\frac{d^2v}{dt^2}\right] = \frac{I_0}{I_2} = 1 \Big/ \frac{Tz}{(z-1)^2} = \frac{(z-1)^2}{Tz}.$$

Because the Z-transform of higher order impulses are not definable, we cannot obtain Z-transforms of 3rd and higher order derivatives under the assumption that $v(t)$ is piecewise-linear.

To obtain Z-transform representations of higher-order derivatives, it is necessary to assume that $v(t)$ is approximated by a set of higher order approximations. For example, if we assume that $v(t)$ is piecewise-quadratic, we can proceed as above using as the input function $u_{-3}(t)$. This will allow us to find a Z-transform form through $Z[d^3v/dt^3]$, but no higher. The results of foregoing are combined in Table 12-2.

TABLE 12-2. Derivative functions.

Derivative	Function and all derivatives linear forward in time	Function Piecewise Linear	Function Piecewise Quadratic
$Z\left[\dfrac{dv}{dt}\right]$	$\dfrac{z-1}{T}V(z)$	$\dfrac{z-1}{T}V(z)$	$\dfrac{2}{T}\dfrac{z-1}{z+1}V(z)$
$Z\left[\dfrac{d^2v}{dt^2}\right]$	$\left(\dfrac{z-1}{T}\right)^2 V(z)$	$\left(\dfrac{z-1}{Tz}\right)^2 V(z)$	$\dfrac{2}{T^2}\dfrac{(z-1)^2}{z+1}V(z)$
$Z\left[\dfrac{d^3v}{dt^3}\right]$	$\left(\dfrac{z-1}{T}\right)^3 V(z)$	not definable	$\dfrac{2}{T^2}\dfrac{(z-1)^3}{z(z+1)}V(z)$

Unless otherwise noted, we shall assume the conditions appropriate to column 2.

Example 12-5.1

Find the solution of the following differential equation by Z-transform methods,

$$\frac{dy}{dt} + 3y = 2e^{-2t} \qquad y(0) = 0.$$

Choose sampling intervals of $T = 0.1$ sec.

Solution. Each term in the differential equation is Z-transformed. The important terms are,

$$Z\left[\frac{dy}{dt}\right] = \frac{z-1}{T}Y(z),$$

$$Z[3y] = 3Y(z),$$

$$Z[2e^{-2t}] = \frac{2z}{2 - e^{-2T}},$$

and the differential equation in Z-transformed form is,

$$\frac{z-1}{T}Y(z) + 3Y(z) = \frac{2z}{z - e^{-2T}}.$$

For $T = 0.1$ this expression reduces to,

$$(10z - 7)Y(z) = \frac{2z}{z - 0.819},$$

so that,

$$Y(z) = \frac{0.2z}{(z-0.7)(z-0.819)} = \frac{0.2z}{z^2 - 1.519z + 0.573}.$$

We proceed in two different ways,
  a. By series expansion, using simple long division methods,

$$
\begin{array}{r}
\frac{0.2}{z} + \frac{0.304}{z^2} + \frac{0.346}{z^3} + \cdots \\
z^2 - 1.519z + 0.573 \overline{)\, 0.2z } \\
0.2z - 0.304 + \frac{0.115}{z} \\
\overline{ 0.304 - \frac{0.115}{z} } \\
0.304 - \frac{0.461}{z} + \frac{0.174}{z^2} \\
\overline{ + \frac{0.346}{z} - \frac{0.174}{z^2}}
\end{array}
$$

The result in this case is,

$$Y(z) = \frac{0.2}{z} + \frac{0.304}{z^2} + \frac{0.346}{z^3} + \cdots.$$

This technique is easily implemented on the digital computer; moreover, it has the advantage that the poles of $Y(z)$ need not be found. However, this long division is sensitive to round-off error and is generally useful only when the poles are well inside the unit circle (large damping).

  b. The procedure now is to find a partial fraction expansion for the function $Y(z)/z$. This is,

$$\frac{Y(z)}{z} = \frac{0.2}{(z-0.7)(z-0.819)} = \frac{A}{(z-0.7)} + \frac{B}{(z-0.819)},$$

where,

$$A = (z-0.7)\left[\frac{0.2}{(z-0.7)(z-0.819)}\right]_{z=0.7} = -1.681,$$

Z-Representation of Differentiation

$$B = (z-0.819)\left[\frac{0.2}{(z-0.7)(z-0.819)}\right]_{z=0.819} = 1.681,$$

so that,

$$Y(z) = 1.681\left(\frac{-z}{z-0.7} + \frac{z}{z-0.819}\right).$$

From Table 12-1, we can deduce the inverse transform. We note that,

$$\frac{1}{T}\ln\frac{1}{0.7} = 10 \times 0.357 = 3.57,$$

$$\frac{1}{T}\ln\frac{1}{0.819} = 10 \times 0.20 = 2.0,$$

and the solution is then,

$$y(t) = 1.681(e^{-2.0t} - e^{-3.57t}).$$

Example 12-5.2

Given the differential equation,

$$\frac{d^2y}{dt^2} + 2\frac{dy}{dt} + 3y = 5 \qquad y(0) = \dot{y}(0) = 0$$

Find solutions by Z-transform methods, assuming piecewise-constant and piecewise-linear functions. Choose $T = 0.1$ sec.

Solution.

a. Piecewise-constant approximation. We write the Z-transformed equation, using the entries in Table 12-2.

$$\left(\frac{z-1}{T}\right)^2 Y(z) + 2\left(\frac{z-1}{T}\right)Y(z) + 3Y(z) = \frac{5z}{z-1}.$$

From this we write,

$$[100(z-1)^2 + 20(z-1) + 3]Y(z) = \frac{5z}{z-1},$$

which is,

$$(100z^2 - 180z + 83)Y(z) = \frac{5z}{z-1}.$$

Solve for $Y(z)$ to find,

$$Y(z) = \frac{5}{100}\frac{z}{(z-1)(z^2-1.8z+0.83)},$$

$$= \frac{0.05z}{z^3 - 2.8z^2 + 2.63z - 0.83}.$$

The series expression is obtained by long division. The result is found to be,

$$Y(z) = \frac{0.05}{z^2} + \frac{0.14}{z^3} + \frac{0.26}{z^4} + \cdots.$$

b. Piecewise-linear function. From the appropriate column in Table 12-2 the Z-transformed equation is,

$$\left(\frac{z-1}{zT}\right)^2 Y(z) + 2\left(\frac{z-1}{T}\right)Y(z) + 3Y(z) = \frac{5z}{z-1},$$

from which we have,

$$[10(z-1)^2 + 20z(z-1) + 3z]Y(z) = \frac{5z^2}{z-1},$$

or,

$$(30z^2 - 37z + 10)Y(z) = \frac{5z^2}{z-1}.$$

Solve for $Y(z)$, to find,

$$Y(z) = \frac{5}{30} \frac{z^2}{(z-1)(z^2-1.233z+0.333)} = \frac{0.1667z^2}{z^3+2.233z^2+1.567z-0.333}.$$

This is written in series form by long division. The result is,

$$Y(z) = \frac{0.167}{z} + \frac{0.373}{z^2} + \frac{0.571}{z^3} + \frac{0.745}{z^4} + \cdots$$

It is observed that the results by the two different function assumptions are rather different.

Example 12-5.3

Consider the differential equation,

$$\frac{d^2y}{dt^2} + 2\frac{dy}{dt} + 3y = 2\frac{du}{dt} + 5u \qquad y(0) = \dot{y}(0) = 0.$$

a. Write $T(s)$, and from this deduce the $T(z)$ function for $T = 0.1$.
b. Find a SFG appropriate to $T(s)$, and then transform it into a z-plane SFG. Write $T(z)$ from this, employing the Mason graph transmittance rule.
c. Find the output $Y(z)$ for $u = u_{-1}(t)$.
Solution. a. $T(s)$ is written directly from an inspection of the differ-

Z-Representation of Differentiation

ential equation. It is,

$$T(s) = \frac{2s+5}{s^2+2s+3} = \frac{\frac{2}{s}+\frac{5}{s^2}}{1+\frac{2}{s}+\frac{3}{s^2}}.$$

To find $T(z)$, make the transformation $T/(z-1)$ for $1/s$. This gives,

$$T(z) = \frac{\frac{2T}{z-1}+5\left(\frac{T}{z-1}\right)^2}{1+\frac{2T}{z-1}+3\left(\frac{T}{z-1}\right)^2} = \frac{2T(z-1)+5T^2}{(z-1)^2+2T(z-1)+3T^2},$$

$$= \frac{2Tz+(5T^2-2T)}{z^2-2z(1-T)+(1-2T+3T^2)}.$$

For $T = 0.1$ (this should be less than the shortest significant time constant),

$$T(z) = \frac{0.2(z-0.75)}{z^2-1.8z+0.83}.$$

b. By the method of Section 8-8a, we deduce the following SFG. (Fig. 12-5.3-1.) In the z-domain this is Fig. 12-5.3-2. But it is noted that a transformation is possible, as per Fig. 12-5.3-3. With this transforma-

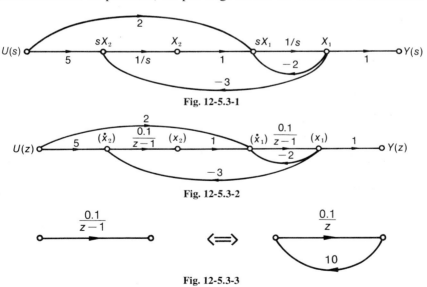

Fig. 12-5.3-1

Fig. 12-5.3-2

Fig. 12-5.3-3

tion, the z-plane SFG appropriate to the given differential equation is that in Fig. 12-5.3-4. The z-plane system function is, using the Mason rule for the transmittance of a SFG,

$$T(z) = \frac{\dfrac{0.2}{z-1} + 5\left(\dfrac{0.1}{z-1}\right)^2}{1 - \dfrac{(-2 \times 0.1)}{z-1} - (-3)\left(\dfrac{0.1}{z-1}\right)^2}.$$

c. For a unit step function, $Z[u_{-1}(t)] = z/(z-1)$, and,

$$Y(z) = T(z)U(z) = \frac{0.2(z-0.75)}{z^2 - 1.8z + 0.83} \frac{z}{(z-1)},$$

$$= \frac{0.2z^2 - 0.15z}{z^3 - 2.8z^2 + 2.63z - 0.83}$$

By long division, the result is,

$$Y(z) = \frac{0.2}{z} + \frac{0.41}{z^2} + \frac{0.62}{z^3} + \frac{0.815}{z^4} + \cdots$$

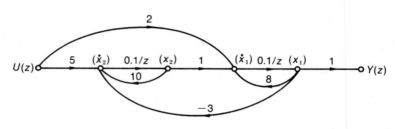

Fig. 12-5.3-4

12-6 DIFFERENCE EQUATIONS AND THE Z-TRANSFORM

We have already noted the close relation between the difference equation and time sampling and the Z-transform. We wish to examine some important results that arise in going from the difference equation to the Z-transform. We begin with the description of the discrete time system in a linear difference equation with constant coefficients, as given in Eq. (2-58), namely,

$$\sum_{j=0}^{k} \xi_j y(n-j) = \sum_{j=0}^{m} \zeta_j u(n-j). \qquad (12\text{-}23)$$

Difference Equations and the Z-Transform 473

The Z-transform of this equation is,

$$(\xi_0 + \xi_1 z^{-1} + \xi_2 z^{-2} + \cdots + \xi_k z^{-k}) Y(z) = (\zeta_0 + \zeta_1 z^{-1} + \cdots + \zeta_m z^{-m}) U(z),$$

from which we write,

$$\frac{Y(z)}{U(z)} = T(z) = \frac{\zeta_0 + \zeta_1 z^{-1} + \cdots + \zeta_m z^{-m}}{\xi_0 + \xi_1 z^{-1} + \cdots + \xi_k z^{-k}} = \frac{\sum_{j=0}^{m} \zeta_j z^{-j}}{\sum_{j=0}^{k} \xi_j z^{-j}}. \quad (12\text{-}24)$$

Here $T(z)$, the Z-transform system function, is a proper rational function of z^{-1}. It is one of several forms for the digital time system function discussed in Section 12-4. A SFG representation is given in Fig. 12-3. This

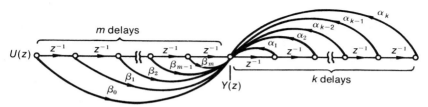

Fig. 12-3. A SFG for the general system described by Eq. (12-24).

expression can be expanded into the ratio of polynomials in z and then written in terms of the roots of the polynomials. This gives $T(z)$ in terms of the poles and zeros of this function,

$$T(z) = H \frac{(z - z_1)(z - z_2) \cdots (z - z_m)}{(z - z_a)(z - z_b) \cdots (z - z_k)}. \quad (12\text{-}25)$$

For a stable system all poles must be included within the unit circle in the z-plane.

In much the same manner as discussed in Section 8-9, $T(z)$ given in Eq. (12-25) can be expanded in partial fraction form to yield an expansion of the form,

$$T(z) = T_1(z) + T_2(z) + \cdots + T_k(z). \quad (12\text{-}26)$$

A graphical portrayal of this system description is given in Fig. 12-4. Other forms for $T(z)$ are also possible, as will be shown below.

There is considerable interest in beginning with Eq. (2-59), namely,

$$y(n) = \sum_{j=0}^{m} b_j u(n-j) - \sum_{j=1}^{k} a_j y(n-j). \quad (12\text{-}27)$$

474 The Z-Transform and Discrete Time Systems

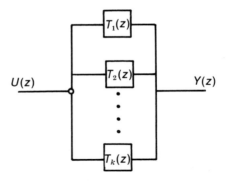

Fig. 12-4. A parallel realization of $T(z)$.

This equation is of importance in the digital processing of signals. It gives a straightforward rule for computing $y(n)$ in terms of the $m+1$ most recent samples of the input $u(n)$ and the k previous values of y, already computed. Once $y(n)$ has been computed, and as soon as a new sample $u(n+1)$ is found, the value of $y(n+1)$ can be found. In such a calculation, the computer must store $m+k$ items of data for the next calculation. The corresponding form for $T(z)$, which is obtained in a manner that exactly parallels that in going from Eq. (12-23) to Eq. (12-24), is,

$$T(z) = \frac{\sum_{j=0}^{m} b_j z^{-j}}{1 + \sum_{j=1}^{k} a_j z^{-j}}. \tag{12-28}$$

As previously noted, from a practical viewpoint of computational demands, there is much to be gained by introducing the auxiliary variable $\eta(n)$, as in Eq. (2-60), and replacing Eq. (12-27) by the pair of equations,

$$\begin{aligned}\eta(n) &= u(n) - \sum_{j=1}^{k} a_j \eta(n-j), \\ y(n) &= \sum_{j=0}^{m} b_j \eta(n-j).\end{aligned} \tag{12-29}$$

The advantage of replacing the single difference equation by the pair of equations is the smaller memory requirement. Now we must save only m or k previous values of $\eta(n)$, depending on which is greater. The essence of this conclusion is exhibited graphically in Fig. 12-5 which shows the

Difference Equations and the Z-Transform

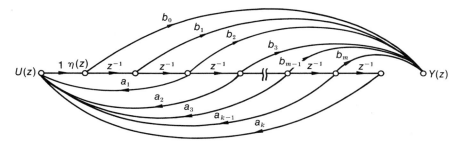

Fig. 12-5. The system of Figure 12-3 with redundant elements eliminated.

description of the system using only k or m delay elements. This SFG should be compared with that in Fig. 12-3 for the same system. In this form, the SFG is very similar to Fig. 8-13 for the continuous system function $T(s)$.

Example 12-6.1

A second order difference equation is given:

$$y(nT) - K_1 y[(n-1)T] - K_2 y[(n-2)T] = x(nT) - Lx[(n-1)T].$$

Find $T(z)$, and give a SFG description of the system.
Solution. The given equation is Z-transformed directly as,

$$(1 - K_1 z^{-1} - K^2 z^{-2}) Y(z) = (1 - Lz^{-1}) X(z),$$

from which we write,

$$T(z) = \frac{Y(z)}{X(z)} = \frac{1 - Lz^{-1}}{1 - K_1 z^{-1} - K_2 z^{-2}}.$$

The SFG is given in the accompanying sketch.

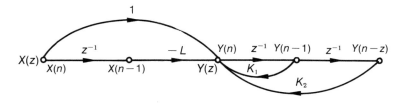

Fig. 12-6.1-1

Example 12-6.2

A system is described by the pair of difference equations,

$$Gu(nT) = y_1(nT) - Ay_1[(n-1)T] - By_2[(n-1)T],$$
$$0 = -Cy_1[(n-1)T] + y_2(nT) - Dy_2[(n-1)T].$$

Write the Z-transform for $Y_1(z)$ and $Y_2(z)$, and draw a system SFG.

Solution. Take the Z-transform of all terms in the equations. This gives,

$$GU(z) = Y_1(z)(1-Az^{-1}) - Bz^{-1}Y_2(z)$$
$$0 = -Cz^{-1}Y_1(z) + Y_2(z)(1-Dz^{-1})$$

These equations are solved for $Y_1(z)$ and $Y_2(z)$ to give,

$$Y_1(z) = \frac{G(1-Dz^{-1})U(z)}{1-(A+D)z^{-1}+(AD-BC)z^{-2}},$$

$$Y_2(z) = \frac{GC\,U(z)}{1-(A+D)z^{-1}+(AD-BC)z^{-2}}.$$

A SFG for this system of equations is given in the accompanying figure.

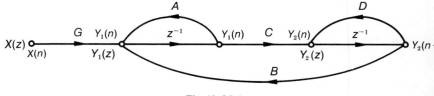

Fig. 12-6.2-1

12-7 SYSTEM DESCRIPTION BY DIFFERENCE EQUATIONS IN NORMAL FORM

The development here parallels that in Section 7-7 which considered continuous time systems which were described by state equations in normal form,

$$\dot{x} = Ax + Bu,$$
$$y = Px + Qu.$$
(12-30)

Now we consider a discrete time system whose difference equation description is the set of equations (*see* Eq. (2-52))

System Description by Difference Equations in Normal Form

$$x(n+1) = Ax(n) + Bu(n),$$
$$y(n) = Px(n) + Qu(n), \qquad (12\text{-}31)$$

where A, B, P, Q, are constant matrixes. The Z-transform is taken of these equations, and yields,

$$zX(z) = AX(z) + BU(z) + zx(0),$$
$$Y(z) = PX(z) + QU(z). \qquad (12\text{-}32)$$

From these we write the equations,

$$X(z) = [zI - A]^{-1}BU(z) + zx(0),$$
$$Y(z) = P[zI - A]^{-1}B + QU(z) + P[zI - A]^{-1}zx(0). \qquad (12\text{-}33)$$

The fundamental matrix $\Phi(z)$ of the discrete system is defined as,

$$\Phi(z) = Z[\Phi(n)] = z[zI - A]^{-1} = [(I - z^{-1}A]^{-1}. \qquad (12\text{-}34)$$

But the fundamental matrix of the discrete time system was previously defined (*see* Eq. (6-48)) as,

$$\Phi(n) = A^n. \qquad (12\text{-}35)$$

It therefore follows that,

$$\Phi(z) = Z[\Phi(n)] = A^n = z[zI - A]^{-1}. \qquad (12\text{-}36)$$

A comparison of this expression with Eq. (7-40) for the continuous system shows that the form of $\Phi(z)$ is similar to $\Phi(s)$ for continuous systems, but differs by the presence of the factor z in the expression for $\Phi(z)$.

In the absence of initial conditions, the relation between the excitation and the response functions is,

$$Y(z) = [P[zI - A]^{-1}B + Q]U(z),$$
$$= [P\Phi(z)z^{-1}B + Q]U(z). \qquad (12\text{-}37)$$

The z-domain system matrix is written,

$$T(z) = P\Phi(z)z^{-1}B + Q, \qquad (12\text{-}38)$$

so that,

$$Y(z) = T(z)U(z), \qquad (12\text{-}39)$$

as in Eq. (12-15), and as is the case for the s-domain models. We note an important feature of these equations, namely, even though there is not an exact analogy between the z-domain and the s-domain forms for the corresponding fundamental matrixes, nevertheless, the extra z in the definition for $\Phi(z)$ disappears in the expression for $T(z)$.

The time sequence vector $y(n)$ corresponding to $Y(z)$ in Eq. (12-33) will require inverting $Y(z)$. But this involves the quantity,

$$[zI - A]^{-1} = \frac{1}{\Delta(z)} \text{adj}\,[zI - A], \tag{12-40}$$

and in the manner discussed in connection with the comparable s-domain form given by Eq. (7-51), this is a matrix whose entries are rational functions of z, and so can be expanded into partial fractions. If $z = z_i$ is a zero of $\Delta(z)$, and $x(0) = 0$, then the expression for $Y(z)$ contains terms of the form,

$$Y_i(z) = \frac{U_i}{z - z_i} U(z), \tag{12-41}$$

where U_i is the residue matrix corresponding to the factor $z - z_i$. By a direct application of the definition of the Z-transform, it can be verified that $z_i(z)$ is the Z-transform of the sequence,

$$y(n) = \begin{cases} U\,C_n{}^{k-1} a^{n-k+1} & (n \geq k-1) \\ 0 & (n < k-1) \end{cases}, \tag{12-42}$$

where,

$$C_n{}^{k-1} = \frac{n!}{(k-1)!(n-k+1)!} = \frac{n(n-1)\cdots(n-k+2)}{(k-1)!}.$$

This specifies the solution to Eq. (12-37).

PROBLEMS

12.1. Determine the Z-transforms of the sequences,

$f(n) = \{nT\} \quad (n = 1, 2, 3, \ldots)$ $\qquad f(n) = \{(n+2)T\} \quad (n = 3, 4, 5, \ldots)$
$f(n) = \sin n\omega h \quad (n = 0, 1, 2, \ldots).$

12.2. Find closed form expressions for the Z-transform of each of the following sequences. Let the first term correspond to $k = 1$ in each case.

a. $0, 1 \times 2, 2 \times 3, 3 \times 4, \ldots,$
b. $2, 2(2)^2, 3(2)^3, 4(2)^4, \ldots.$

12.3. a. Show that the Z-transform of $f(n) = a^n$ is $Z[a^n] = (1/1 - az^{-1})$ for $|z| > |a|$.

b. Let $a = e^{j\omega}$ and then find $Z[e^{j\omega}]$.
c. Equate real and imaginary parts to find $Z[\sin n\omega]$, $Z[\cos n\omega]$.

12.4. Solve the differential equation,

$$\dot{x} + 0.1x = 3 - 3e^{-2t} \qquad x(0) = 0.$$

a. Using a finite difference approximation.
b. Z-transform techniques, using a sampling period $T = 1$.
c. Compare the results obtained under a and b.

12-5. A function $\{f(t)\}$ is zero for $t < 0$ and has a Laplace transform $(s+b)/s(s+a)$.
a. Find the Z-transform of $\{f(kT - nT)\}$, n a positive integer.
b. What is the region of convergence (the range over which is finite and positive).

12.6. A continuous time system is characterized by the Laplace transform $10/s(s+5)$. There is applied the impulse function series,

$$2e^{-2kT} \quad T = 0.01 \text{ sec} \qquad (k = 0, 1, 2, \ldots).$$

What is the output at $t = kT$; at $kT/3$?

12.7. Determine the Z-transforms of the time sequences that would be obtained by sampling at $t = nT$ the time functions that possess the following Laplace transforms,

$$\frac{K}{s(s+a)} \qquad \frac{k}{(s+a)(s+b)} \qquad \frac{K(s+a)}{s(s+b)(s+c)}.$$

12.8. Find the time sequence $f(n)$ corresponding to the following z-domain functions,

$$\frac{z}{(z-1)(z-2)} \qquad \frac{z^2}{(z-1)(z-2)^2} \qquad \frac{z+2}{(z-1)^2} \qquad \frac{3z^4 - 8z^3 + 15z^2 + 7z + 15}{z(z^2 - 10z + 25)}.$$

12.9. Find the inverse Z-transforms by: a. inversion formulas, b. power series expansion, of the following,

$$\frac{1}{(1-z^{-1})(1-0.5z^{-1})} \qquad \frac{z^{-1}}{(1-z^{-1})^2} \qquad \frac{z^{-1}+z^{-2}}{(1-2z^{-1})^2}.$$

12-10. A discrete time system function is the following,

$$T(z) = 1 + 2z^{-1} - 17z^2,$$

which is defined for all z except $|z| = 0$. The input is,

$$u(n) = (\tfrac{1}{2})^n \qquad (n = \text{integer}).$$

Find the output $y(n)$ for all n.

480 The Z-Transform and Discrete Time Systems

12-11. A discrete time system function is given by,

$$T(z) = \frac{3(1-z^{-1})}{1-z^{-1}-2z^{-2}}.$$

The input consists of,

$$u(0) = 0; \quad u(1) = 1; \quad u(2) = 2; \quad u(3) = 0.$$

Determine the output $y(n)$ for $n \geq 0$.

12-12. A discrete system function is,

$$T(z) = \frac{3(2-z^{-1})}{1-10z^{-1}+25z^{-2}} \quad (|z| \leq 5).$$

The input is: $u(0) = 1$; $u(1) = 1$; $u(2) = -1$; $u(3) = -1$. Determine $y(n)$ for $n \geq 0$.

12-13. The SFG of a specified sampled data system is shown.

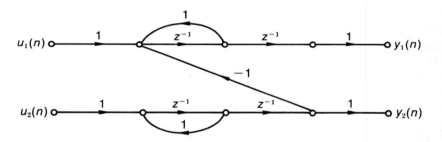

a. Write the difference equations governing the system.
b. Determine the system transfer matrix.

12-14. A system is described by the difference equation,

$$y(n+2) + 2.5y(n+1) + 4y(n) = u(n),$$

with initial conditions $y(0)$ and $y(1)$, and with an input $u(n) = 2^n$ at $n = 0$. Find the values of $y(n)$ for $n \geq 2$.

12-15. A system is described by the difference equation,

$$y(n) - 3y(n-1) + 5y(n-2) - 7y(n-3) = u(n-1) - 2u(n-2),$$

subject to the input $u = 3^n$. For specified $y(0)$, first forward difference $\Delta y(0)$ and second forward difference $\Delta^2 y(0)$.
  a. Represent the system in terms of state equations.
  b. Draw a SFG of the system.
  c. Determine $y(0)$ appropriate to a.
  d. Determine $y(2), y(3)$.

12-16. A system is described by the difference equation,
$$y(n+3) - 5y(n+2) + 3y(n+1) - 9y(n) = u(n+2),$$
with $u(n) = 0$ for $n \geq 0$, and with $y(1) = 0$; $y(2) = 1$; $y(2) = 5$. Determine the output $y(n)$ for $n \geq 1$.

12-17. Solve the system of difference equations by means of the Z-transform,
$$\begin{bmatrix} x_1(n+1) \\ x_2(n+1) \end{bmatrix} = \begin{bmatrix} 1 & 0.5 \\ 0 & -2 \end{bmatrix} \begin{bmatrix} x_1(n) \\ x_2(n) \end{bmatrix} + \begin{bmatrix} u(n) \\ 0 \end{bmatrix},$$
for
$$\begin{matrix} u(n) = 0 & (n < 0) \\ u(n) = 1 & (n > 0) \end{matrix} \qquad \begin{bmatrix} x_1(0) \\ x_2(0) \end{bmatrix} = \begin{bmatrix} 1 \\ 0 \end{bmatrix}.$$

12-18. A system is described by the difference equations
$$x_1(n+1) = -2x_1(n) - x_2(n),$$
$$x_2(n+1) = 3x_1(n) + x_2(n) + u(n),$$
$$y(n) = 2x_1(n) + x_2(n) + u(n).$$

a. This system is to be described by a difference equation of the form,
$$y(n+2) + \alpha_1 y(n+1) + \alpha_0 y(n) = \beta_2 u(n+2) + \beta_1 u(n+1) + \beta_0 u(n).$$
Find the appropriate α and β factors.

b. Suppose that the input is,
$$\begin{matrix} u(n) = 1 & (n = 0), \\ = 1 & (n = 1), \\ = 0 & (n = 2, 3, \ldots). \end{matrix}$$
Find the output $y(n)$ by the use of the Z-transform.

13

Stability

Stability is a time domain concept, and implies that a system, when initially at rest, will remain at rest in the absence of a disturbance, and further, when subject to a transient disturbance, the system response will tend toward zero, or at most, to some bounded or limited response. One or the other of these ideas is involved in the study of the stability of a system.

13-1 POLE LOCATIONS AND STABILITY

The Routh-Hurwitz and the Nyquist methods are very important techniques for studying the stability of a linear system. Both of these involve considerations of the natural modes of the system. These methods begin with the fact that the transient behavior of a system is intimately related to the system function $T(s)$, which will be the ratio of two polynomials in s, in the forms,

$$T(s) = \frac{P(s)}{Q(s)} = \frac{b_m s^m + b_{m-1} s^{m-1} + \cdots + b_0}{a_n s^n + a_{n-1} s^{n-1} + \cdots + a_0},$$

$$= H \frac{(s-s_a)(a-s_b) \cdots (s-s_m)}{(s-s_1)(s-s_2) \cdots (s-s^n)}, \qquad (13\text{-}1)$$

$$= \frac{A_1}{s-s_1} + \frac{A_2}{s-s_2} + \cdots + \frac{A_k}{s-s_k} + \cdots + \frac{A_n}{s-s_n}.$$

When expressed in the second form in terms of their roots, since the basic controlling differential equation has real coefficients, the roots of both

$P(s)$ and $Q(s)$ either are real, or if complex, will occur in complex conjugate pairs. Furthermore, the location in the s-plane of the poles of the system function will represent the natural modes of the system. Specifically, a system with simple poles s_k that is subjected to a sudden excitation function of any sort will contain the sum of exponential terms of the form $A_k e^{s_k t}$, with A_k being a finite real or complex number, and with s_k denoting one of the natural modes of the system. The nature of the response depends on the location of s_k in the s-plane.

The following three important general cases exist that depend intimately on the location of s_k in the s-plane,

1. The point representing s_k lies in the left half of the s-plane.
2. The point representing s_k lies on the $j\omega$-axis.
3. The point representing s_k lies in the right half of the s-plane.

We examine each of these cases.

Case 1. s_k lies in the left half of the s-plane. In this case, s_k will be of the form $s_k = \sigma_k + j\omega_k$, with $\sigma_k < 0$. The response term will then be of the form,

$$\text{response} = A_k e^{(\sigma_k + j\omega_k)t} = A_k e^{\sigma_k t + j\omega_k t}. \tag{13-2}$$

But since $\sigma_k < 0$, then as t increases, the value of $e^{\sigma_k t}$ decreases. After a lapse of time, the response due to terms of this type will disappear. This is also the situation when the roots are real and simple, where now $\omega_k = 0$.

Suppose that there are one or more pairs of complex conjugate roots. Now, corresponding to each root s_k there will also be a root s_k^*. For each such pair of roots the response will contain terms of the form,

$$\text{response} = A_k e^{s_k t} + A_k^* e^{s_k^* t}, \tag{13-3}$$

where A_k and A_k^* specify the appropriate amplitude factors. The response terms may be combined, noting that in general $A_k = a + jb$, so that,

$$\text{response} = (a+jb)e^{(\sigma_k + j\omega_k)t} + (a-jb)e^{(\sigma_k - j\omega_k)t}.$$

In the manner of combination discussed in Section 9-2, this expression will reduce to the form,

$$\text{response} = 2\sqrt{a^2 + b^2}\, e^{\sigma_k t} \cos(\omega_k t + \beta_k),$$

where, $\tag{13-4}$

$$\beta_k = \tan^{-1} \frac{b}{a}.$$

This response term is a damped sinusoid, which has the form shown in Fig. 13-1. As a general conclusion, therefore: systems with simple natural modes in the left half plane are stable.

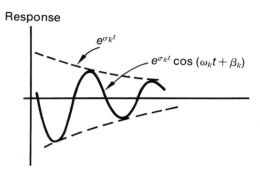

Fig. 13-1. Transient response due to conjugate poles.

Case 2. s_k lies on the imaginary axis. This is a special case under Case 1, but now $\sigma_k = 0$. The response for conjugate poles is, from Eq. (13-4),

$$\text{response} = 2\sqrt{a^2 + b^2}\cos(\omega_k t + \beta_k). \tag{13-5}$$

Observe that there is no damping, and the response is thus a sustained oscillatory function. Such a system is bounded, and is defined to be stable, even though it is oscillatory.

Case 3. s_k lies in the right half plane. The response function will be of the form,

$$\text{response} = Ae^{\sigma_k t}, \tag{13-6}$$

for real roots, and will be of the form,

$$\text{response} = 2\sqrt{a^2 + b^2}\, e^{\sigma_k t}\cos(\omega_k t + \beta_k), \tag{13-7}$$

for conjugate complex roots. Because both functions increase in time without limit, the system for which these are roots is said to be unstable.

The conclusions that result from these three cases are: a system with simple poles is unstable if one or more of its natural modes appear in the right half plane. Conversely, a system with simple poles is stable when all of its natural modes are in the left half plane or on its boundary. In fact, the distance of the points representing the natural modes from the

Pole Locations and Stability

imaginary axis gives a measure of the decay rate of the response with time.

We now wish to reexamine the situation when multiple order poles exist. We must again consider the three cases.

Case 1. Multiple real poles in the left half plane. As previously discussed (*see* Section 7-6), a second order real pole (two repeated roots) gives rise to the response function,

$$\text{response} = (A_{k1} + A_{k2}t)e^{\sigma_k t}. \tag{13-8}$$

For negative values of σ_k, the exponential time function decreases faster than the linearly increasing time factor. The response ultimately dies out, the rapidity of decay depending on the value of σ_k. The system with such poles is stable.

Case 2. Multiple poles on the imaginary axis. The response function is made up of the response due to each pair of poles, and is,

$$\text{response} = (A_{k1} + A_{k2}t)e^{j\omega_k t} + (A_{k1}^* + A_{k2}^* t)e^{-j\omega_k t}. \tag{13-9}$$

This result can be written, following the procedure discussed above for the simple complex poles,

$$\text{response} = 2\sqrt{a^2 + b^2} \cos(\omega_k t + \beta_k) + 2\sqrt{c^2 + d^2}\, t \cos(\omega_k t + \gamma_k). \tag{13-10}$$

The first term on the right is a sustained oscillatory function. The second term is a time-modulated oscillatory function which increases with time. Clearly, the system in this case is unstable.

Case 3. Multiple roots in the right half plane. The solution in this case will be, for real roots,

$$\text{response} = (A_{k1} + A_{k2}t)e^{\sigma_k t}, \tag{13-11}$$

and for complex roots,

$$\text{response} = e^{\sigma_k t}(2\sqrt{a^2 + b^2} \cos(\omega_k t + \beta_k) + 2\sqrt{c^2 + d^2}\, t \cos(\omega_k t + \gamma_k)). \tag{13-12}$$

In both cases, owing to the factor $e^{\sigma_k t}$, the response increases with time, and each system is unstable.

The foregoing considerations can be summarized as follows: A system with multiple poles is unstable if one or more of its normal modes appear on the j-axis or in the right half plane. Conversely, when all of the natural modes of the system are confined to the left half plane, the system is stable.

13-2 PROPERTIES OF DRIVING POINT FUNCTIONS

On the basis of the foregoing section, and based also on our previous studies relative to networks of passive linear elements, we can summarize the properties of driving point *immittance* functions. The term immittance function is to denote either the impedance or the admittance function. The properties are, since such functions for passive networks must be stable; i.e., they do not produce an output in the absence of an input.

1. It is the ratio of polynomials with real positive coefficients.
2. The degree of the numerator polynomial cannot exceed the degree of the denominator polynomial by more than unity.
3. No poles or zeros may exist in the right half s-plane.
4. Poles and zeros on the imaginary axis must be simple.
5. Complex poles and zeros must occur in conjugate pairs.

These are necessary conditions, but they are not sufficient to insure that $T(s)$ is the transform of a driving point immittance. To insure sufficiency, it is required that,

6. $T(s)$ is real when s is real.
7. $Re\ T(s) \geq 0$ when $Re\ s = 0$.
8. $Re\ T(s) > 0$ when $Re\ s > 0$.

Functions which satisfy these conditions are called *positive real functions*, which are usually written as *prf* functions.

As an example of the latter conditions, we examine the function,

$$T(s) = \frac{s+4}{s^2+2s+5}.$$

This function satisfies all conditions except 7. It is noted, for example, that when $s = j5$, hence $Re\ s = 0$,

$$T(j\omega) = -0.06 - j0.28.$$

Clearly, since a negative resistance implies power generation, then this particular $T(s)$ is not a suitable function for a passive network.

A large body of literature has been developed over the years that is concerned with the synthesis of passive networks having prescribed driving point functions. In all cases the basic requirement for such studies is that the specified functions to be synthesized are *prf* functions.

13-3 ROUTH-HURWITZ TEST

Return now to Section 13-1. With the stability of a linear system precisely defined in terms of the location of the poles of $T(s)$, it is a simple matter to establish the stability of a given system when the denominator polynomial is known in factored form. As has already been discussed, finding these can become a very tedious chore, especially for a complicated interconnected system with feedback. We should like to ascertain stability without being compelled to factor the denominator polynomial. Several simple tests do exist, especially for certain special cases. Important tests are: A system is unstable if,

1. All terms of the characteristic equations are not of the same sign.
2. All terms of the characteristic equation are not present in descending powers.

Either of these conditions can be applied directly to the differential equation that describes the system. Observe specifically that these tests are not adequate (the sufficiency condition) when all terms of the characteristic equation are of the same sign, and all terms are present in descending powers. More elaborate tests are now necessary.

The work of Routh in 1875 and independently of Hurwitz in 1895 provides a means for deciding whether a polynomial has zeros with positive real parts, without the necessity for factoring the polynomial. That is, the Routh-Hurwitz test allows one to determine whether any roots of an algebraic equation lie in the right hand s-plane. If the coefficients of the equation are known only in literal form, the test yields a set of inequality conditions for stability. If numerical coefficients are available, the test permits a check on stability.

The Routh-Hurwitz test requires developing a schedule from the coefficients of the algebraic equation. Consider the general algebraic equation,

$$a_n s^n + a_{n-1} s^{n-1} + \cdots + a_1 s + a_0 = 0. \qquad (13\text{-}13)$$

We now develop a schedule of the form shown in Fig. 13-2.
The entries in this schedule are deduced as follows,

Row 1. Alternate coefficients of the given equation,

$$a_n, a_{n-2}, a_{n-4}, \ldots$$

Row 2. The remaining coefficients of the original equations,

$$a_{n-1}, a_{n-3}, a_{n-5}, \ldots$$

$$\begin{vmatrix} a_n & a_{n-2} & a_{n-4} & a_{n-6} & a_{n-8} \\ a_{n-1} & a_{n-3} & a_{n-5} & a_{n-7} & \\ b_{n-1} & b_{n-3} & b_{n-5} & b_{n-7} & \\ c_{n-1} & c_{n-3} & c_{n-5} & & \\ d_{n-1} & d_{n-3} & d_{n-3} & & \\ e_{n-1} & e_{n-3} & & & \\ f_{n-1} & f_{n-3} & & & \\ g_{n-1} & & & & \end{vmatrix}$$

Fig. 13-2. Schedule for the Routh-Hurwitz test.

Row 3. The b-factors are deduced from Rows 1 and 2,

$$b_{n-1} = \frac{-1}{a_{n-1}} \begin{vmatrix} a_n & a_{n-2} \\ a_{n-1} & a_{n-3} \end{vmatrix}$$

$$b_{n-3} = \frac{-1}{a_{n-1}} \begin{vmatrix} a_n & a_{n-4} \\ a_{n-1} & a_{n-5} \end{vmatrix}$$

$$b_{n-5} = \frac{-1}{a_{n-1}} \begin{vmatrix} a_n & a_{n-6} \\ a_{n-1} & a_{n-7} \end{vmatrix} \text{ etc.}$$

Row 4. These are deduced from Rows 2 and 3, as follows,

$$c_{n-1} = \frac{-1}{b_{n-1}} \begin{vmatrix} a_{n-1} & a_{n-3} \\ b_{n-1} & b_{n-3} \end{vmatrix}$$

$$c_{n-3} = \frac{-1}{b_{n-1}} \begin{vmatrix} a_{n-1} & a_{n-5} \\ b_{n-1} & b_{n-5} \end{vmatrix}$$

$$c_{n-5} = \frac{-1}{b_{n-1}} \begin{vmatrix} a_{n-1} & a_{n-7} \\ b_{n-1} & b_{n-7} \end{vmatrix} \text{ etc.}$$

Row 5. etc. Repeat the procedure shown for Rows 3 and 4 until all the elements of a row are zero.

The Routh-Hurwitz test states: The number of roots of Eq. (13-13) that lie in the right half s-plane equals the number of sign changes of the elements of the first column of the schedule of Fig. 13-2.

Example 13-3.1

The characteristic polynomial of a system is given as,

$$s^5 + 2s^4 + 3s^3 + 7s^2 + 14s + 22 = 0.$$

Employ the Routh-Hurwitz test to ascertain whether the system is stable.

Solution. Prepare the Routh-Hurwitz schedule. The result is that shown,

$$
\begin{array}{ccc}
1 & 3 & 14 \\
2 & 7 & 22 \\
-\tfrac{1}{2} & 3 & \\
19 & 22 & \\
68/19 & & \\
22 & &
\end{array}
$$

This schedule shows one sign change in going from the second to the third row, and another sign change in going from the third row to the fourth row. Consequently, two roots of the given equation lie in the right half s-plane.

While the Routh-Hurwitz test does provide useful information when the algebraic equation is known, there are several important features of the test that are not fully satisfactory. Clearly, while it is certainly helpful to know whether a system is stable or not, it would be most desirable, and often it would be necessary to be able to answer the following,

a. What is the degree of instability?
b. What might one do to stabilize the system?

While it is not a failure of the test, nevertheless the test is not useful in those real experimental situations when one cannot find the coefficients of the characteristic polynomial. This situation can be particularly true for closed loop systems.

13-4 THE NYQUIST CRITERION

The Nyquist test provides a rather different approach for testing the stability of a linear system. In particular, it provides a test of whether any roots of the equation,

$$1 + L(s) = 0, \qquad (13\text{-}14)$$

lie in the right half plane, where for closed loop systems $L(s)$ is the open loop response function for the system.

Before proceeding with the details of the test, we shall consider the class of systems for which this test will be of particular importance. Let us consider the simple feedback loop shown in Fig. 13-3. As we know,

490   Stability

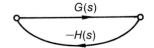

Fig. 13-3. A simple feedback signal flow graph.

the total gain (transmittance) of this system will be,

$$T(s) = \frac{G(s)}{1+G(s)H(s)}. \qquad (13\text{-}15)$$

In this function, $G(s)$ is the forward system function, and $H(s)$ is the feedback system function. Observe moreover, that the total system function of the combined forward and feedback path, but with the feedback connection open, is $G(s)H(s)$. Clearly, right half-plane poles in the denominator of Eq. (13-15) denotes an unstable system. However, right half-plane poles of Eq. (13-15) are precisely the poles of the function $G(s)H(s)$ itself. This means, therefore, that one may infer the stability of a closed loop system from a study of the open loop characteristics of the system.

A property that is a necessary preliminary to the Nyquist test relates to mappings of contours in the s-plane for poles and zeros. We first consider a contour in the clockwise direction, say Γ, that encloses a zero in the s-plane, as illustrated in Fig. 13-4a. A map or image of the contour in the Γ-plane, in accordance with the discussion in Section 9-3, is a closed loop, also in the clockwise direction, that encloses the origin, as shown in Fig. 13-4b. If the closed contour in the s-plane were to enclose two zeros in the clockwise direction, then the corresponding mapping in the Γ-plane would encircle the origin twice in the clockwise direction. This situation is illustrated in Fig. 13-5.

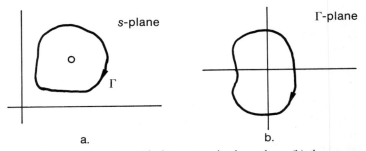

Fig. 13-4. (a) A closed contour enclosing a zero in the s-plane, (b) the corresponding mapping in the Γ-plane.

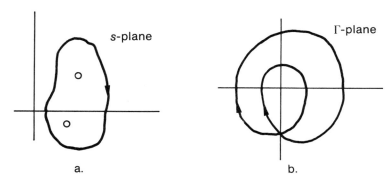

Fig. 13-5. (a) A closed contour enclosing two zeros in the *s*-plane, (b) the corresponding mapping in the Γ-plane.

Now consider the case when the closed contour in the clockwise direction encloses a pole in the *s*-plane. The mapping in the Γ-plane encloses the origin in the counter-clockwise direction, as shown in Fig. 13-6. Consistent with the foregoing discussion is the fact that if a closed contour in the counter-clockwise direction in the *s*-plane encloses neither poles nor zeros, then the mapping in the Γ-plane will be a closed contour also in the counter-clockwise direction, but which does not enclose the origin. Likewise, a closed contour in the counter-clockwise direction in the *s*-plane that encloses an equal number of poles and zeros will map into a closed contour in the counter-clockwise direction in the Γ-plane, but not enclosing the origin.

We use these ideas of mapping for the function $G(s)H(s)$ of Eq. (13-15) which, in the general case, is the ratio of two polynomials of the form given in Eq. (13-1). Moreover, we are concerned with roots in the right half plane that might produce instability, and these are the poles of $G(s)H(s)$. To include right half plane poles, if any, we choose the contour in the *s*-plane extending from $-j\infty$ to $+j\infty$ which is closed by a semicircle at infinity in the clockwise direction. We can write, in general,

$$N = Z - P, \qquad (13\text{-}16)$$

where N = the total number of clockwise encirclements of the origin in the $G(s)H(s)$ plane; Z = the number of clockwise encirclements of the origin due to zeros with real parts in the right half plane; P = the number of counter-clockwise encirclements of the origin due to poles. For no zeros in the right half plane $Z = 0$, the number of counter-clockwise encirclements about the origin will equal the number of poles with positive

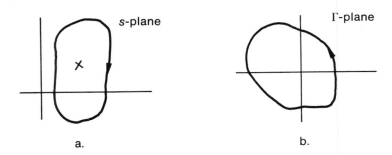

Fig. 13-6. (a) A closed contour enclosing a pole in the s-plane, (b) the corresponding mapping in the Γ-plane.

real parts,

$$N = -P. \qquad (13\text{-}17)$$

From this we conclude that: for no zeros and no poles in the right half plane, a system is stable if it does not encircle the origin.

To adapt the foregoing to the closed loop system, we consider Eq. (13-15) which shows that the system is unstable if the denominator is zero, that is, if,

$$G(s)H(s) = -1. \qquad (13\text{-}18)$$

We thus write the Nyquist stability criterion, "A closed-loop feedback system is stable if and only if its open loop transfer locus does not enclose the point $(-1,0)$ and if the number of clockwise encirclements about $(-1,0)$ equals the number of poles of $G(s)H(s)$ with positive real parts." According to the discussion above, the procedure is to find the image of the imaginary axis for the function $G(s)H(s)$, as shown in Fig. 13-7. Here, since the contour in the $G(j\omega)H(j\omega)$ plane is clockwise (no right half plane poles) and since the contour does not encircle the point $(-1,0)$ the system is stable. Figure 13-7b is typical of systems that have no right half plane poles, an important case. In this case the Nyquist stability criterion can be simplified to, "A feedback system with no right half plane poles is stable if and only if $G(j\omega)H(j\omega) < 1$ when arg $G(j\omega)H(j\omega) = 180°$."

Special attention must be given to poles of $G(s)H(s)$ that lie on the imaginary axis or at the origin of the s-plane. These points are avoided by detouring around them with very small semicircular contours to the right. By this procedure, all of the above discussion remains pertinent.

In summary, the following data are required in the Nyquist procedure:

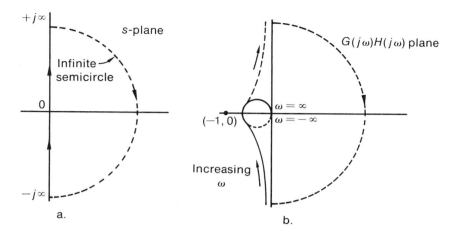

Fig. 13-7. (a) Contour in the s-plane for the Nyquist test, (b) an assumed form for $G(j\omega)H(j\omega)$ for a system under study.

1. The magnitude and phase angle of $G(j\omega)H(j\omega)$ for ω from $-\infty$ to $+\infty$.
2. The behavior of $G(s)H(s)$ at the poles that lie on the imaginary axis or at the origin of the s-plane.
3. The number of $G(s)H(s)$ right half plane poles.

The Nyquist diagram provides a qualitative indication of the sensitivity of the system stability to changes in system parameters. That is, since instability is evidenced by the $G(j\omega)H(j\omega)$ locus covering or encircling the point $(-1,0)$, then it is reasonable to expect that the greater the distance of the locus from the point $(-1,0)$ the smaller the likelihood that the system will become unstable with small changes in the system parameters.

13-5 DISCRETE TIME SYSTEMS

The Nyquist criterion for stability is directly applicable for discrete time systems for which the transmittance is of the form,

$$T(z) = \frac{G(z)}{1 + G(z)H(z)}. \qquad (13\text{-}19)$$

Here, instead of the region to the right of the imaginary axis in the s-plane, the region outside of the unit circle in the z-plane is involved in the stability studies. Now the unit circle in the z-plane is mapped onto the

$G(z)H(z)$ plane. If the map encloses the point $(-1,0)$, then $T(z)$ has no poles outside of the unit circle. If $T(z)$ does have poles outside of the unit circle, then Eq. (13-17) is used in ascertaining the number of poles external to the unit circle, as a step in the application of the Nyquist criterion. Figure 13-8 shows the situation for the discrete time system that roughly parallels that in Fig. 13-7 for the continuous time system.

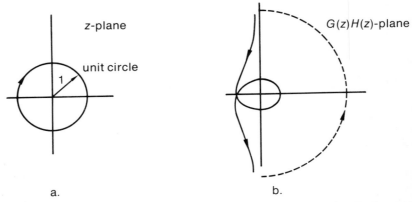

Fig. 13-8. (a) Contour in the z-plane for a discrete time system (b) the mapping on the $G(z)H(z)$ plane.

13-6 CONTROLLABILITY AND OBSERVABILITY OF LINEAR SYSTEMS

It has been assumed in all of the foregoing discussion that the input to the system will excite all of the natural modes, i.e., the response of the unforced system has been assumed to be the superposition of all of the modes of the system. The question of controllability of a given mode is concerned with the presence of coupling of a given mode to the input variable. The question of observability is concerned with the existence of coupling of a given mode to the output variable. For example, refer to Fig. 13-9. A simple calculation will show that the state variables of the system, v_c and i_L, will be affected by the input excitation, hence the modes are controllable. However, denoting the current i as the output, we find that $Y(s) = 1$, owing to the cancellation of the zero and the pole of this network. As a result, the modes are not coupled to the output, and these are not observable.

The concept of controllability and observability of linear systems plays

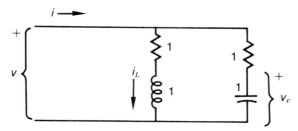

Fig. 13-9. A network with unobservable modes.

a very important role in modern control theory. Clearly, if there are unobservable modes in a feedback system, and if the unobservable states are unstable, they will not be detected in the output. Likewise, if the system is not completely controllable, a control signal may be unable to affect the unstable modes. Evidently, the feedback loop will not be able to stabilize a system, if the design has assumed complete observability and controllability.

To establish the conditions for the excitation of all natural modes, we shall assume that the eigenvalues are distinct. Repeated roots may be included by assuming that they are separated by negligible amounts, thereby being distinct. Now we may use the Jordan matrix discussed in Section 7-7, or the partial fraction expansion method discussed in Section 8-8c to uncouple each response mode. Thus by introducing the linear matrix transformation,

$$x = WX, \qquad (13\text{-}20)$$

into the state equations,

$$\dot{x} = Ax + Bu, \qquad (13\text{-}21)$$

there results, following the methods in Section 7-7,

$$\dot{X} = \Lambda X + W^{-1}Bu. \qquad (13\text{-}22)$$

Since Λ, the Jordan matrix, is diagonal, each mode or transformed state variable X_i is uncoupled from the other state variables X_j. If we write the equation for the first state variable as,

$$\dot{X}_1 = \lambda_1 X_1 + g_{11} u_1 + g_{12} u_2 + \cdots, \qquad (13\text{-}23)$$

where $G = W^{-1}B$, then to control this mode, there must be at least one element of the first row of the G matrix that is nonzero. In general, we can write that a necessary and sufficient condition for a linear time-invariant system to be controllable is that each row of G contain at least one nonzero element.

Now let us consider the output vector y, given by,

$$y = Px + Qu. \tag{13-24}$$

We write this in terms of the transformed uncoupled state variables X. This requires that we combine Eq. (13-24) with Eq. (13-20). This gives,

$$y = PWX + Qu = RX + Qu, \tag{13-25}$$

where $R = PW$. But for the unforced system, $u = 0$, and in this case

$$y = [r_1 \quad r_2 \quad r_3 \cdots r_n] \begin{bmatrix} X_1 \\ X_2 \\ \cdot \\ \cdot \\ \cdot \\ X_n \end{bmatrix}, \tag{13-26}$$

where R is the row matrix $[r]$. Then,

$$y = r_1 X_1 + r_2 X_2 + \cdots + r_n X_n. \tag{13-27}$$

To be specific, consider a third order system which is described by,

$$\begin{bmatrix} y_1 \\ y_2 \\ y_3 \end{bmatrix} = \begin{bmatrix} r_{11} & r_{12} & r_{13} \\ r_{21} & r_{22} & r_{23} \\ r_{31} & r_{32} & r_{33} \end{bmatrix} \begin{bmatrix} X_1 \\ X_2 \\ X_3 \end{bmatrix}. \tag{13-28}$$

It is evident that in order to determine the state y_1 at least one element of the first column of R must be nonzero. In general for a system to be observable, it is necessary and sufficient for at least one element of each column of $R = PW$ to be nonzero.

13-7 OBSERVING THE STATE OF A SYSTEM

In many cases it is necessary that the states (26) of the system be known. Often, however, only input and output information is available. There is interest in being able to deduce the unmeasurable states of the system. Fortunately it is possible to deduce the otherwise unmeasurable states by means of a system modeling procedure. In developing a model that provides equivalent states of the system, the input information u and the output information y, which are assumed available, are used as inputs to the model.

Refer to Fig. 13-10, which shows the system and the model that is to be developed. The system is described by the state equations,

$$\dot{x} = Ax + Bu,$$
$$y = Px + Qu, \qquad (13\text{-}29)$$

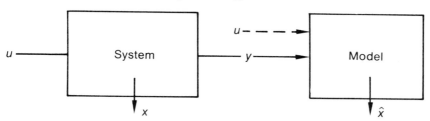

Fig. 13-10. System modeling for state determination.

and the model is described by,

$$\dot{\hat{x}} = A'\hat{x} + B'y, \qquad (13\text{-}30)$$

with A' and A being supposed to have different characteristics. It is noted that if y in Eq. (13-29) does not include u explicitly, then u will be available as an input to the model. Equation (13-30) is combined with Eq. (13-29) to yield,

$$\dot{\hat{x}} = A'\hat{x} + B'(Px + Qu),$$
$$= A'\hat{x} + B'Px + B'Qu. \qquad (13\text{-}31)$$

Now consider the difference between the states of the system and the model. From Eqs. (13-29) and (13-31) we have that,

$$\dot{x} - \dot{\hat{x}} = (A - B'P)x - A'\hat{x} + (B - B'Q)u. \qquad (13\text{-}32)$$

Suppose that we now require that $x = \hat{x}$; in this case the states of the model are the same as those of the original system. This requires that,

$$A - B'P - A' = 0,$$
$$B - B'Q = 0. \qquad (13\text{-}33)$$

Clearly, therefore, if the model is prescribed by,

$$A' = A - B'P,$$
$$B' = BQ^{-1}, \qquad (13\text{-}34)$$

the entire system state vector is made available. Of course, since some of the system's state variables are available by direct measurement, then the model may be used to deduce the unmeasurable ones.

13-8 STABILITY IN THE SENSE OF LIAPUNOV

The physical idea of stability, when one proceeds from a state equation formulation of the system behavior, is closely related to a bounded system response to a sudden disturbance or input. If the system is disturbed and is displaced slightly from the equilibrium state, several different behaviors are possible. If the system remains near the equilibrium state, the system is said to be stable. If the system tends to return to the equilibrium state, it is said to be *asymptotically stable*. An equilibrium state x_e is said to be asymptotically stable *in the large* if it is asymptotically stable for any initial state vector x_0 such that every motion converges to x_e as $t \to \infty$. The Liapunov Second Method (27) which is also known as the Liapunov Direct Method, allows one to ascertain the stability of a system precisely. A feature of this method, in common with the Routh-Hurwitz and the Nyquist tests, is that it permits a study of the stability of the dynamic system without requiring the solution of the vector differential equations of state.

To discuss the general features of the Liapunov Second or Direct Method, suppose that an equilibrium state exists at the origin, whence $x(0) = 0$. It may be necessary to effect a transformation of coordinates to refer the equations to the origin, but this can always be done. The Euclidean length of the vector from the origin, often called the norm, may be written,

$$\|x\| = (x_1^2 \times x_2^2 + \cdots + x_n^2)^{1/2}. \tag{13-35}$$

Now let $S(R)$ be a spherical region of radius $R > 0$ about the origin. Clearly, $S(R)$ will contain those points x which satisfy the condition $\|x\| < R$. Now consider a second region $S(r)$ within $S(R)$, as illustrated in Fig. 13-11. The origin is said to be stable if, corresponding to each $S(R)$, there is an $S(r)$ such that a solution starting in $S(r)$ does not leave $S(R)$. Under these conditions, with $x(0)$ in $S(r)$, $x(t)$ will remain within $S(R)$ for all $t > 0$. If, in addition, there is a neighborhood $S(R_0)$ such that every solution that starts in $S(R_0)$ approaches the origin as $t \to \infty$, the system is asymptotically stable. If R_0 is arbitrarily large, then the solution is asymptotically stable in the large. Actually, R_0 is not limited for a linear system, but the nature of the nonlinearity governs the size of the region $S(R_0)$.

13-9 THE DIRECT METHOD OF LIAPUNOV

The essence of the direct method of Liapunov is best understood by considering the total energy of a system. It is a generalization of the idea

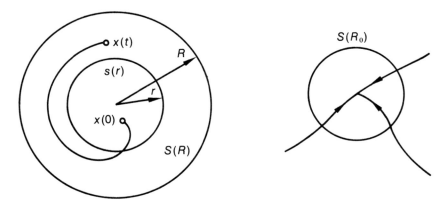

Fig. 13-11. Stability regions in state space.

that the total energy of a system is always decreasing in the neighborhood of an equilibrium state. To establish the ideas a little more firmly, consider the case of a series RL circuit, the solution for which is discussed in Section 5-6. The current is given by the equation,

$$i = \frac{V}{R}(1 - e^{-Rt/L}),$$

which shows that the system is stable. Let us consider the magnetic coenergy stored in the magnetic field of the inductor. This is,

$$W_m = \tfrac{1}{2}Li^2 = \frac{L}{2}\left[\frac{V}{R}(1 - e^{-Rt/L})\right]^2,$$

which is a positive quantity, always. Let us also examine the time derivative of this function. It is,

$$\frac{dW_m}{dt} = \frac{L}{2} 2\left[\frac{V}{R}\right]^2 (1 - e^{-Rt/L})\left(\frac{R}{L}e^{-Rt/L}\right),$$

$$= \left[\frac{Re^{-Rt/L}}{1 - e^{-Rt/L}}\right]W_m.$$

This shows that the time derivative of the energy is always decreasing.

The Liapunov generalization is essentially the following: Suppose that within some neighborhood $S(R)$ of the origin a scalar function $V(x)$ of the state $x(t)$ can be constructed, such that $V(x)$ has continuous first partials and such that $V(x) = 0$ when x equals the equilibrium state. We require that $V(x) > 0$ for all x other than the equilibrium state, and that the time derivative of $V(x)$ shall be negative. Then the system represented by $V(x)$ is asymptotically stable. These results are best given by the Liapunov Stability theorem:

Given $\dot{x} = X(x)$, where $\dot{x}_i$ is continuous in the state variable x_j for all $i, j = 1, 2, \ldots, n$. If there exists a $V(x)$ such that,

   a. $V(x)$ has continuous first partial derivatives with respect to x_i,
   b. $V(x)$ is positive definite† for all $x_i > 0$,
   c. $V(x) \to \infty$ for $|x| \to \infty$,

then:

1. The system is stable with respect to x_i if there is a region $S(R)$ defined by $0 < \|x_i\| < S(R)$, where $S(R)$ is some real positive constant, such that in this region $-\dot{V}(x)$ is positive semidefinite, that is, for all $(x_i, t > 0)$,

$$-\dot{V} = \sum_{i=1}^{n} \frac{\partial V}{\partial x_i} \dot{x}_i \geq 0.$$

2. The system is asymptotically stable with respect to x_i, if in $S(R)$ $-\dot{V} > 0, x \neq 0$ (i.e., $-\dot{V}$ is positive definite).

3. The system is asymptotically stable in the large if condition 2 is satisfied and if $S(R)$ is the whole of state space.

It is important to bear in mind that this theorem gives a set of sufficiency conditions for stability; thus if one is able to find a positive definite function which is nonincreasing, the origin is thereby proved to be stable. If, however, some arbitrarily selected positive definite function $V(x)$ is increasing for some motion of the state vector, no conclusion can be drawn. That is, if a proposed Liapunov function fails to satisfy the requirements of the stability theorem, no valid conclusions are possible, since another choice of Liapunov function may satisfy the stability requirements. A difficulty in the use of the Liapunov stability theorem is in the construction of a suitable $V(x)$ for a given system. Formal methods exist for generating suitable Liapunov functions for linear systems. Certain classes of nonlinear problems lend themselves to formal procedures in finding suitable $V(x)$ functions, but generating suitable $V(x)$ functions in the general nonlinear case is not yet possible.

†Given a quadratic form $X^T a X$ associated with a real symmetrix matrix a. If $X^T a X$ is nonnegative for all real values of the variables X_i and is zero only if each of these n variables is zero, the quadratic form is positive definite. It is customary to say that the matrix is positive definite. A necessary condition for a matrix to be positive definite is that each element on the diagonal of a symmetric matrix must be positive. The Sylvester rule for a matrix to be positive definite is that the determinant of all principal minors of the matrix must be positive.

Example 13-9.1

A simple nonlinear is described by the differential equation,

$$\ddot{x} + a\dot{x} + bx + cx^2 = 0 \quad (a > 0, b > 0, c > 0).$$

Examine the system stability in the neighborhood of $x = 0$.
a. From considerations of the differential equation.
b. By the Liapunov theorem.

Solution. a. Refer to the differential equation in the neighborhood of $x = 0$. The differential equation now approximates to,

$$\ddot{x} + a\dot{x} + bx = 0,$$

which is a simple second order linear equation, which we know to be asymptotically stable.

b. Transform the differential equation to normal form. This becomes,

$$\dot{x} = y,$$
$$\dot{y} = -ay - bx - cx^2.$$

Define as a possible Liapunov function,

$$V(x,y) = \tfrac{1}{2}(bx^2 + y^2 + 2\beta xy) \quad (\beta > 0).$$

$V(x,y)$ is positive definite, as required. From this we write

$$\dot{V} = \frac{\partial V}{\partial x}\dot{x} + \frac{\partial V}{\partial y}\dot{y} = (bx + \beta y)\dot{x} + (y + \beta x)\dot{y},$$
$$= (bx + \beta y)y - (y + \beta x)(ay + bx + cx^2),$$
$$= -(a - \beta)y^2 - a\beta xy - cyx^2 - b\beta x^2 - c\beta x^3.$$

Clearly, for $\beta < a$, $\dot{V}(x,y) < 0$. Thus $V(x,y)$ is a Liapunov function, and by the Liapunov theorem, the origin is asymptotically stable.

13-10 GENERATING LIAPUNOV FUNCTIONS

As noted above, an explicit method is available for generating a Liapunov function for a linear system. To see how this is accomplished, consider the system that is described by,

$$\dot{x} = Ax. \qquad (13\text{-}36)$$

Let us select as a Liapunov function,

$$V(x) = x^T E x. \qquad (13\text{-}37)$$

The time derivative of V is,

$$\dot{V} = x^T E \dot{x} + \dot{x}^T E x. \tag{13-38}$$

Combine this equation with Eq. (13-36) to get,

$$\dot{V} = x^T E A x + (Ax)^T E x.$$

But since $(Ax)^T = x^T A^T$, then,

$$\dot{V} = x^T (EA + A^T E) x. \tag{13-39}$$

Since $V(x)$ was defined as positive, then for an asymptotically stable system $\dot{V}(x)$ must be negative. If the vector F is defined such that,

$$\dot{V} = -x^T F x, \tag{13-40}$$

then,

$$-F = EA + A^T E, \tag{13-41}$$

and the system is stable. That is, for asymptotic stability for a linear system, it is sufficient that F be positive definite. The necessity portion of the proof relies on several theorems that we have not discussed (28).

Example 13-10.1

Consider a second order system,

$$\frac{d}{dt}\begin{bmatrix} x_1 \\ x_2 \end{bmatrix} = \begin{bmatrix} a_{11} & a_{12} \\ a_{21} & a_{22} \end{bmatrix} \begin{bmatrix} x_1 \\ x_2 \end{bmatrix}.$$

Specify the conditions on A and E to satisfy Eq. (13-41).

Solution. Since F is an arbitrary symmetric positive definite matrix, it is chosen as the unity matrix, $F = I$. Equation (13-41) becomes,

$$\begin{bmatrix} e_{11} & e_{12} \\ e_{21} & e_{22} \end{bmatrix} \begin{bmatrix} a_{11} & a_{12} \\ a_{21} & a_{22} \end{bmatrix} + \begin{bmatrix} a_{11} & a_{21} \\ a_{12} & a_{22} \end{bmatrix} \begin{bmatrix} e_{11} & e_{12} \\ e_{21} & e_{22} \end{bmatrix} = \begin{bmatrix} -1 & 0 \\ 0 & -1 \end{bmatrix}.$$

This matrix equation is expanded and rearranged to the following form,

$$\begin{bmatrix} 2a_{11} & 2a_{21} & 0 \\ a_{12} & a_{11} + a_{22} & a_{21} \\ 0 & 2a_{12} & 2a_{12} \end{bmatrix} \begin{bmatrix} e_{11} \\ e_{12} \\ e_{22} \end{bmatrix} = \begin{bmatrix} -1 \\ 0 \\ -1 \end{bmatrix}.$$

These three equations can be solved simultaneously to yield the elements of the matrix E,

$$\begin{bmatrix} e_{11} & e_{12} \\ e_{21} & e_{22} \end{bmatrix} = \frac{-1}{2(Tr A)\Delta(A)} \begin{bmatrix} \Delta(A) + a_{21}^2 + a_{22}^2 & -(a_{12}a_{22} + a_{21}a_{11}) \\ -(a_{12}a_{22} + a_{21}a_{11}) & \Delta(A) + a_{11}^2 + a_{12}^2 \end{bmatrix}.$$

where $Tr A = a_{11} + a_{22}$, the sum of the diagonal terms of the matrix A. The matrix E is positive definite if each element on the diagonal (the principal minors of E) is positive. This requires that,

$$e_{11} = \frac{\Delta(A) + a_{21}^2 + a_{22}^2}{2(Tr A)\Delta(A)} > 0; \quad e_{22} = -\frac{\Delta(A) + a_{11}^2 + a_{12}^2}{2(Tr A)\Delta(A)} > 0,$$

and,

$$\Delta(E) = \frac{(a_{11} + a_{22})^2 + (a_{12} - a_{21})^2}{2(Tr A)^2 \Delta(A)} > 0; \quad e_{12} = \frac{a_{12}a_{22} + a_{21}a_{11}}{2(Tr A)\Delta(A)}.$$

This latter expression implies, since the numerator of the inequality is always positive, that,

$$\Delta(A) = a_{11}a_{22} - a_{12}a_{21} > 0,$$

and consequently,

$$Tr A = a_{11} + a_{22} < 0.$$

These two expressions give the required conditions for the stability of the second order system.

The problem of establishing necessary and sufficient conditions for the stability of nonlinear systems is still only partially resolved. Some progress has been made in obtaining construction algorithms for Liapunov functions, at least to establish sufficiency conditions. But the formulation of satisfactory Liapunov functions is still the subject of research. The formulation of such Liapunov functions is largely a matter of trial and error, and the process of evaluation V and $\dot{V}$ throughout the state space often becomes a tedious algebraic process. Because of this, the use of the digital computer is becoming more widespread in these studies (29). Often, in fact, even the desired form of the Liapunov function is not known. Lur'e and Letov (30) have shown that for particular classes of nonlinear systems for which the system equations can be expressed in one of two canonical forms, a suitable form of a Liapunov function is known. This is an important step, since many control systems problems can be arranged in the prescribed canonical form. Other important methods exist for generating Liapunov functions and for studying the stability of nonlinear systems (31). The general problem has not yet been solved.

PROBLEMS

13-1. Apply the Routh-Hurwitz test to determine whether the systems specified by the following characteristic equations are stable:

$$4s^4 + 3s^3 + 2s^2 + s + 1 = 0$$
$$s^5 + 2s^4 + 13s^3 + 11s^2 + 17s + 8 = 0$$
$$s^6 + 2s^5 + 3s^4 + 4s^3 + 5s^2 + 6s + 7 = 0$$
$$6s^6 + 5s^5 + 4s^4 + 3s^3 + 2s^2 + s + 1 = 0$$
$$4s^4 + 3s^3 - 2s^2 - s + 1 = 0$$

13-2. Determine, using the Routh-Hurwitz test, the range of K, if any, for which the systems illustrated are stable.

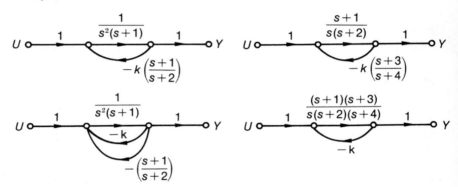

13-3. Consider the general fourth order characteristic equation,

$$a_4 s^4 + a_3 s^3 + a_2 s^2 + a_1 s + a_0 = 0.$$

All the a-s are positive. Develop a set of rules, using the Routh-Hurwitz algorithm, that insures that all roots of the characteristic equation will represent a stable system.

13-4. Consider the system functions,

$$\frac{H}{s(s+1)} \qquad \frac{H(s-1)}{s+1} \qquad \frac{H}{s^2+3s+5} \qquad \frac{H}{s(s^2-9)}$$

For each of these functions,
 a. Plot $T(j\omega)$ in the range $0 < \omega < \infty$, with $H = 1$.
 b. Determine by the Nyquist test the range of H for a stable system when simple feedback is applied, so that $T(s) - 1 = 0$.

13-5. Plot the function $T(s)$ on the $T(s)$-plane for the values of s along the paths designated,

 a. $T(s) = \dfrac{s+2}{s(s-1)}$ b. $T(s) = \dfrac{1}{s^2+2s+3}$.

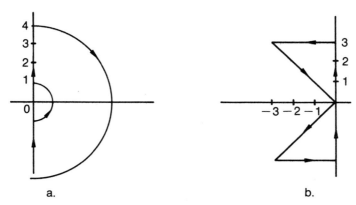

a. b.

13-6. Given are the frequency plane diagrams of a number of systems. Give the conditions on the number of poles of $GH(s)$ with positive real parts for a stable closed loop system.

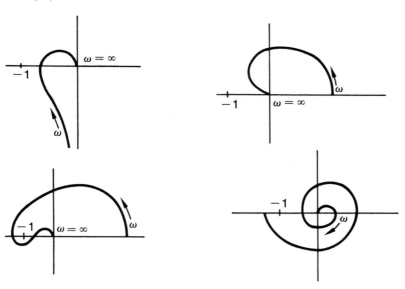

13-7. Given feedback systems which are described by the open loop functions $GH(s)$,

$$\frac{1000}{s(s+1)} \qquad \frac{500}{s(s^2+2s+3)}.$$

Investigate the stability of each by the Nyquist test.

13-8. Consider the feedback control system function,

$$GH(s) = \frac{H}{s\left(\frac{s}{50}+1\right)\left(\frac{s}{10}+1\right)}.$$

Plot the Nyquist diagram for $H = 5, 20, 50$. Determine the gain H at which the system becomes unstable. Check your results using the Routh-Hurwitz test.

13-9. Consider the closed loop system function,

$$T(s) = \frac{G(s)}{1+G(s)H(s)},$$

with

$$GH(s) = \frac{K}{(1+0.1s)(1+0.3s)(1+s)}.$$

Determine by means of the Routh-Hurwitz test,
   a. The range of K for which the system is stable.
   b. Draw the Bode plots for the function when $K = 10$.
   c. From b construct the Nyquist diagram. Does this check your conclusions under a?

13-10. The open loop transmittance functions for some physical systems are given below. Use the Nyquist method to investigate the stability of the systems represented by these functions, with $K = 0$.

$$\frac{K(1+\tau_1 s)}{(1+\tau_2 s)(1+\tau_3 s)} \quad \text{(with } \tau_1 > \tau_2 > \tau_3 > 0\text{)} \quad \frac{K}{s(1+s)(1+0.5s)(1+0.2s)}$$

$$\frac{K(1+2s)}{s(s+1)(1+0.05s)(1+0.02s)}.$$

13-11. Repeat Problem 13-2 using the Nyquist criterion.

13-12. An unforced, initially excited, system is characterized by the matrixes,

$$A = \begin{bmatrix} 0 & 1 \\ 2 & -1 \end{bmatrix} \quad P = [P_1 \quad P_2].$$

   a. Is the system stable?
   b. Find an expression for $y(t)$. Is it stable?
   c. If $P_1 = -P_2$, is the output stable?

13-13. An initially relaxed system is characterized by the matrixes,

$$A = \begin{bmatrix} 1 & 1 \\ 0 & -1 \end{bmatrix} \quad B = \begin{bmatrix} 1 \\ -2 \end{bmatrix}.$$

   a. Is the relaxed system stable?
   b. Now consider the response to a unit impulse. Is the excited system stable?

Relate this result to the excitation of the natural modes by the excitation.

13-14. A system is described by the state equations,

$$\dot{x}_1 = x_2$$
$$\dot{x}_2 = -ax_2 - bx_1 - x_1^2 \quad (a > 0, b > 0).$$

Define as a possible Liapunov function,

$$V(x_1,x_2) = \tfrac{1}{2}[bx_1^2 + 2\beta x_1 x_2 + x_2^2] \quad (\beta > 0).$$

Is the origin stable? If so, establish the level of stability.

13-15. A system is described by the state equation,

$$\frac{d}{dt}\begin{bmatrix} x_1 \\ x_2 \end{bmatrix} = \begin{bmatrix} 0 & 1 \\ -1 & 0 \end{bmatrix}\begin{bmatrix} x_1 \\ x_2 \end{bmatrix} - a\begin{bmatrix} x_1 & x_1 \\ x_2 & x_2 \end{bmatrix}\begin{bmatrix} x_1^2 \\ x_2^2 \end{bmatrix},$$

with $a > 0$. A possible Liapunov function is

$$V(x_1,x_2) = x_1^2 + x_2^2.$$

Is the system stable? If so, establish the level of stability.

13-16. The SFG of a simple feedback system that is known to be asymptotically stable is given in the accompanying figure.

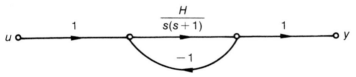

a. Set up state equations of the system.
b. Two possible Liapunov functions are suggested,

$$V(x_1,x_2) = x_1^2 + x_2^2,$$
$$V(x_1,x_2) = Hx_1^2 + (x_2 + ax_1)^2.$$

Ascertain whether either is a suitable Liapunov function.

13-17. A system is described by the state equation,

$$\frac{d}{dt}\begin{bmatrix} x_1 \\ x_2 \end{bmatrix} = \begin{bmatrix} 1 & -1 \\ -1 & 1 \end{bmatrix}\begin{bmatrix} x_1 \\ x_2 \end{bmatrix} + \begin{bmatrix} u \\ 0 \end{bmatrix}.$$

Ascertain whether the system is stable.

13-18. Investigate the stability of the system described by the state equation,

$$\frac{d}{dt}\begin{bmatrix} x_1 \\ x_2 \end{bmatrix} = \begin{bmatrix} 0 & 1 \\ -\alpha & -\beta \end{bmatrix}\begin{bmatrix} x_1 \\ x_2 \end{bmatrix} + \begin{bmatrix} u \\ 0 \end{bmatrix}.$$

Select as a possible Liapunov function,

$$V(x_1,x_2) = x_1^2 + x_2^2.$$

13-19. An input free system is characterized by the matrix,

$$A = \begin{bmatrix} 0 & 1 & 0 \\ 0 & 0 & 1 \\ 1.6 & -0.8 & 3 \end{bmatrix}.$$

Apply the Liapunov test to determine whether or not this system is stable.

13-20. Determine whether a system having the following characteristic equation is stable,

$$|sI - A| = 4s^3 - 21s^2 + 3s - 3 = 0.$$

13-21. A discrete time system is characterized by the equations,

$$\begin{bmatrix} x_1(n+1) \\ x_2(n+1) \end{bmatrix} = \begin{bmatrix} 1 & -\frac{1}{2} \\ -2 & 1 \end{bmatrix} \begin{bmatrix} x_1(n) \\ x_2(n) \end{bmatrix} + \begin{bmatrix} u(n) \\ 0 \end{bmatrix}.$$

From considerations of the characteristic polynomial, ascertain whether the system is stable.

13-22. Determine whether the system described by the system function,

$$T(z) = \frac{2z^2 - 3z + 1}{z(8z^3 + 4z^2 + 2z + 4)}.$$

13-23. Determine the stability of the systems specified by the characteristic equations,

$$7z^5 - 5z^4 + 11z^3 - 9z^2 - 3z + 1 = 0,$$
$$7z^5 + 5z^4 + 11z^3 + 9z^2 + 3z + 1 = 0,$$
$$5z^6 + 11z^5 + 14z^4 + 8z^3 + 2z^2 = 0.$$

References

1. S. Seely, "Electromechanical Energy Conversion," McGraw Hill, New York, 1962.
2. S. Mason and H. Zimmerman, "Electronic Circuits, Signals and Systems" Wiley, New York, 1960.
3. F. B. Hildebrand, "Introduction to Numerical Analysis," McGraw Hill, New York, 1956.
4. F. H. Branin, G. R. Hogsett, R. L. Lunde, L. E. Kugel, "ECAP-II—An Electronic Circuit Analysis Program" *IEEE Spectrum* **8**, 14, June 1971.
5. D. A. Calahan, "Numerical Solution of Linear Systems with Widely Separated Time Constants," *Proc. IEEE*, **55**, 2016, November 1967.
   R. V. Stineman, "Digital Time-Domain Analysis of Systems with Widely Separated Poles," *J. ACM*, **12**, 186, April 1965.
   P. M. Russo, "On the Time Domain Analysis of Linear Time Invariant Networks with Large Time-Constant Spreads by Digital Computer," *IEEE Trans. on Circuit Theory*, CT 18, 194, January 1971.
6. R. C. Hamming, "Numerical Methods for Scientists and Engineers," McGraw Hill, New York, 1962.
7. M. L. Liou, "A Novel Method of Evaluating Transient Response," *Proc. IEEE* **54**, 20, Jan. 1966.
   W. E. Thomson, "Evaluation of Transient Response," *Proc. IEEE* (Letters) **54**, 1584, Nov. 1966.
   M. L. Liou, "Evaluation of Transition Matrix," *Proc. IEEE* (Letters) **55**, 228, Feb. 1967.
   W. Everline, "On the Evaluation of e^{AT} by Power Series," *Proc. IEEE* (Letters) **55**, 413, March 1967.
   T. A. Bickart, "Matrix Exponential: Approximation by Truncated Power Series," *Proc. IEEE* (Letters) **56**, 872, May 1968.
   E. J. Mastascusa, "A Relation Between Liou's Method and the Fourth-order Runge-Kutta Method for Evaluating Transient Response," *Proc. IEEE* (Letters) **57**, 802, May 1969.
   P. N. Robinson, "Steady-State Solution of Linear Time Invariant Systems," *Proc. IEEE* (Letters) **57**, 2079, Nov. 1969.

L. W. Cahill, "Computation of the Transient Response of Linear Systems to Step and Ramp Inputs," *Proc. IEEE* (Letters) **58**, 1169, July 1970.

G. Fronza and S. Rinaldi, "On the evaluation of the Transition Matrix," *Proc. IEEE* (Letters) **58**, 1951, Dec. 1970.

8. E. R. Cohen, "Some Topics in Reactor Kinetics," Proc. 2nd U.N. International Conference on Peaceful Uses of Atomic Energy," **11**, 302–309, Geneva, Switzerland, 1958.

   E. R. Cohen and H. P. Flatt, "Numerical Solution of Quasi-Linear Equations," Proc. of Seminar on Codes for Reactor Computations, pp. 461–484, Vienna, Austria, 1960.

9. J. Certaine, "The Solution of Ordinary Differential Equations with Large (small) Time Constants," Chapter 11, Mathematical Methods for Digital Computers, A. Ralston and H. S. Wolf, (Eds.) Wiley, New York, 1960.

10. D. A. Callahan, "Numerical Solution of Linear Systems with Widely Separated Time Constants," *Proc. IEEE*, **55**, 2016, November 1967.

11. E. J. Davison and H. W. Smith, "A Computational Technique for Determining the Steady-State Output of a Linear System with Periodic Input," *IEEE Trans. on Circuit Theory*, CT-18, 181, January 1971.

12. W. R. LePage, "Complex Variables and the Laplace Transform for Engineers," McGraw Hill, New York, 1961.

13. P. A. McCollum and B. F. Brown, "Laplace Transform Tables and Theorems" Holt, Rinehart and Winston, New York, 1965.

14. J. M. Souriau, "Une methode pour la decomposition spectrale a l'inversion de matrices," Compt. rend., 227, 1010–1011, 1948.

    J. S. Frame, "Matrix Functions and Applications," part IV, *IEEE Spectrum* **1**, 123–131, June 1964.

    L. A. Zadeh and C. De Soer, "Linear Circuit Theory," McGraw Hill, New York, 1963.

15. F. H. Branin, Jr., "Computer Methods of Network Analysis," *Proc. IEEE* **55**, 1787, Nov. 1967.

16. G. V. S. S. Raju, "The Routh Canonical Form," *IEEE Trans. on Automatic Control* (C) **12**, 463, Aug. 1967.

17. N. Balabanian, "Network Synthesis," Prentice Hall, Englewood Cliffs, N.J., 1958.

    D. Hazony, "Elements of Network Synthesis," Reinhold, New York, 1963.

    E. A. Guillemin, "Synthesis of Passive Networks," Wiley, New York, 1957.

    M. E. Van Valkenburg, "Introduction to Modern Network Synthesis, "Wiley, New York, 1960.

18. P. M. Chirlian, "Integrated and Active Network Analysis and Synthesis," Prentice Hall, Englewood Cliffs, N.J., 1967.

    G. S. Moschytz, "Inductorless Filters: A Survey" *IEEE Spectrum*, 7, 63, Sept. 1970.

    L. Huelsman, "Active Filters," McGraw Hill, New York, 1970.

19. G. R. Steber and R. J. Krneger, "On a Completely tunable active filter using fixed RC elements," *Proc. IEEE* (Letters), **57**, 727, April 1969.

20. S. Seely, "Realizing System Functions with Operational Amplifiers," *Proc. IEEE (Letters)*, **57**, 1665, Sept. 1969.

21. B. D. H. Tellegen, "A General Network Theorem with Applications," *Philips Research Rep.* **7**, 259, (1952).

22. H. S. Carslaw, "Introduction to the Theory of Fourier's Series and Integrals," 3d. rev. ed., Dover, New York, 1930.

23. J. W. Cooley and J. W. Tukey, "An algorithm for the machine calculation of complex Fourier series," Math. of Computers **19**, 297, April 1965.

24. W. T. Cochran, et al., "What is the Fast Fourier Transform," *Proc. IEEE*, **55**, 1664–1674, October 1967.

    E. O. Brigham and R. E. Morrow. "The Fast Fourier transform," *IEEE Spectrum*, **4**, 63–70, Dec. 1967.

    B. Gold and C. M. Rader, "Digital Processing of Signals," Chap. 6, McGraw Hill, New York, 1969.

25. J. R. Ragazzini and G. F. Franklin, "Sampled Data Control Systems," McGraw Hill, New York, 1958.

    E. I. Jury, "Sampled Data Control Systems," Wiley, New York, 1958.

26. M. L. Bakker and C. J. Triska, "Determination of the state variable A matrix by Direct Measurement," *IEEE Trans on Automatic Control* (C) **11**, 610, July 1966.

27. A. Liapunov, "Probleme general de la stabilite du movement," Annals of Mathematical Studies No. 17, Princeton University Press, Princeton, N.J., 1949.

    J. P. La Salle and S. Lefschetz, "Stability by Liapunov's Direct Method with Applications," Academic Press, New York, 1961.

28. R. E. Kalman and J. E. Bertram, "Control Systems Analysis and Design via the Second Method of Liapunov," *J. Basic Engrg., Trans. ASME*, 371, June 1960.

    S. Lefschetz, "On Automatic Control," *IRE Trans.* PGCT-7, 474, 1962.

29. J. J. Rodden, "Numerical Applications of Liapunov Stability Theory," *Proc. of J.A.C.C.*, 261–268, June 1964.

30. A. M. Letov, "Stability in Nonlinear Control Systems," Princeton University Press, Princeton, N.J. 1961.

    A. I. Lur'e, "Some Nonlinear Problems in the Theory of Automatic Control," Her Majesty's Stationary Office, London, 1957.

    A. I. Lur'e and E. N. Rozenvasser, "On Methods of Constructing Liapunov Functions in the Theory of Nonlinear Control Systems," *Proc. IFAC* vol. 2, Butterworth's Scientific Publications, London, 1960.

31. N. N. Krasovskii, "On the Global Stability of a System of Nonlinear Differential Equations," *Prikladnaja Matematika i Mekanika*, **18**, 735–737, 1954.

    D. G. Schultz and J. E. Gibson, "The Variable Gradient Method of Generating Liapunov Functions," *AIEE Trans.*, Part II, pp. 203–210, Sept. 1962.

Appendix A Matrixes

Matrixes are used extensively in this text for representing a system of equations with both continuous and with discrete variables. It is the purpose of this appendix to summarize a number of the important properties and mathematics of matrixes.

To introduce the study, consider the following set of linear algebraic equations,

$$
\begin{aligned}
a_{11}x_1 + a_{12}x_2 + \cdots + a_{1n}x_n &= u_1 \\
a_{21}x_1 + a_{22}x_2 + \cdots + a_{2n}x_n &= u_2 \\
&\vdots \\
a_{n1}x + a_{n2}x + \cdots + a_{nn}x_n &= u_n.
\end{aligned}
\tag{A-1}
$$

Here the u-s may be excitation functions for a given system, and the x-s may denote the state variables. The a-s are linear coefficients which are defined by the system. In matrix form, this set of equations is written as,

$$
\begin{bmatrix}
a_{11} & a_{12} & \cdots & a_{1n} \\
a_{21} & a_{22} & \cdots & a_{2n} \\
\vdots & \vdots & & \vdots \\
a_{n1} & a_{n2} & \cdots & a_{nn}
\end{bmatrix}
\begin{bmatrix} x_1 \\ x_2 \\ \vdots \\ x_n \end{bmatrix}
=
\begin{bmatrix} u_1 \\ u_2 \\ \vdots \\ u_n \end{bmatrix}.
\tag{A-2}
$$

This is usually written more simply as,

$$AX = U,$$

where

$$A = \begin{bmatrix} a_{11} & a_{12} & \cdots & a_{1n} \\ a_{21} & a_{22} & \cdots & a_{2n} \\ \vdots & \vdots & & \vdots \\ a_{n1} & a_{n2} & \cdots & a_{nn} \end{bmatrix}; \quad X = \begin{bmatrix} x_1 \\ x_2 \\ \vdots \\ x_n \end{bmatrix}; \quad U = \begin{bmatrix} u_1 \\ u_2 \\ \vdots \\ u_n \end{bmatrix}. \quad \text{(A-3)}$$

It is evident that the matrixes as defined here are arrays of coefficients; there is no numerical value associated with the array. The matrix A as defined comprises the coefficients a_{ij}, where i refers to the row and j to the column; hence a_{ij} is the coefficient belonging to the ith row and the jth column.

Clearly from Eq. (A-3), matrixes may be square, they may be rectangular, they may be columnar, or they may be row. The relationship between the number of rows and columns in matrixes is important in certain operations, and this matter will receive attention at the appropriate point in our study.

A-1 DEFINITIONS

It is customary to define a matrix according to the character of the array, the elements of the array, or some special property of the array. The important definitions follow:

a. Order of a matrix: a matrix of m rows and n columns is called an $m \times n$ matrix.

b. Square matrix: a matrix of order $m \times m$.

c. Column matrix: a matrix containing only one column, hence it is of order $m \times 1$.

d. Row matrix: a matrix containing only one row; it is of order $1 \times m$.

e. Real matrix: all elements of the matrix are real.

f. Complex matrix: some elements of the matrix are complex.

g. Zero or null matrix: all elements of the matrix are zero.

h. Sparse matrix: many of the elements of the matrix are zero (no quantitative measure is given).

i. Diagonal matrix: a square matrix having nonzero elements along the diagonal.

j. Scalar matrix: a diagonal matrix whose diagonal elements are equal.

k. Unit matrix: a scalar matrix whose diagonal elements are unity.

l. Transposed matrix: A transposed matrix A^T has rows and columns which are interchanged with the rows and columns of the original matrix A, i.e., if $A = [a_{ij}]$, then $A^T = [a_{ji}]$.

m. Complex conjugate of a matrix. A^* is the complex conjugate of matrix A if each element of A^* is the complex conjugate of the corresponding element of A. Clearly for real matrices $A = A^*$.

Additional definitions will be introduced as details of matrix operations are discussed.

A-2 MATRIX ALGEBRA

Some of the more important algebraic operations involving matrixes are now to be discussed.

a. Equality of matrixes: two matrixes are equal if and only if each of their corresponding elements are equal. This prescribes that for two matrixes to be equal, they must contain the same number of rows and columns.

b. Addition of matrixes. The sum of two matrixes A and B, both of the same order, is,

$$A + B = [a_{ij} + b_{ij}] \qquad \begin{array}{l}(i = 1, 2, \ldots, m) \\ (j = 1, 2, \ldots, n).\end{array} \qquad \text{(A-4)}$$

c. Subtraction of matrixes: (a corollary of b.) The difference of two $m \times n$ matrixes is another $m \times n$ matrix, the elements of which are the difference of the two corresponding elements in the two matrixes being subtracted.

d. Matrix multiplication: The product of a matrix of m rows and n columns and a matrix B of n rows and p columns is a matrix C of m rows and p columns, with elements obtained as follows,

$$C = AB,$$

where,

$$c_{ij} = \sum_{k=1}^{m} a_{ik} b_{kj} \qquad \begin{array}{l}(i = 1, 2, \ldots, m) \\ (j = 1, 2, \ldots, p).\end{array} \qquad \text{(A-5)}$$

As a simple example, consider the product,

$$\begin{bmatrix} 1 & 0 & 3 \\ -2 & 3 & 1 \\ 6 & 5 & 0 \end{bmatrix} \times \begin{bmatrix} 0 & 3 \\ 1 & 4 \\ 2 & -6 \end{bmatrix} = \begin{bmatrix} (0+0+6) & (3+0-18) \\ (0+3+2) & (-6+12-6) \\ (0+5+0) & (18+20+0) \end{bmatrix} = \begin{bmatrix} 6 & -15 \\ 5 & 0 \\ 5 & 38 \end{bmatrix}.$$

e. Conformable matrixes: two matrixes which can be multiplied together are said to be conformable. This means that the matrix AB

might exist, whereas the matrix BA might not exist, i.e., AB might satisfy condition d, whereas BA might not do so. In general $AB \neq BA$ (e.g., in the simple numerical example given in d, AB is defined but BA does not exist.) When $AB = BA$, the matrixes are said to commute.

f. Matrix division: Matrix division is not a defined operation. The equivalent operation involving the inverse of a matrix is defined.

g. Inverse of a matrix: The inverse A^{-1} of a square matrix A is defined by the relation,

$$A^{-1}A = AA^{-1} = I \quad \text{(the unit matrix).} \tag{A-6}$$

Based on this, a solution may be written to Eq. (A-3). Now begin with Eq. (A-3),

$$AX = U.$$

Premultiply by A^{-1} to get,

$$A^{-1}AX = X = A^{-1}U. \tag{A-7}$$

It has been assumed that given a square matrix A, that A^{-1} exists. When the inverse does not exist, A is said to be a singular matrix. The inverse of a matrix is obtained as follows:

given

$$A = [a_{ij}],$$

then,

$$A^{-1} = [a_{ij}]^{-1} = \frac{\text{transpose cofactor } a_{ij}}{\Delta(A)} = \frac{\text{cofactor } a_{ji}}{\Delta(A)}, \tag{A-8}$$

where $\Delta(A)$ is the determinant of the matrix (the determinant that has the same elements as A). If $\Delta(A) = 0$, the matrix is singular and the inverse does not exist.

h. Partitioning of matrixes: When matrixes have a large number of rows and columns, it may be desirable to partition them into smaller sections, called submatrixes. Partitioning is best illustrated by an example. Refer to the product matrix below,

$$\begin{bmatrix} a_{11} & a_{12} \\ a_{21} & a_{22} \\ \hline a_{31} & a_{32} \end{bmatrix} \begin{bmatrix} b_{11} & b_{12} & b_{13} & b_{14} \\ b_{21} & b_{22} & b_{23} & b_{24} \end{bmatrix}$$

$$= \begin{bmatrix} (a_{11}b_{11}+a_{12}b_{21}) & (a_{11}b_{12}+a_{12}b_{22}) & (a_{11}b_{13}+a_{12}b_{23}) & (a_{11}b_{14}+a_{12}b_{24}) \\ (a_{12}b_{11}+a_{22}b_{21}) & (a_{21}b_{12}+a_{22}b_{22}) & (a_{21}b_{13}+a_{22}b_{23}) & (a_{21}b_{14}+a_{22}b_{24}) \\ (a_{31}b_{11}+a_{32}b_{21}) & (a_{31}b_{12}+a_{32}b_{22}) & (a_{31}b_{13}+a_{32}b_{23}) & (a_{31}b_{14}+a_{32}b_{24}) \end{bmatrix}.$$

Now suppose that matrix A is partitioned into two parts, as per the broken line shown, and let B be partitioned into three parts, also as shown. The resulting matrix may then be written,

$$\begin{bmatrix} A_1 \\ A_2 \end{bmatrix} \begin{bmatrix} B_1 & B_2 & B_3 \end{bmatrix} = \begin{bmatrix} A_1B_1 & A_1B_2 & A_1B_3 \\ A_2B_1 & A_2B_2 & A_2B_3 \end{bmatrix}. \tag{A-9}$$

It is observed that the matrix product AB may be evaluated as though the submatrixes were ordinary matrix elements. Clearly, for partitioning, the submatrixes must be conformable.

i. Inverse of product of matrixes: Consider the product of two conformable nonsingular matrixes. From the fact that,

$$(AB)(AB)^{-1} = I.$$

and also from the fact that we can write,

$$ABB^{-1}A^{-1} = AA^{-1} = I,$$

then it follows that

$$(AB)^{-1} = B^{-1}A^{-1}. \tag{A-10}$$

That is, the inverse of the product of two nonsingular matrixes of the same order is the product of the inverse matrixes in reverse order.

j. Transpose of a product: Consider two conformable matrixes A and B. The typical term of their product is,

$$[AB]^T = [ab]^T_{ij} = [ab]_{ji},$$
$$= \sum_k a_{jk}b_{ki} = \sum_k b^T_{ik}a^T_{kj} = [b^T a^T]_{ij}.$$

It is seen that,

$$AB^T = B^T A^T. \tag{A-11}$$

That is, the transpose of a matrix product is the product of their transposes in reverse order.

A-3 QUADRATIC FORMS

Consider the function,

$$B = A^T P A. \tag{A-12}$$

The quantity on the right is known as a congruent transformation of P.

Specifically, suppose that the matrixes are,

$$A = \begin{bmatrix} a_1 \\ a_2 \\ a_3 \end{bmatrix} \quad P = \begin{bmatrix} p_{11} & p_{12} & p_{13} \\ p_{21} & p_{22} & p_{23} \\ p_{31} & p_{32} & p_{33} \end{bmatrix} \quad A^T = [a_1 \; a_2 \; a_3].$$

Upon expansion of Eq. (A-12) there results,

$$B = p_{11}a_1^2 + p_{22}a_2^2 + p_{33}a_3^2 + 2p_{12}a_1a_2 + 2p_{32}a_2a_3 + 2p_{31}a_3a_1. \quad (A\text{-}13)$$

This expression is homogeneous of the second degree in the variables $a_1, a_2, a_3,$.

When $B = A^T P A$ is greater than zero for $A \neq 0$, the quadratic form is called positive definite (it is customary to say that P is positive definite in this case). If B is less than zero for $A \neq 0$, the quadratic form is negative definite. If B is essentially positive but may become negative, it is called positive semidefinite. For example,

$$\begin{aligned} x_1^2 + 2x_1x_2 + 2x_2^2 & \quad \text{positive definite,} \\ -(x_1^2 + 2x_1x_2 + 2x_2^2) & \quad \text{negative definite,} \\ x_1^2 + 2x_1x_2 + x_2^2 & \quad \text{positive semidefinite,} \\ -(x_1^2 + 2x_1x_2 + x_2^2) & \quad \text{negative semidefinite.} \end{aligned}$$

A-4 EIGENVALUES AND EIGENVECTORS

We consider the following set of algebraic equations,

$$AX = \lambda X, \quad (A\text{-}14)$$

where A = a known real square matrix of order $n \times n$,
X = an unknown column vector,
λ = a scalar parameter (often the complex number s).

Clearly, we have that the vectors λX and X have the same direction. Eq. (A-14) poses the question—are there any directions that are left invariant by the transformation defined by A? A nontrivial solution of Eq. (A-14) exists only if the scalar parameter λ assumes certain specific values. To find these, we rewrite Eq. (A-14) in the form,

$$(A - \lambda I)X = 0. \quad (A\text{-}15)$$

This expression has a nontrivial solution X if and only if,

$$\det |A - \lambda I| = 0. \quad (A\text{-}16)$$

This determinant is written explicitly,

$$\begin{bmatrix} a_{11}-\lambda & a_{12} & \cdots & a_{1n} \\ a_{21} & a_{22}-\lambda & & a_{2n} \\ \multicolumn{4}{c}{\dotfill} \\ a_{n1} & a_{n2} & & a_{nn}-\lambda \end{bmatrix} = 0.$$

It is clearly of degree n in λ. To expand this, it is noted that the terms in λ^n, λ^{n-1}, etc. occur only in the product of all the terms in the leading diagonal. The term independent of λ is obtained by setting $\lambda = 0$ in the determinant, and is thus $\Delta[A] \equiv |A|$. Multiplying by $(-1)^n$, the equation for λ has the form,

$$\lambda^n - (a_{11} + a_{22} + \cdots + a_{nn})\lambda^{n-1} + \cdots + (-1)^n |A| = 0. \qquad \text{(A-17)}$$

This equation is called the *characteristic equation* of the matrix A. The roots of this equation are called the *eigenvalues* of matrix A. An equation of degree n has n roots, some of which may be repeated, and may be real or complex. If the coefficients are all real, the complex roots occur in conjugate pairs. Let the eigenvalues of A be $\lambda_1, \lambda_2, \ldots, \lambda_k$; the determinant may thus be written,

$$|A - \lambda I| = (\lambda - \lambda_1)^{r_1}(\lambda - \lambda_2)^{r_2}, \ldots, (\lambda - \lambda_k)^{r_k}, \qquad \text{(A-18)}$$

where the r_i specify the order of root λ_i. It is noted that when A is symmetric, the eigenvalues are all real.

Corresponding to each eigenvalue there exists a nonzero column vector X satisfying Eq. (A-15). X is known as the (column) *eigenvector* corresponding to the eigenvalue λ.

A-5 TRANSFORMATION TO DIAGONAL FORM

Calculations are sometimes greatly simplified if a given matrix is first transformed to diagonal form. Such transformations are possible for a large class of matrixes. The simplest case to consider is the matrix A for which all eigenvalues are distinct, since in this case the eigenvectors are linearly independent. Now it is possible to find a nonsingular matrix W of the same order as A such that the product (the *similarity transformation*— the eigenvalues are unchanged in the transformation) $W^{-1}AW$ is a diagonal matrix. The proof proceeds by considering A to be of order $n \times n$, with eigenvalues $\lambda_1, \lambda_2, \ldots, \lambda_n$, and with the corresponding eigenvectors $W^{(1)}, W^{(2)}, \ldots, W^{(n)}$. It is here noted that matrix W whose columns are $W^{(1)}, W^{(2)}, \ldots, W^{(n)}$ is nonsingular, and consequently W^{-1} exists. Now we write,

$$AW^{(1)} = \lambda_1 W^{(1)}; \quad AW^{(2)} = \lambda_2 W^{(2)}, \ldots, AW^{(n)} = \lambda_n W^{(n)}.$$

These n equations are equivalent to the single matrix equation,

$$AW = W\Lambda, \qquad (A\text{-}19)$$

where Λ is the $n \times n$ diagonal matrix,

$$\begin{bmatrix} \lambda_1 & 0 & \cdots & 0 \\ 0 & \lambda_2 & \cdots & 0 \\ \multicolumn{4}{c}{\dotfill} \\ 0 & 0 & \cdots & \lambda_n \end{bmatrix}.$$

Multiply Eq. (A-19) on the left by W^{-1}, which yields,

$$W^{-1}AW = W^{-1}W\Lambda = \Lambda. \qquad (A\text{-}20)$$

Clearly, this similarity transformation reduces the matrix A to diagonal canonical form Λ.

It is noted that the proof of Eq. (A-20) depends only on the fact that the eigenvectors $W^{(1)}, W^{(2)}, \ldots, W^{(n)}$ are linearly independent. This is always the case when the eigenvalues are distinct. This proof remains valid when the eigenvalues of the matrix are not all distinct.

Index

abcd parameters, 397
Active element, 51
Active network, 401
  equivalent networks, 402
Admittance, 312
  driving point, 313
  indicial, 246
  operator, 68
  relation to impedance, 68
Admittance matrix of two-port, 389
Admittance parameters, 313
Amplifier, operational, 222
Amplitude, of complex number, 321
  of sinusoid, 344
Analogs, 29, 55
Analog computer, function generation, 227
  initial conditions, 230
  operational amplifier, 222
    differentiator, 226
    integrator, 225
    summing, 223
  programming, 228
  scaling, time and magnitude, 231
  simulation by digital computer, 233
Analysis, loop, 77, 146
  node, 69, 144
  state variable 82ff., 149ff.
Angle, of complex number, 321
Approximation, of differential equation by difference equation, 96

  incremental, 5
  least squares of Fourier series, 429
Asymptotic stability, 498
  existence of Liapunov function for, 498
Autonomous system, 85
Average power, definition, 353

Bandwidth, 365
Bilateral, 6
Bode plot, 369
Branch, 133
  current, 69
  equations, 71
  voltage, 69
Branch parameter matrix, 141
Butterworth filter, 380

Canonic *LC* network, 149
Capacitance, electrical, 13
  fluid, 41
  gas, 44
  thermal, 48
Cascade connection of two-ports, 398
Cascade parameters, 396
Cascaded systems, 397
Characteristic equation, 300
Characteristic matrix, 297
Characteristic polynomial, 300
  algorithm for computing coefficients, 300
Characteristic roots, 175, 300

522   Index

of A matrix, 300
  determining stability, 482
Circuit, elements (*see* Elements, system)
  linear, 5
  lumped, 5
  nonlinear, 5
  two terminal, 486
Coefficient, admittance, 72
  impedance, 79
  mutual inductance, 21
  self induction, 17
Coenergy function, 15
Complementary function, 174
Complete solution, 173
Complex Fourier series, 423
Complex frequency, 276
Complex number, 321
  complex conjugate, 321
  magnitude, 321
  phase, 321
  polar representation, 321
Complex plane, 321
Components of state vector, 83
Computer, analog (*see* Analog computer)
  digital, 233
Conductance, 23
Connection matrix, 134
Conservation, electric charge, 186
  energy, 69
  flux linkages, 187
Continuous time systems, 65ff.
  differential equations for, 169
  impulse response of, 249, 326
  periodic inputs, 265
  sampled inputs, 263
  solution of differential equations, 255
    numerical methods, 261
  stability criterion (*see* Stability)
  superposition integral, 246
Controllability, criterion for, 495
Controlled source, 402
Convolution, integral, 249
  computation of, 253
  graphical interpretation, 251
  matrix, 259
  property, Discrete Fourier Transform, 445
    Fourier transform, 453

Laplace transform, 289
Z-transform, 463
Convolution summation, 254
Coupled inductors, 19
Cramer's rule, 293, 316
Criterions for stability (*see* Stability)
Critical frequencies, poles and zeros, 310
Critically damped, 200
Current, 13
  branch, 69
  loop, 78
  reference direction, 24
Current, complex amplitude, 354
  conduction, 14
  displacement, 14
  effective, 346
Current source, dependent, 402
  independent, 24
Cutset, 133
  matrix, 137

D'Alembert's principle, 70
Dash pot, 32
Damping constant, 32
D-c transients, 187
Decibel, 370
Decomposition of signals, (*see* Fourier series, Singularity functions)
Delta connection, 378, 395
Delta function, 244
Dependent source, 402
Determinant, 316
  expansion, Gauss reduction, 318
    Laplace, 317
Diagonal matrix, 513
Difference equations, approximation of differential equations, 95, 212
  characterization of systems by, 97
  for discrete time systems, 262
    solutions of, 262
  in normal form, 99
  Z-transform of, 472
  simultaneous, 107
Differences, forward, 240
  Z-transform of, 463
Differential equations, 169ff.
  approximation by difference equation, 95, 212

complementary equation, 172, 174
constant coefficients, 91
 linear, 91, 169
 nonhomogeneous, 170
 nth order, 91
 normal form of, 90, 171
 numerical solution of, 210
  Euler method, 210, 219
  explicit method, 213
  implicit method, 213
  Predictor-Corrector, 221
  Runge-Kutta, 220
 simultaneous, 71, 112
 solution, particular, 172, 178
Digital computer, simulation languages, 233
Dirichlet conditions, in Fourier transform, 435
 in Fourier series, 415
 in Laplace transform, 277
Discrete Fourier Transform, 443 (*see* Fourier, Discrete transform)
Discontinuity, in time domain analysis, 246
Discrete spectrum, 425
Discrete time convolution property, 254
Discrete time system, 95, 262
 difference equation of, 97
 equivalent, for continuous time system, 264
  with sampled input, 96
 fundamental matrix, 262
 signal flow graph, 125
 solution of difference equations for, 262
 stability of, 493
 superposition summation, 271
Discrete time transfer function, for systems in normal form, 478
 Z-transform, 464
Doublet function, 245
Driving point admittance (*see* Admittance)
Driving point impedance (*see* Impedance)
Dual graphs, construction of, 159
Dual networks, 161
Duality, 28, 159
Dynamic system, description of, 1

e^{At} (*see* Fundamental matrix)
Effective value, 346

Eigenvalue, 175, 300 (*see also*, Natural modes)
Eigenvector, 301
Elements, active, 51
 bilateral, 6
 linear, 5
 lumped, 5
Element of matrix, 83
Elements, system, electrical, 13ff.
 fluid, 39
 gas, 43
 mechanical, translational, 29ff.
 mechanical, rotational, 35
 thermal, 45
 transducers, 49
Energy density spectrum, 438
Envelope of spectrum, 426
Equilibrium equations, loop, 77, 146
 node, 72, 144
 state, 82, 149
 through-across, 69
Equilibrium state, 85
 stable (*see* Stability)
Equivalent networks, of active networks, 402
 Norton, 382
 T and Π, 395
 Thévenin, 382
Euclidean norm, 498
Euler's identity, 177
Euler method in solving differential equations, 210, 219
Evaluation of integration constants, 185
Even function, 419

Fast Fourier Transform, 446
Feedback, 77, 117, 490
Final value theorem, Laplace transform, 288
First order system, 187
Flow graph, signal (*see* Signal flow graph)
Fluid flow, laminar, 39
 turbulent, 39
Flux linkage, 16
Forced response, 178
Forward difference, operator, 463
 Z-transform of, 463
Four terminal networks (*see* Two-port networks)

Fourier, Discrete Transform, 443ff.
   evaluation of, 445
      fast Fourier transform, 446
      relation to, Fourier integral, 445
      Z-transform, 462
   integral, 437
   integral transform, 435ff.
      convolution, 453
      inversion of, 438
      Parseval's theorem, 437
      relation to Discrete transform, 445
      response of RC network to pulse, 453
      sampling theorem, 439
   law of heat conduction, 46
   series, 414ff.
      Dirichlet conditions, 415
      exponential, 423
      impulse train, 426
      least squares approximation, 429
      line spectrum, 425
      numerical calculation of coefficients, 431
      properties of, 428
         Parseval's theorem, 428
      rapidity of convergence, 417
      symmetry, effects of, 419
      transform pairs, 437
Frequency, angular, 35, 344
   resonant, 361
   3 dB cutoff, 365
Frequency domain analysis, 275ff.
Frequency response plane, 368
Frequency spectrum, 425
Friction, 32
Function, sampling, 426
   singularity, 243
   transfer (*see* System function)
   unit impulse (*see* Impulse)
   unit step, 243
Fundamental solution matrix, 255
   calculation of, 255
   discrete time system, 262
   Laplace transform of, 297
   physical significance of, in continuous time system, 256
   in discrete time system, 262
   relation to impulse response, 328
   Z-transform of, 472

Gain, current, 313
   signal flow graph, 121
   voltage, 313
Gauss elimination, 318
General circuit parameters, two-port (*see* Two-port networks)
General solution, 173
Generating function for Z-transform, 461
Global stability, 498
Graph, 130ff.
   connected, 130
   dual, 159
   mappable, 159
   oriented, 131
   planar, 159
   separate parts, 132
   signal flow, 108ff.
   topological, 131
Graphical methods for stability, in discrete time systems, 493
   Nyquist criterion, 492
Green's function, 249

h-parameters, 52
Half-power bandwidth, 365
Half-power frequency, 365
Heat conduction, 46
   convection, 47
Homogeneous equation, 170
Hurwitz test, 487
Hybrid parameters, 52, 395
Hysteresis, 18

Ideal diode, 7
Image diagrams, 369
Impedance, driving point, 313
   locus, 368
   operator, 68
   parallel combination, 314
   passive two-port, 400
   properties of passive one-port, 486
   poles and zeros (*see* System function)
   relation to admittance, 68
   series combination, 314
Impedance matrix of two-port, 391
Improper fraction, 294
Impulse function, 244
   replacement for initial conditions, 281

Index 525

resolution of signal into, 459
strength of, 459
train, Fourier series of, 426
Impulse response, relation to network function, 327
   relation to step response, 249
Indicial response, 246
Incremental operation, 5
Inductance, mutual, 19
   self, 17
Inductor, 16
   energy stored in, 17
   linear, 17
   nonlinear, 18
Initial conditions, 185
   in analog computers, 230
   Laplace transform of, 281
   replacement of, in charged capacitors, 283
   in fluxed inductors, 281
   vector of, 303
Initial state, 303
Initial value theorem, Laplace transform, 287
Input-output relation (*see* System function)
Instability conditions (*see* Stability)
Instantaneous power, 23, 353
Integral, convolution, 249
   shifting, 284
   superposition, 246
Integration constants, evaluation of, 185
Integrator, 225
   initial conditions, 231
   for determining elements of fundamental matrix, 256
Interpolating function, 441
Inverse matrix, 515
Inverse network, 161
Inverse of $(sI - A)$, 297
Inversion integral, Fourier, 438
   Laplace, 290
   Z-transform, 461

Jordan matrix, 302

Kirchhoff current law, KCL, 69
   cutset analysis, 144
   node pair analysis, 69
Kirchhoff formulation, relation to state description, 82, 122
Kirchhoff voltage law, KVL, 69
   cutset analysis, 146
   loop analysis, 77

Laminar flow, 39
Laplace transform, 275ff.
   convolution theorem, 289
   defining integral, 276
   of elementary functions, 277
   expansion theorem, 293
   of fundamental matrix, 297
   inverse, 290
   one sided, 276
   problem solving using, 291
   properties of, 279
      convolution, 289
      differentiation, 280
      final value, 288
      initial value, 287
      linearity, 280
      time translation, 283
   of square wave, 285
   shifting theorem, 284
   state equations, 297
   table of, 279
Laurent series, expansion of Z-transform into, 461
Least squares approximation of Fourier series, 429
Liapunov function, 500
   generating, 501
   stability theorem, 500
Linear differential equation (*see* Differential equations)
Linear systems, definition of, 170
Linearity of Laplace transform, 280
Links, 133
Long division method of inverting Z-transform, 460
Loop, 77, 133
   current, 78
   fundamental, 133
   impedance matrix, 79
Loop analysis, 77, 146
Loop currents, 78
   related to fundamental circuit matrix, 138
Lumped elements, 5

Magnetic field, 16
Magnetically coupled systems, 19
Magnitude and phase plots, 369
Mappable graph, 159
Matrix, 512ff.
  addition of, 514
  adjoint, 300
  algebra, 514
  branch parameter, 141
  chain, 397
  characteristic polynomial, 300, 518
  characteristic value of, 518
  conformable, 514
  connection, 134
  cutset, 137
  definitions, 513
  diagonal, 513
  discrete time transfer, 255
  dominantly diagonal, 260
  eigenvalue, 517
  eigenvector, 517
  functions of (*see* Fundamental solutions matrix)
  fundamental circuit, 138
  fundamental cutset, 137
  hybrid, for two-port network, 395
  identity, 137
  impedance, 148
  inverse, 515
  Jordan, 302
  loop impedance, 148
  node admittance, 72
  order of, 513
  partitioning, 137, 145, 147, 515
  positive definite, 517
  quadratic form, 516
  singular, 94
  square, 513
  state transition, 256
  tie set, 138
  trace of, 301
  transfer function, 328
  transpose of, 513
  transformation, similarity, 302, 518
  unity, 137, 297, 513
  y-system, 389
  z-system, 392
Matrix system function, 328
  characterized by simultaneous difference equations, 477
Maximum power transfer, 385
Memoryless system, 84
Mesh (*see* Loop)
Models and modeling, 3ff.
  electrical elements, 13ff.
  mechanical elements, 29ff.
  fluid, 39
  thermal, 45
  transducers, 49
Model approximations, 7
Moment of inertia, 35
Mutual inductance, 19
  dot convention, 19
  sign of M, 20

Natural frequencies, 175, 300 (*see also*, Natural modes)
  relation to network functions, 294
Natural modes, 175, 300
  conditions for excitation, 495
  observation of, 495
Negative definite function, 517
Network, active, 401
  determinant, 317
  dual, 161
  graph (*see* Graph)
  reciprocal, 161
  stable (*see* Stability)
  topology (*see* Graph)
Network behavior, calculation of fundamental matrix, 255
  convolution integral, 249
  excited by piecewise constant input, 247
  principle of superposition in, 246
  RLC, 198
  step and impulse response, 326
Network function, 310
  cascaded networks, 399
  general properties, 323
  magnitude, 324
  phase, 324
  phasor locus, 368
Network theorems, 382ff.
  independent loop variables, 139
  independent node pair variables, 139
  maximum power, 385

Norton, 382
  source transformation, 388
  superposition, 170
  T and Π, 395
  Tellegen, 407
  Thévenin, 382
Newton, law of cooling, 47
Newton-Raphson method, 209
Node analysis, 69, 144
  equilibrium conditions, 69
  variables, 139
Nonlinear system, definition of, 5
  solutions, 216
Nonrecursive process, 98
Norm, Euclidean, 498
Normal form of differential equations, 90
  procedure for determining, 90
Normal modes, 175, 300 (*see* Natural modes)
Norton theorem, 382
Numerical integration, 219
Numerical methods, 207ff.
Nyquist criterion, stability, 492
  for discrete time system, 493
Nyquist diagram, 369

Observability, criterion for, 496
Odd function, 419
Ohm, 23
One-sided Laplace transform, 276
One sided Z-transform, 461
Open circuit impedance, two-port networks, 392
Operational amplifier, 222
  used as integrator, 225
  in system function realization, 336
Order of system, 90
Ordinary differential equations (*see* Differential equations)
Oriented graph (*see* Graph)
Origin, equilibrium of, 498
  stability of, 498
Orthogonal set of functions, 415
Output (*see* Response)
Output port, 389
Overdamped case, 200

Parallel connection of admittances, 314

Parameter matrix, 141
Parseval's theorem, 428, 437
Partial fraction expansion, 294
Particular solution, 178
Partitioning of matrix, 137, 515
Periodic function, Fourier series, 414
  Laplace transform of, 278, 285
Phase, of driving point impedance, 352
  of network function, 325
  of sinusoid, 345
Phasor, 348
  addition of, 349
  diagrams, 355
Π equivalent network, 395
Piecewise constant signal, 9
  applied to RC network, 239
  in sampled data systems, 458
  and superposition integral, 248
Planar network, 159
Polarity of voltage, 24
Poles, 294, 323
  dynamic interpretation, 294
  multiple order, 294
  of network function, 322
  location and stability, 482
  simple, 294
Pole-zero configuration, 323
Polynomial, characteristic, 300
Positive definite function, 517
Power, average, 353
  instantaneous, 23, 353
  maximum transfer, 385
Power factor, 355
Predictor-Corrector method, 221
Problem oriented languages, 234
Proper fraction, 294
Properties of, Fourier series, 428
  Fourier transform, 437
  Laplace transform, 279
  Z-transform, 462
Properties of general solution, 173
  solution to singularity functions, 326
Pulse response of RC network, 453
Pulse train, Fourier series of, 425
  least square approximation, 431
  spectrum of, 425

Q of a resonant circuit, 362

definition, 362, 366
  relation to bandwidth, 365
Quadratic form, 516
  positive definite, 517

Radius of convergence, Z-transform, 461
Ramp function, 243
Rational function, 311
  inverse Laplace transform of, 293
  proper, 295
RC circuit, pulse response of, 453
Reciprocity conditions, 389
Reciprocal network, 161
Recursive process, 98, 212
Reference conditions, 24
Residue, 294
Resistor, 22
Resolution of continuous time waveform, 248
  into impulses, 458
  into steps, 458
  of discrete time signal, into Z-transformed form, 479
Resonance, series, 361
Resonant circuit, parallel, 367
  series, 361
  Q of parallel circuit, 367
  series circuit, 362
Resonant frequency, 361
Response, damped sinusoid, 201
  forced, 172
  general periodic excitation, 414ff.
  general nonperiodic excitation, 435
  impulse, 249
  indicial, 246
  state, discrete time, 262
    force free, 256
    forced, 259
    initial state, 303
  steady state, 178
  step, 187
  sinusoidal, 206
  transient, 172, 174
    impulse, 326
    step function, 187, 326
Response function (*see* System function)
Root mean square, 346
Routh-Hurwitz test, 487

Runge-Kutta method, 220

$(s\mathbf{I}-A)^{-1}$ evaluation, 297
s-plane, 310ff.
  graphical representation in, 321
Sampled data system, 95
  treatment by Z-transform, 457
Sampled signal, 95
  as input to continuous time system, 95
  Laplace transform of, 459
  Z-transform of, 459
Sampler, types of, 458
Sampling, reduction to impulse modulation, 459
Sampling function, 426
Sampling interval, 458
Sampling theorem, 439
Scale factor, 231
Scaling in analog computers, 231
Second method of Liapunov, 498
Second order system, 198
Separate parts, 132
Series combination of impedances, 314
Shifting formula, 284
Signals, 10
  average value, 345
  discrete time, 96 (*see also*, Sampled signal)
  Fourier integral representation, 437
  Fourier series representations, 414
  piecewise constant, 9
  root mean square value, 346
  stepwise approximation, 9
  singularity representations, 266
  truncated, 429
Signal flow graph, 108ff.
  algebra of, 116
  computer program, 228
  gain equation, 121
  properties of, 108
  rules for drawing, 109
  sampled data systems, 471
  simultaneous differential equations, 112
  state equations, 122, 328
  transmittance, 108
Simulation, analog, 228
  use of CSMP for, 233
Simultaneous differential equations, 112

(see also, Differential equations)
Singularity functions, 243
Sinor, 348
Sinusoid, 344
 amplitude, 344
 effective value, 346
 full cycle average, 345
 phase, 345
 phasor, 348
 summation, 347
Sinusoidal steady-state analysis, 350
Sinusoidal steady-state response, 206, 356
Solution of differential equations, analytic, 243ff.
 computer, 221
 numerical, 207
Sources, dependent, 402
 independent, 24
 transformation of, 27, 388
Space, state, 83
Spectrum, amplitude, 426
 envelope of, 426
 phase, 426
Square wave, Laplace transform of, 285
Stability, 482ff.
 asymptotic, 498
 bounded input bounded output, 498
 criterions for (see Stability, determination)
 eigenvalue location, 482
 global, 498
 graphical method for determining, 489
 in the large, 498
 in Liapunov sense, 498
 Nyquist criterion for determining, 492
 of origin, 498
 relation to pole location, 482
Stability, determination, Liapunov, 500
 Nyquist, 489
 Routh-Hurwitz, 487
State, characteristic matrix, 297
 equations, 85, 149
 autonomous, 85
 solutions, Laplace transform, 297
  numerical, 261
  time domain, 255
 formulation, 82, 149
  differential equation, 331
  signal flow graph, 122

 system function, 328
 fundamental matrix, 256
 Liapunov function, 500
 residue matrix, 303
 space, 83
 stability, 498
  Liapunov theorem, 500
 transition matrix, 256
 variables, 83
  selection of, 85
   by analog computer program, 241
   measurement of unobservable, 496
 vector, 83
  initial, 303
State, equilibrium, 85
Step function, unity, 243
Step response, 326
 of first order circuits, 187
 relation to impulse response, 249
 of second order circuit, 198
Strength of impulse, 244, 459
Summing amplifier, 223
Superposition, in sinusoidal steady state, 350
 in terms of network function, 424
Superposition integral, 246
 in terms of step response, 248
 in terms of impulse response, 249
Superposition and linearity, 170
Superposition summation, 271
System, 65ff.
 causal, 173
 characterized by differential equations, 65ff.
 continuous time (see, Continuous time systems)
 controllable, 495
 difference equations for, 97
 discrete time (see Discrete time system)
 dynamic, 65
 elements (see Elements, system)
 feedback, 77, 117, 490
 inverse of, 161
 lumped, 5
 memory of, 83
 observable, 496
 order of, 90
 signal flow graph for, 108

stable (*see* Stability)
Systems equations, linear form of, 90
  normal form of, 90, 171
System function, 310
  continuous time system, 311
  fundamental matrix, 256
  graphical evaluation, 322
  locus diagrams, 368
  matrix, 328
  realization using operational amplifiers, 336
  poles and zeros of, 322
  relation to impulse response, 329
  roots of, 322
  signal flow graph, 121
  and state models, 328
System of simultaneous differential equations, 72
  determinant, 316
System response, frequency domain, 275ff.
  time domain, 243ff.
System transfer function (*see* System function)

T equivalent network, 394
T and Π transformation, 394
Tellegen theorem, 407
Thermal conduction (*see* Heat conduction)
Thévenin theorem, 382
3 dB bandwidth, 365
Three phase, 378
Through variable, 13
Through-across equilibrium law, 69
Tie set, 133
  matrix, 138
Time average of signal, 345
Time constant, 190
  barrier in transient solutions, 215
Time domain analysis, 169ff.
Time functions, 10
Time translation property, Laplace transform, 283
  Z-transform, 463
$T(j\omega)$ plane, 368
Topological definitions, 133
Topology, 130
Trace of matrix, 301, 503
Train of impulses, Fourier series of, 426

use in sampled data systems, 459
Transducers, 49
Transfer function (*see* System function)
Transfer matrix, 328
Transform, Discrete Fourier (*see* Fourier transform, Discrete)
  Fourier (*see* Fourier transform, integral)
  Laplace (*see* Laplace transform)
  Z- (*see* Z-transform)
Transform pairs, Laplace, 279
  Z-, 462
Transformation, impedance (T and Π), 394
  source, 26
Transformer, 19
Transient response, classical methods, 174
  a-c excitation, 206
  impulse and step function, 326
  initial conditions, effects of, 173
  pole-zero location, 482
  state space, 255
    numerical methods, 261
  transform methods, 292
    expansion theorem, 293
Transition matrix, 255 (*see also,* Fundamental matrix)
Translation property, Laplace transform, 283
  Z-transform, 463
Tree, 133
  proper, 151
Tree branches, 133
Tree complement, 133
Tree link, 133
Two-port networks, 388ff.
  $abcd$ matrix, 397
  hybrid matrix, 395
  impedance of, input, 400
    output, 401
    transfer, 401
  T-Π equivalent, 395
  y-system matrix, 389
  z-system matrix, 391
Two-sided, Laplace transform, 276
  Z-transform, 461
Two terminal element (*see* Elements)

Underdamped case, 201
Unforced fixed linear system, stability

Index 531

criterion for, 484
Unit delay operator, 284
Unit impulse, 245
Unit step, 326
  approximations to, 265
Unit step response, 326
Unity coupled transformer, 20
Unstable system, 484

Van der Pol, 90
Variables, across, 13
  loop, 139
  node pair, 139
  through, 13
Variation of parameters, method of, 182
Vector, state, 83
Vector space 83
Voltage, 13
  branch, 69
  node to datum, 13
  reference polarity of, 25
Voltage law, Kirchhoff KVL, 69
Voltage source, dependent, 402
  independent, 24

Waveform (*see* Signals)
Weighting function, 415
Weighting sequence, 255

Y-delta transformation, 394
y-system matrix, admittance parameters, 389

z-parameters, impedance parameters, 391
Zero of network function, 322
Z-transform, 457ff.
  convergence of, 461
  convolution, 463
  definition of, 461
  differentiation, 465
  difference equations, 472
    normal form, 477
  Discrete Fourier transform, 462
  forward difference, 463
  inversion of, 461
  one-sided, 461
  properties of, convolution, 463
    difference, 462
  system function, 473
  table of, 462
  translated function, 463
  two-sided, 461
  unit response, 464
Zeros, network response function, 322
Zero memory system, 84

OTHER TITLES IN THE PERGAMON UNIFIED ENGINEERING SERIES

Vol. 1. W. H. DAVENPORT/D. ROSENTHAL – *Engineering: Its Role and Function in Human Society*
Vol. 2. M. TRIBUS – *Rational Descriptions, Decisions and Designs*
Vol. 3. W. H. DAVENPORT – *The One Culture*
Vol. 4. W. A. WOOD – *The Study of Metal Structures and Their Mechanical Properties*
Vol. 5. M. SMYTH – *Linear Engineering Systems: Tools and Techniques*
Vol. 6. L. M. MAXWELL/M. B. REED – *The Theory of Graphs: A Basis for Network Theory*
Vol. 7. W. R. SPILLERS – *Automated Structural Analysis: An Introduction*
Vol. 8. J. J. AZAR – *Matrix Structural Analysis*
Vol. 9. S. SEELY – *An Introduction to Engineering Systems*
Vol. 10. D. T. THOMAS – *Engineering Electromagnetics*
Vol. 12. S. J. BRITVEC – *The Stability of Elastic Systems*
Vol. 13. A. R. M. NOTON – *Modern Control Engineering*
Vol. 14. B. MORRILL – *An Introduction to Equilibrium Thermodynamics*
Vol. 15. R. PARKMAN – *The Cybernetic Society*